Progress in Computer Science
No. 2

Edited by
E. Coffman
R. Graham
D. Kuck

Springer Science+Business Media, LLC

Applied Probability–Computer Science: The Interface Volume I

Sponsored by Applied Probability Technical Section
College of the Operations Research Society of America
The Institute of Management Sciences
January 5-7, 1981
Florida Atlantic University
Boca Raton, Florida

Ralph L. Disney,
Teunis J. Ott,
editors

1982 Springer Science+Business Media, LLC

Editors:

Ralph L. Disney
Department of Industrial Engineering
and Operations Research
Virginia Polytechnic Institute
and State University
Blacksburg, Virginia 24061, USA

Teunis J. Ott
Bell Laboratories
Hölmdel, New Jersey 07733, USA

Applied probability-computer science.
(Progress in computer science ; no. 2-3)
1. Electronic data processing--Congresses.
2. Electronic digital computers--Programming--
Congresses. 3. Probabilities--Congresses.
4. Queuing theory--Congresses. I. Disney, Ralph L.,
1928- . II. Ott, Teunis J. III. Applied
Probability Technical Section-College of the
Operations Research Society of America, the Institute
of Management Sciences. IV. Series.
QA75.5.A66 1982 0001.64 82-18506

ISBN 978-0-8176-3116-1

CIP-Kurztitelaufnahme der Deutschen Bibliothek

Applied probability-computer science, the
interface / sponsored by Applied Probability
Techn. Sect., College of the Operations Research
Soc. of America, The Inst. of Management Sciences.
- Boston ; Basel ; Stuttgart : Birkhäuser

1981. January 5 - 7, 1981, Florida Atlantic
University, Boca Raton, Florida.
Vol. 1 (1982).
(Progress in computer science ; Vol. 2)
ISBN 978-0-8176-3116-1 ISBN 978-1-4899-4975-2 (eBook)
DOI 10.1007/978-1-4899-4975-2

NE: Florida Atlantic University Boca Raton, Fla.
GT

Originally published by Birkhäuser Boston, inc., in 1982
Softcover reprint of the hardcover 1st edition 1982

ISBN 978-0-8176-3116-1

TABLE OF CONTENTS

NETWORKS OF QUEUES, II
Jean Walrand, Chairman

QUEUEING MODELS IN PERFORMANCE ANALYSIS, II
Benjamin Melamed, Chairman

TABLE OF CONTENTS BY AUTHORS

PREFACE

These two volumes are the Proceedings of the first special interest meeting instigated and organized by the joint Technical Section and College in Applied Probability of ORSA and TIMS. This meeting, which took place January 5-7, 1981 at Florida Atlantic University in Boca Raton, Florida, had the same name as these Proceedings: Applied Probability-Computer Science, the Interface. The goal of that conference was to achieve a meeting of, and a cross fertilization between, two groups of researchers who, from different starting points, had come to work on similar problems, often developing similar methodologies and tools.

One of these groups are the applied probabilists, many of whom consider their field an offspring of mathematics, and who find their motivation in many areas of application. The other is that group of computer scientists who, over the years, have found an increasing need in their work for the use of probabilistic models.

The most visible area of common methodology between these two groups is networks of queues, which by itself could have been the theme of an entire conference. Functional areas which are, or are becoming, sources of exciting problems are computer performance analysis, data base analysis, analysis of communication protocols, data networks, and mixed voice-data telephone networks. The reader can add to this list by going through the papers in these Proceedings.

At the conference, 55 papers were presented. Some authors committed their work for publication elsewhere. 45 papers are included in these Proceedings. The first eight papers correspond with the 1 hour long talks, often of a survey nature, that were given. The other papers were presented in 20 minute talks. We attempted to give these Proceedings, as much as possible, an archival character and for that reason have included, where possible, an account of the formal discussions and author's responses to their discussions.

The meeting drew about 150 participants from the U.S., Canada, England, France, East- and West Germany and Poland. More importantly, the meeting has every indication of having been successful in cross fertilization and in stimulating interest. One indication of this is that a similar meeting is being planned for January 1983 at INRIA, Paris, France.

The efforts of a large number of people were necessary for the success of the meeting. We thank ORSA and TIMS for sponsoring the meeting and for making available the competent assistance of Mrs. Julie Eldridge. Without the help of Florida Atlantic University and our local organizer, Prof. Robert Cooper, the conference could not have become the success it was.

We also thank Bell Telephone Laboratories for allowing one of the editors the time the conference needed.

Virginia Polytechnic Institute and State University, the Department of Industrial Engineering and Operations Research and Dr. Robert D. Dryden, Head of that department have been sympathetic to our needs and have supported this effort superbly. We thank those people.

Ms. Lauren Klein of Birkhäuser-Boston has been a constant source of information and liaison between the publisher and the editors. The task would have been considerably more difficult without her help.

Mr. Jagadeesh Chandramohan, Peter Kiessler, and Ms. Georgia Ann Klutke read portions of papers herein and proof read final typing. This was certainly an imposition on the time of three graduate students.

Authors, discussants and especially reviewers of these papers must be thanked. They lived with tight deadlines, harassment by hurried editors and produced well and in a timely fashion.

Finally, we must thank Ms. Paula L. Kirk. Not only did she retype these nearly 1000 pages on camera ready paper with an eye-appeal that is stunning, she redrew several of the diagrams so as to make them amenable to photo-copying. She was the liaison between author and editor when questions arose about manuscripts. Without her tireless, professional, dedicated effort, these two volumes would never have appeared.

Ralph L. Disney
Teunis J. Ott

Blacksburg, Virginia
Holmdel, New Jersey

ORGANIZING COMMITTEE

Teunis Ott
Bell Labs

Edward Coffman
Bell Labs

Robert Cooper
Florida Atlantic University

Ralph Disney
Virginia Polytechnic Institute
and State University

Carl Harris
Center for Management and Policy Research

Steven Lavenberg
IBM

Robert Stark
University of Delaware

ATTENDEES*

Sven Axsäter
Graduate Program in Operations Research
North Carolina State University
Box 5511
Raleigh, North Carolina 27650

John Bartholdi
School of Industrial and Systems Engineering
Georgia Institute of Technology
Atlanta, Georgia 30332

Vaclav E. Beneš
Bell Laboratories
Murray Hill, New Jersey 07901

Jon L. Bentley
Department of Computer Science
Carnegie-Mellon University
Pittsburgh, Pennsylvania 15213

Menachem Berg
2729 West Jarvis
Chicago, Illinois 60645
University of Illinois, Chicago

Narayan Bhat
Southern Methodist University
Perkins Administration Building
Room 221
Dallas, Texas 75275

Howard Blum
305 East 86th Street, Apt. 7FW
New York, New York 10028
Rutgers University

Nancy Boynton
Department of Mathematics and Computer Science
Michigan Technological University
Houghton, Michigan 49931

Theodore Brown
Department of Computer Science
Queens College of the City of New York
Flushing, New York 11367

Steven C. Bruell
123A Lind Hall
Computer Science Department
University of Minnesota
Minneapolis, Minnesota 55455

John Bruno
Department of Computer Science
University of California
Santa Barbara, California 93106

W. Bulgren
2810 Trail Road
Lawrence, Kansas 66044
University of Kansas

David Burman
Room WB1G302
Bell Laboratories
Holmdel, New Jersey 07733

M. Jeya Chandra
Department of Industrial and Management Systems Engineering
Pennsylvania State University
207 Hammond Building
University Park, Pennsylvania 16802

Alain Chesnais
LRI Bat 490
Université Paris Sud
Orsay 91405, FRANCE

Stavros Christodoulakis
Department of Computer Science
University of Toronto
CANADA

A. Bruce Clarke
College of Arts and Sciences
Western Michigan University
Kalamazoo, Michigan 49008

E. G. Coffman
Bell Laboratories
600 Moutain Avenue
Murray Hill, New Jersey 07974

Arthur Cohen
22 Cold Forge Drive
Warren, New Jersey 07060
Rutgers University

Robert Cooper
Department of Computer Systems
and Management Science
Florida Atlantic University
Boca Raton, Florida 33431

P. J. Courtois
Philips Research Laboratory
2, Av. E. van Becelaere
1170, Brussels BELGIUM

Ralph L. Disney
Department of IEOR
302 Whittemore Hall
Virginia Polytechnic Institute
and State University
Blacksburg, Virginia 24061

Bharat T. Doshi
HP 1B323
Bell Laboratories
Holmdel, New Jersey 07733

Larry Dowdy
Department of Computer Science
University of Maryland
College Park, Maryland 20742

Peter J. Downey
7201 Sabino Vista Drive
Tucson, Arizona 86715
University of Arizona

Isaac Dukhovny
335 2nd Avenue, #5
San Francisco, California 94118
Bank of America

Adrian E. Eckberg, Jr.
P. O. Box 316
Holmdel, New Jersey 07733
Bell Laboratories

Mark I. Farber
Miami Laboratory
National Marine Fisheries
75 Virginia Beach Drive
Miami, Florida 33149

Fayolle
INRIA
Domaine de Voluceau Rocquencourt
B.P. 105 - 78150 Le Chesnay
FRANCE

Awi Federgrun
Columbia University
Uris Hall
New York, New York 10027

Martin Fischer
2503 Pinoak Lane
Reston, Virgnia 22091

Leopold Flatto
3116 Arlington Avenue
Bronx, New York 10463
Bell Laboratories

R. D. Foley
Department of IEOR
302 Whittemore Hall
Virgnia Polytechnic Institute
and State University
Blacksburg, Virginia 24061

C. Ed Ford
234 Alhambra Road
Oak Ridge, Tennessee 37830
Union Carbide Nuclear Division

Ernest Forman
1438 Ironwood Drive
McLean, Virginia 22101
George Washington University

John Franco
3164 Ludlow Road
Shaker Heights, OH 44120
Case Western Reserve University

Peter Franken
Sektion Mathematik
Humbold Universität
1006 Berlin, PSF 1297
German Democratic Republic

Richard L. Franks
HP 1D315
Bell Laboratories
Holmdel, New Jersey 07733

A. A. Fredericks
29 Yellow Brook Drive
Colts Neck, New Jersey 07733
Bell Laboratories

Richard Gail
Computer Science Department
University of California
Los Angeles, California 90024

Donald P. Gaver
26780 Paseo Robles
Carmel, California 93921
Naval Postgraduate School

Ilya B. Gertsbakh
Department of Mathematics
Ben Gurion University of the Negev
P.O. Box 653
Beersheva, ISRAEL

Thomas Giammo
Department of the Air Force
FEDSIM/CA
Washington, DC 20330

Jerren Gould
2440 North Bradley Avenue
Claremont, California 91711
Hughes Aircraft Company

Winfried Grassmann
Department of Computational Sciences
University of Saskatchewan
Saskatoon, Saskatchewan
CANADA S7N OWO

Linda Green
Graduate School of Business
Columbia University
New York, New York 10027

Irwin Greenberg
6619 Midhill Place
Falls Church, Virginia 22043
George Mason University

Donald Gross
Department of Operations Research
George Washington University
Washington, DC 20002

Shlomo Halfin
15 Daniel Drive
Little Silver, New Jersey 07739
Bell Laboratories

Carl M. Harris
Center for Management and Policy Research
1625 I Street, NW
Washington, DC 20006

Philip Heidelberger
IBM Research Center
Yorktown Heights, New York 10598

David Heimann
U.S. Department of Transportation
Transportation Systems Center
55 Broadway
Cambridge, Massachusetts 02142

Daniel P. Heyman
Building WB, Room 1G-311
Bell Laboratories
Holmdel, New Jersey 07733

J. M. Holtzman
Room HO, 2G500
Bell Laboratories
Holmdel, NJ 07733

Dr. A. Hordijk **
Department of Mathematics
University of Leiden
P.O. Box 9512
2300 RA Leiden
THE NETHERLANDS

David Houck
7 Tory Court
Colts Neck, New Jersey 07722
Bell Laboratories

David Hunter
Room 31-138
IBM, T. J. Watson Research Center
P.O. Box 218
Yorktown Heights, New York 10598

Donald L. Iglehart
Department of Operations Research
Stanford University
Stanford, California 94305

Patricia A. Jacobs
Operations Research Department
Naval Postgraduate School
Monterey, California 93940

F. Jelinek
IBM, Box 218
Yorktown Heights, New York 10598

** Paper presented; author not in attendance.

S. S. Katz
Building HP Room 1B-304
Bell Telephone Laboratories
Holmdel, New Jersey 07733

Leonard Kleinrock
Computer Science Department
(3732BH)
University of California
Los Angeles, California 90024

J. Keilson
Graduate School of Management
University of Rochester
Rochester, New York 14627

F. P. Kelly
Statistical Laboratory
University of Cambridge
16 Mill Lane
Cambridge, UNITED KINGDOM

Alan G. Konheim
IBM, T. J. Watson Research Center
P. O. Box 218
Yorktown Heights, New York 10598

V. G. Kulkarni
2524 Shallowford Road NE #85
Atlanta, Georgia 30345
Georgia Institute of Technology

Guy Latouche
Université Libre de Bruxelles
Faculté des Sciences - C.P. 212
Lab. d'Informatique Theorique
Boulevard du Triomphe
1050 Brussels BELGIUM

Stephen S. Lavenberg
IBM, T. J. Watson Research Center
Yorktown Heights, New York 10598

John P. Lehoczky
Department of Statistics
Carnegie-Mellon University
Pittsburgh, Pennsylvania 15213

Austin J. Lemoine
Systems Control, Inc.
1801 Page Mill Road
Palo Alto, California 94304

Peter A. W. Lewis
Naval Postgraduate School
Monterey, California 93940

Melvin Gail Linnell
166 Statesir Place
Red Bank, New Jersey 07701
Bell Laboratories

E. H. Lipper
Building HP 1C322
Bell Telephone Laboratories
Holmdel, New Jersey 07733

C. L. Liu
Department of Computer Science
University of Illinois
Urbana, Illinois 61801

David M. Lucantoni
Department of Mathematical Sciences
University of Delaware
Newark, Delaware 19711

George S. Lueker
ICS Department
University of California
Irvine, California 92717

Michael Magazine
University of Waterloo
Waterloo, Ontario
CANADA N2L 3G1

J. B. Major
5 Place Ville Marie
Montreal, Quebec
CANADA H3B 2G3
IBM, Canada

Gerald Marazas
IBM Corporation
2000 NW 51st Street
Boca Raton, Florida 33432

David F. McAllister
205 Furches Street
Raleigh, North Carolina 27607
North Carolina State University

James McKenna
Bell Laboratories
600 Moutain Avenue
Murray Hill, New Jersey 07974

Robert A. McLaren
121 Normandy Road
Oak Ridge, Tennessee 37830
Union Carbide Corporation

Edward P. McMahon
CIA
8224 Iverness Hollow Terrace
Potomac, Maryland 20854

B. Melamed
Department of IE/MS
Northwestern University
Evanston, Illinois 60201

Haim Mendelson
Graduate School of Management
University of Rochester
Rochester, New York 14627

Douglas R. Miller
Department of Operations Research
George Washington University
Washington, DC 20052

D. Mitra
Bell Telephone Laboratories
Murray Hill, New Jersey 07974

I. Mitrani
Computing Laboratory
University of Newcastle
Newcastle Upon Tyne, UK

George E. Monahan
College of Management
Georgia Institute of Technology
Atlanta, Georgia 30332

John A. Morrison
28 Ashwood Road
New Providence, New Jersey 07974
Bell Laboratories

Eginhard J. Muth
Department of Industrial and Systems Engineering
University of Florida
Gainesville, Florida 32611

A. Nakassis
8107 Fallow Drive
Caithersburg, Maryland 20760
NIA

Marcel F. Neuts
Department of Mathematical Sciences
University of Delaware
Newark, Deleware 19711

Teunis Ott
Building HP, Room 1A-332
Bell Telephone Laboratories
Holmdel, New Jersey 07733

Linda M. Ottenstein
Box 333
Michigan Technical University
Houghton, Michigan 49931

Michael Pinedo
Instituto Venezolano de Investigaciones Cientificas
Apartado 1827
Caracas 101 VENEZUELA

Brigitte Plateau
LRI Bat 490 Université d'Orsay
Université de Paris Sud
91405 Orsay, FRANCE

Loren Platzman
ISYE - Georgia Institute of Technology
Atlanta, Georgia 30332

Guy Pujolle
Ecole Nationale Superieure des Telecommunications
46, Rue Barrault
75634 Paris Cedex 13
FRANCE

Marty Reiman
Bell Telephone Laboratories
Murray Hill, New Jersey 07901

Martin Reiser
IBM Research Laboratory
Saumerstr 4, CH-8803
Ruschlikon, Zurich
SWITZERLAND

Raymond Rishel
Department of Mathematics
University of Kentucky
Lexington, Kentucky 40506

Tomasz Rolski
Wroclaw University
Mathematics Institute
pl. Grunwaldzki 2/4
50-384 Wroclaw
POLAND

Walter A. Rosenkrantz
Department of Mathematics and Statistics, GRC Tower
University of Massachusetts
Amherst, Massachusetts 01003

Charles H. Sauer
IBM, T. J. Watson Research Center
P.O. Box 218
Yorktown Heights, New York 10598

R. Schassberger
Fachbereich 3
T. U. Berlin
Str. D17 Juni 135
Berlin, WEST GERMANY

T. Schonfeld
45 Tanglewood Drive
Livingston, New Jersey 07039
Bell Laboratories

Lee W. Schruben
Cornell University
Ithaca, New York 14850

Paul Schweitzer
Graduate School of Management
University of Rochester
Rochester, New York 14627

Herbert D. Schwetman
Computer Science Department
Purdue University
West Lafayette, Indiana 47906

Richard Serfozo
Bell Telephone Laboratories
Holmdel, New Jersey 07733

Kenneth C. Sevcik
CSRG, University of Toronto
121 St. Joseph Street
Toronto, Ontario
CANADA M5S 1A4

Diane Sheng
WB1H314
Bell Telephone Laboratories
Holmdel, New Jersey 07733

Martin L. Shooman
Polytechnic Institute of New York
Route 110
Farmingdale, New York 11735

S. H. Sim
Department of Industrial Engineering
University of Toronto
Toronto, Ontario
CANADA M5S 1A4

Burt Simon
4E337
Bell Telephone Laboratories
Holmdel, New Jersey 07733

Donald R. Smith
WB 1H316
Bell Telephone Laboratories
Holmdel, New Jersey 07733

Kimming So
IBM, T. J. Watson Research Center
P.O. Box 218
Yorktown Heights, New York 10598

Matthew J. Sobel
College of Management
Georgia Institute of Technology
Atlanta, Georgia 30332

W. J. Stewart
Department of Computer Science
North Carolina State University
Raleigh, North Carolina 27650

Shaler Stidham, Jr.
Department of Industrial Engineering, Box 5511
North Carolina State University
Raleigh, North Carolina 27650

G. B. Swartz
28 Sand Sprint Drive
Eatontown, NJ 07724

Andre Tchen
102-D121
Exxon Corporation
P.O. Box 153
Florham Park, New Jersey 07932

Yiu Kwok Tham
59 St. George Street
Toronto, Ontario
CANADA M5S 2E6
University of Toronto

Henk Tijms
Department of Actuarial Sciences
Free University
De Boelelaan 1081
1081 VN Amsterdam
THE NETHERLANDS

David Trutt
Room 5B109
Bell Telephone Laboratories
Murray Hill, New Jersey 07974

Edwin C. Tse
197 Michael Drive
Middletown, New Jersey 07701
Bell Telephone Laboratories

Percy Tzelnic
Wang Laboratories, Inc. MIS 1379
One Industrial Avenue
Lowell, Massachusetts 01851

Victor L. Wallace
Computer Science Department
116 Stong Hall
University of Kansas
Lawrence, Kansas 66045

Jean C. Walrand
1434 Hanshaw Road
Ithaca, New York 14850
Cornell University

Glenn Weber
Christopher Newport College
50 Shoe Lane
Newport News, Virginia 23606

Peter Welch
85 Croton Avenue
Mt. Kisco, New York 10549
IBM

J. Wessels
Prins Clauslaan 18
5582 Jr Waalre
THE NETHERLANDS
Eindhoven University of Technology

Ward Whitt
Bell Telephone Laboratories
Holmdel, New Jersey 07733

Dan E. Willard
114B Eaton Crest Drive
Eatontown, NJ 07724
Bell Telephone Laboratories

Daniel Wood
Modular Computer Systems
P.O. Box 6099
Ft. Lauderdale, Florida 33309

C. Murray Woodside
Department of Systems/Computer Engineering
Carleton University
Ottawa K1S 5B6
CANADA

*Addresses as of time of registration.

MAJOR SPEAKERS

F. P. Kelly
M. F. Neuts
P. W. Glynn & D. L. Iglehart
V. E. Beneš

J. L. Bentley
P. Franken
P. J. Courtois
C. H. Sauer

NETWORKS OF QUASI-REVERSIBLE NODES

F. P. Kelly

Summary

Many analytical results are available for a network of queues when the nodes in the network have a simplifying property. This property, called here quasi-reversibility, was first identified by Muntz and has since been investigated by a number of authors. A closely related concept, partial balance, has been central to the investigation of insensitivity begun by Matthes.

Here we describe the concept of quasi-reversibility, provide new examples of quasi-reversible nodes, discuss the range of arrival rates for which a node remains quasi-reversible, and analyse a model of a communication network insensitive to patterns of dependence more general than have previously been considered.

1. Introduction

A great number of analytical results are available concerning the equilibrium behaviour of a queueing network when the nodes in the network have a certain simplifying property. This property, called here quasi-reversibility, was first identified by Muntz [17] and various examples of quasi-reversible nodes have been presented by Baskett, Chandy, Muntz and Palacios [2] and Kelly ([7], [8]).

Important aspects of the equilibrium behaviour of some quasi-reversible nodes are insensitive to the precise specification of the some initial states this procedure generates an infinite number of jumps in a finite time with positive probability. When this is not so, i.e., in the non-explosive case, the procedure can be used to construct a Markov process defined on $[0,\infty)$ with an arbitrary initial distribution. In any event a discrete time Markov process defined on $(0,1,2,\ldots)$ can be constructed by using the same procedure but with the holding periods in each state set to one: call this process the jump chain associated with Q.

Suppose now that the transition rates Q admit a positive invariant measure $(\pi(x), x \in S)$, taken here to mean that $(\pi(x), x \in S)$ is a collection of positive numbers satisfying

$$\pi(x)q(x) = \sum_{x' \in S} \pi(x')q(x',x) \quad x \in S . \tag{1}$$

Define $q'(x,x')$ by

$$\pi(x)\, q'(x,x') = \pi(x')\, q(x',x), \tag{2}$$

let

$$Q' = (q'(x,x'),\ x,x' \in S)$$

and let

$$q'(x) = \sum_{x' \in S} q'(x,x').$$

Observe that equation (1) implies

$$q(x) = q'(x). \tag{3}$$

If $(x(t), t \in \mathbb{R})$ is a non-explosive stationary Markov process with transition rates Q and stationary distribution π then the transition rates Q' are precisely those of the reversed process $(x(-t), t \in \mathbb{R})$.

Assume now that certain transitions are identified with arrivals or departures of customers (or items) of class $c \in C$, for C a finite or countable collection of possible customer classes. Specifically suppose that for each $(c,x) \in C \times S$ there are subsets $S^a(c,x)$, $S^d(c,x) \subset S$

satisfying

$$S^a(c_1,x) \cap S^a(c_2,x) = \phi \quad S^d(c_1,x) \cap S^d(c_2,x) = \phi \quad \forall c_1 \neq c_2 \tag{4}$$

$$\{(x,x'): x' \in S^a(c_1,x),\ x \in S^d(c_2,x')\} = \phi \quad \forall c_1,c_2 \tag{5}$$

with the interpretation that a transition from x to $x' \in S^a(c,x)$ signals the arrival of a customer of class c, and a transition from $x' \in S^d(c,x)$ to x signals the departure of a customer of class c. Condition (4) ensures that customers arrive singly and that they depart singly. This condition is essential in what follows. Condition (5) rules out the possibility that a single transition may signal both an arrival and a departure, and is made for convenience of exposition (cf. [8]). Since customer classes are essentially defined in terms of subsets of the state space S it will be natural in what follows to suppose that the symbol C fixes the family

$$((S^a(c,x), S^d(c,x)): (c,x) \in C \times S).$$

Call the node (Q,π,C) <u>quasi-reversible</u> if there exists a collection $(\alpha(c),\ \beta(c),\ c \in C)$ such that

$$\sum_{x' \in S^a(c,x)} q(x,x') = \alpha(c) \tag{6}$$

$$\sum_{x' \in S^d(c,x)} q'(x,x') = \beta(c) \tag{7}$$

for all $(c,x) \in C \times S$. If $(x(t),\ t \in \mathbb{R})$ is a non-explosive stationary Markov process with transition rates Q and equilibrium distribution π then quasi-reversibility reduces to the property that the state of the process at time $t, x(t)$, is independent of:

(i) the arrival times of class c customers, $c \in C$, subsequent to time t;

(ii) the departure times of class c customers, $c \in C$, prior to time t.

This property in turn implies that:

(i) arrival times of class c customers, for $c \in C$, form independent Poisson processes;

(ii) departure times of class c customers, for $c \in C$, form independent Poisson processes.

An especially simple example of a quasi-reversible node is the following system:

$$S = \{0,1,2\} \qquad C = \{1\}$$

$$q(x,x') = \alpha \qquad x' = x+1 \pmod 3$$

$$= \beta \qquad x' = x-1 \pmod 3$$

$$\pi(x) = 1 \qquad x \in S$$

$$S^a(1,x) = x+1 \pmod 3 \qquad S^d(1,x) = x-1 \pmod 3.$$

This node can be viewed as acting as a source of rate α and a sink of rate β.

We shall now discuss how a number of quasi-reversible nodes can be linked together to form a network. Let $((Q_j, \pi_j, C), j = 1,2,\ldots,J)$ be a finite collection of quasi-reversible nodes, and use the subscript j generally to identify entities associated with the jth node. Thus arrivals and departures are defined for node j in terms of the family

$$((S_j^a(c,x_j), S_j^d(c,x_j)): (c,x_j) \in C \times S_j)$$

and the collection of customer classes C is the same for each node. Let

$$(\xi,k): C \times \{1,2,\ldots,J\} \to C \times \{1,2,\ldots,J\}$$

be a bijection, with the interpretation that when a customer of class c departs from node j he transmutes into a customer of class $\xi(c,j)$ who then arrives at node $k(c,j)$. Assume that

$$\alpha_{k(c,j)}(\xi(c,j)) = \beta_j(c), \tag{8}$$

a requirement that will emerge as necessary to match departures of class c customers from node j with arrivals of class $\xi(c,j)$ customers at node $k(c,j)$.

Now define a collection of transition rates $Q = (q(x,x'),\ x,x' \in S)$ on the state space $S = S_1 \times S_2 \times \cdots \times S_J$ as follows. If

$$x = (x_1, x_2, \ldots, x_j', \ldots, x_k, \ldots, x_J) \tag{9}$$

and

$$x' = (x_1, x_2, \ldots, x_j, \ldots, x_k', \ldots, x_J) \tag{10}$$

where

$$x_j' \in S_j^d(c, x_j),\ k = k(c,j),\ x_k' \in S_k^a(\xi(c,j), x_k) \tag{11}$$

put

$$q(x,x') = q_j(x_j', x_j)\, \frac{q_k(x_k, x_k')}{\alpha_k(\xi(c,j))}\ ;$$

if

$$x = (x_1, x_2, \ldots, x_j, \ldots, x_J) \tag{12}$$

and

$$x' = (x_1, x_2, \ldots, x_j', \ldots, x_J) \tag{13}$$

where

$$x_j' \notin \bigcup_{c \in C} S_j^a(c, x_j),\ x_j \notin \bigcup_{c \in C} S_j^d(c, x_j') \tag{14}$$

put

$$q(x,x') = q_j(x_j, x_j');$$

otherwise put $q(x,x') = 0$. The transition rates Q are thus defined in the obvious way: a node behaves as it would in isolation except that arrivals are triggered exogenously, by departures from other nodes, rather than by an endogenous mechanism.

Theorem. The transition rates Q admit a positive invariant measure

$$\pi(x) = \prod_{j=1}^{J} \pi_j(x_j) \qquad x \in S$$

Proof. Define a collection of transition rates Q' in terms of $(Q'_j,\ j = 1,2,\ldots,J)$ as follows. If relations (9), (10) and (11) hold put

$$q'(x',x) = q'_k(x'_k,x_k)\,\frac{q'_j(x_j,x'_j)}{\beta_j(c)}\ ;$$

if relations (12), (13) and (14) hold put

$$q'(x',x) = q'_j(x'_j,x_j);$$

otherwise put $q'(x',x) = 0$. The definition (2) of Q'_j, $j = 1,2,\ldots,J$, and the equalities (8) imply that

$$\pi(x)q(x,x') = \pi(x')q'(x',x) \qquad x,x' \in S \tag{15}$$

Since

$$q(x) = \sum_{j=1}^{J} [q_j(x_j) - \sum_{c\in C} \alpha_j(c)]$$

and

$$q'(x) = \sum_{j=1}^{J} [q'_j(x_j) - \sum_{c\in C} \beta_j(c)]$$

it follows from equations (3) and (8) that

$$q(x) = q'(x) \qquad x \in S$$

This and equation (15) establish the result:

$$\sum_{x\in S} \pi(x)q(x,x') = \pi(x')q(x') \qquad x \in S \quad \square$$

The generality of our approach has resulted in some simplicity in the statement and proof of the Theorem, since we have been able to postpone asking whether Q is explosive, reducible or positive recurrent. In applications, however, we usually seek a unique stationary distribution rather than just a particular invariant measure. This final step must be justified by appeal to specific properties of the network under consideration, but the general line of argument usually proceeds as follows ([9], [10]). If $S^* \subset S$ is a closed communicating class then $(\pi(x), x \in S^*)$ is a positive invariant measure for the transition rates $Q^* = (q(x,x'), x,x' \in S^*)$. If the Markov process constructed from Q^* is non-explosive then it has a stationary distribution if and only if

$$B^{-1} \triangleq \sum_{x \in S^*} \pi(x) < \infty , \tag{16}$$

and when this condition is satisfied $(B\pi(x), x \in S^*)$ is the unique stationary distribution. Observe that condition (16) may be satisfied even if some or all of the measures $(\pi_j, j = 1,2,\dots,J)$ are not summable.

Sometimes interest is focused not directly on the Markov process constructed from Q or Q^*, but on chains embedded in this process - for example we may be interested in the state observed at times immediately following a particular sort of transition. To obtain results for such chains it is useful to consider the transition probabilities $P = (p((x,y), (y,z)), x,y,z \in S)$ where

$$p((x,y), (y,z)) = q(y,z)/q(y).$$

These are just the transition probabilities of the Markov chain $((x(n), x(n+1)), n = 0,1,\dots)$ formed by taking each successive pair of states of the jump chain $(x(n), n = 0,1,\dots)$. It is immediately verified that an invariant measure for P is $(\pi(x)q(x,y), (x,y) \in S^2)$, and from this invariant measures for chains embedded in the sequence $((x(n), x(n+1)), n = 0,1,\dots)$ can be readily deduced. If $(x(t), t \in \mathbb{R})$ is a non-explosive, irreducible stationary Markov process with transition rates Q^* and stationary distribution $(\pi(x), x \in S^*)$ then $(\pi(x)q(x,y), x,y \in S^*)$ is the unique stationary distribution for the chain $((x(n), x(n+1)), n = 0,1,\dots)$ with state space $S^* \times S^*$. In this case $\pi(x)q(x,y)$ has an interpretation as the probability flux from state x to state y [8].

Often the embedded chain of interest has itself an invariant measure of product form but the appropriate closed communicating class to which the measure should be restricted differs in some respect from the state space S* of the original process. For example, results contained in [4], [8], [14] and [21] are concerned with a chain embedded at certain arrival times in a closed network and the appropriate class is isomorphic to the state space of a closed network with one less customer. As another example [4] if a closed network with homogenous customers is observed at just the times when the number in a particular queue increases from $n-1$ to n and if the state of the particular queue is deleted from the observation then the appropriate class for the resulting chain is isomorphic to the state space of a closed network with n less customers and one less queue.

3. A Many Server Queue

The examples of quasi-reversible nodes presented by Baskett, Chandy, Muntz and Palacios [2] and Kelly [7] are widely known. Here and in Section 5 we describe two simple examples not covered in those papers.

The first example is a queue with s servers at which customers of a single class arrive in a Poisson stream of rate α. The servers may differ in efficiency: specifically, a customer's service time at server i is exponentially distributed with parameter μ_i, for $i = 1,2,\ldots,s$. Define the state of the queue to be the vector $x = (n, i_1,i_2,\ldots,i_{s-n})$, read as (n) when $n \geq s$, where n is the number of customers at the queue and $i_1,i_2,\ldots,i_{s-n}$ is a list of the free servers arranged in order according to the length of time they have been free. Suppose that if a customer arrives to find the queue in state $x = (n, i_1,i_2,\ldots,i_{s-n})$ with $n < s$ he is allocated to server i_r with probability $p(r,s-n)$,

$r = 1,2,\ldots,s-n$. For example if $p(1,m) = 1$ for $m = 1,2,\ldots,s$ then a customer is always allocated to the server who has been idle for the longest time. If the customer arrives to find $n \geq s$ he waits in line. It is elementary to check that an invariant measure for the resulting transition rates Q is

$$\pi(x) = \prod_{r=1}^{s-n} \frac{\mu_{i_r}}{\alpha} \qquad n < s$$

$$= \left(\frac{\alpha}{\sum_{i=1}^{s} \mu_i}\right)^{n-s} \qquad n \geq s$$

The transition rates Q' defined by equation (2) are easily calculated and can be regarded as describing a similar s-server queue with a slightly different method of handling idle servers. It then follows that, with the obvious transitions signalling arrivals and departures of customers of the single class, the queue is quasi-reversible, with $\alpha(1) = \beta(1) = \alpha$. If $\Sigma\mu_i < \alpha$ the invariant measure π can be normalized to give the unique stationary distribution. In equilibrium the service time of a customer is distributed as a mixture of exponential distributions. The convex combination defining the mixture depends on the arrival rate α as well as on $\mu_1,\mu_2,\ldots,\mu_n$, and the service times of successive customers are dependent.

Note that if customers leaving the queue who have been served by server i are assigned class i then the queue is not quasi-reversible. In constrast if customers of class c, $c \in C$, arrive in independent Poisson streams of rate $\alpha(c)$, where $\alpha = \Sigma\alpha(c)$, and if a customer's class neither changes nor affects his progress as he passes through the queue then the queue is quasi-reversible. To show this the state of the process must be expanded from x to $(x,\underline{c})$ where $\underline{c} = (c_1,c_2,\ldots,c_n)$ determines the class of each customer in each possible position in the system: an invariant measure is then

$$\pi(x,\underline{c}) = \pi(x) \prod_{r=1}^{n} \frac{\alpha(c_i)}{\alpha}$$

and the conditions for quasi-reversibility are readily verified with $\beta(c) = \alpha(c)$ for all $c \in C$.

The above discussion shows that the queue is quasi-reversible for all values of the arrival rates $\alpha(c)$, $c \in C$, satisfying $\Sigma\alpha(c) < \infty$. Chandy, Howard and Towsley [5] have observed that symmetric queues also have this property. The next Section derives the property in a more general setting.

4. Varied Arrival Rates

Consider a quasi-reversible node (Q,π,C) with arrival and departure rates $\alpha(c)$, $\beta(c)$, $c \in C$. Define a new collection of transition rates $Q^+ = (q^+(x,x'),\ x,x' \in S)$ by

$$q^+(x,x') = \frac{\alpha^+(c)}{\alpha(c)}\, q(x,x') \text{ if } x' \in S^a(c,x)$$

$$= q(x,x') \quad \text{otherwise}$$

where $\alpha^+(c) = 0$ if and only if $\alpha(c) = 0$, and $\Sigma\alpha^+(c) < \infty$. The interpretation here is that the arrival rate of class c customers has been altered from $\alpha(c)$ to $\alpha^+(c)$, for $c \in C$. The next result gives sufficient conditions for the altered node to be quasi-reversible.

<u>Proposition</u>. If

(a) there exists a function $n\colon C \times S \to \mathbb{Z}$ such that

$$x' \in S^a(c,x) \cup S^d(c,x) \iff n(c,x') = n(c,x)+1$$

$$x' \notin S^a(c,x) \cup S^d(c,x) \iff n(c,x') = n(c,x)$$

(b) $\alpha(c) = \beta(c) \quad c \in C$

then the node (Q^+,π^+,C) is quasi-reversible, where

$$\pi^+(x) = \pi(x) \prod_{c\in C} \left[\frac{\alpha^+(c)}{\alpha(c)}\right]^{n(c,x)} .$$

Remark. The integer n(c,x) can be regarded as the number of class c customers in the node when its state is x. Conditions (a) and (b) can then be interpreted as a requirement that the node be customer conserving for each class $c \in C$. Condition (b) can be deduced from condition (a) when the Markov process constructed from Q is positive recurrent. All the nodes considered in [2] and [7] and the networks formed from these nodes can be formulated so that they satisfy conditions (a) and (b).

Proof. Let $\underline{n} = (n(c),\ c \in C) \in \mathbb{Z}^C$. Consider a collection of transition rates Q_Ψ defined on the state space $\{\underline{n}: \Sigma n(c) < \infty\}$ by

$$q(\underline{n},\ T_c\underline{n}) = \alpha(c)\ \frac{\Psi(T_c\underline{n})}{\Psi(\underline{n})}$$

$$q(T_c\underline{n},\underline{n}) = \alpha(c)$$

where

$$T_{c'}\underline{n} = (n(c) + I[c=c'],\ c \in C)$$

with all other transition rates zero. An invariant measure for Q_Ψ is clearly Ψ. If we identify a transition from $\underline{n}$ to $T_c\underline{n}$ as a <u>departure</u> of a customer of class c and a transition from $T_c\underline{n}$ to $\underline{n}$ as an <u>arrival</u> of a customer of class c then the node (Q_Ψ,Ψ,C) is quasi-reversible, with $\beta(c) = \alpha(c)$ for all $c \in C$.

Now form a network from the nodes (Q,π,C) and (Q_Ψ,Ψ,C) by having a departure of a class c customer from one node trigger the arrival of a class c customer at the other node, for each $c \in C$. Observe that those network states in which the state of the second node $(n(c),\ c \in C)$ corresponds precisely to the list $(n(c,x),\ c \in C)$ derived from the state x of the first node form a closed class and so an invariant measure over this class is

$$\Psi(n(c,x),\ c \in C)\pi(x)$$

The choice

$$\Psi(\underline{n}) = \prod_{c\varepsilon C} \left(\frac{\alpha^+(c)}{\alpha(c)}\right)^{n(c)} \tag{17}$$

and the bijection

$$((n(c,x),\ c \ \varepsilon\ C),\ x) \leftrightarrow x$$

establish that π^+ is a positive invariant measure for Q^+. Substitution into equations (6) and (7) then shows that (Q^+,π^+,C) is quasi-reversible, with arrival and departure rates $\alpha^+(c)$ for customers of class c, c ε C. □

Choices more general than expression (17) can be made for the function Ψ, and some of these produce nodes quasi-reversible with respect to customer classifications less fine than C ([8], see also [22]). Indeed any quasi-reversible node with state space $\{\underline{n},\ \Sigma n(c) < \infty\}$ and arrival and departures rates $\alpha(c)$, c ε C, can be joined with the node (Q,π,C) to form a network, and the outcome viewed as a variation of the arrival rates at the node (Q,π,C). The result of Lam [13] can be interpreted in this way.

Examples of quasi-reversible nodes which do not satisfy condition (a) can be constructed from the reversible migration processes introduced by Kingman [10] (see [8]). While these nodes do not conserve customers of each class c, c ε C, they can be regarded as conserving customers unidentified by class. It is possible to show that if the arrival rates $\alpha(c)$, c ε C, at such a node are all multiplied by the same factor the resulting node is quasi-reversible. In the next Section we shall discuss a node which does not conserve even unclassified customers, but first we shall describe an example at the other extreme where the arrival rates can depend on more than the information contained in $(n(c,x),\ c\ \varepsilon\ C)$.

Let x be the state of a series of first come first served M/M/1 queues at which arrivals of customers of class c form a Poisson process of unit rate for $c \in C$, where C is a finite set. Suppose that a customer's class neither changes nor affects his progress as he moves through the series of queues. The resulting node (Q,π,C) is quasi-reversible. Let $\underline{c}(x) = (c_1,c_2,\ldots,c_n)$ be the classes of the n customers in the series of queues arranged in order of their arrival at the first queue in the series so that, for example, c_1 is the class of the customer who has been in the node the least time. Observe that if at some point in time $\underline{c}(x)$ is given, its future evolution can be tracked by a simple updating procedure applied whenever an arrival at or departure from the node occurs. By joining the node (Q,π,C) to another quasi-reversible node it is possible to show that if the arrival rate of class c customers is altered to

$$\frac{\Psi(c,\underline{c}(x))}{\Psi(\underline{c}(x))}$$

when the state of the node is x then an invariant measure for the resulting system is

$$\Psi(\underline{c}(x))\ \pi(x) \qquad x \in S$$

provided

$$\sum_{c\in C} \Psi(\underline{c},c) = \sum_{c\in C} \Psi(c,\underline{c}) \ .$$

For example if

$$\Psi(\underline{c}) = p^{M(\underline{c})}$$

where

$$M(\underline{c}) = \#\{i:\ c_i = c_{i+1},\ 1 \le i \le n-1\}$$

then an arriving customer of class c is lost with probability p when $c = c_1$.

5. A Clustering Node

We shall now discuss in detail a quasi-reversible node at which the arrival rates $\alpha(c)$, $c \in C$, cannot be varied independently without loosing quasi-reversibility. Let $S = \mathbb{N}^2$ and let the non-zero transition rates from the collection Q be given by

$$q((n_1,n_2),\ (n_1+1,n_2)) = \alpha(1)$$
$$q((n_1,n_2),\ (n_1,n_2+1)) = \alpha(2)$$
$$q((n_1,n_2),\ (n_1-1,n_2)) = \mu_1 n_1$$
$$q((n_1,n_2),\ (n_1,n_2-1)) = \mu_2 n_2$$
$$q((n_1,n_2),\ (n_1-2,n_2+1)) = \gamma_{12} n_1 (n_1-1)$$
$$q((n_1,n_2),\ (n_1+2,n_2-1)) = \gamma_{21} n_2$$

Provided that

$$\delta_1^2\ \gamma_{12} = \delta_2\ \gamma_{21}$$

where (18)

$$\alpha(1) = \delta_1 \mu_1 \qquad \alpha(2) = \delta_2 \mu_2$$

the rates Q admit an invariant measure

$$\pi(n_1,n_2) = \frac{\delta_1^{\ n_1}}{n_1!}\ \frac{\delta_2^{\ n_2}}{n_2!}$$

With $C = \{1,2\}$ and

$$s^a(1,\ (n_1,n_2)) = s^d(1,\ (n_1,n_2)) = (n_1+1,n_2)$$

$$s^a(2,\ (n_1,n_2)) = s^d(2,\ (n_1,n_2)) = (n_1,n_2+1)$$

the node (Q,π,C) is quasi-reversible. It is not difficult to show that if the arrival rates $\alpha(1),\alpha(2)$ are altered to $\eta_1\ \alpha(1)$, $\eta_2\ \alpha(2)$ respectively the resulting system is quasi-reversible if and only if $\eta_2 = \eta_1^2$. Condition (a) of the preceding Section does not hold: all that is conserved is the sum $n_1 + 2n_2$.

The presence in a network of a quasi-reversible node without property (a) can give rise to interesting phenomena not observed in the networks considered in [2] and [7]. For example, suppose we form a network from the above node labelled 1, and two single-class quasi-reversible nodes, labelled 2 and 3, satisfying conditions (a) and (b). Link the nodes as indicated in Figure 1: the important point to notice here is that an item leaving node 1 will eventually return as an item of a different class.

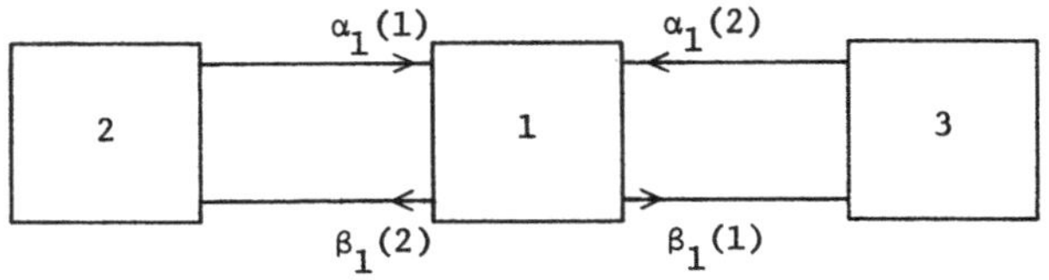

Figure 1. Network With Clustering Node

In addition to equations (8) we must now satisfy the non-linear constraint (18): a solution exists, given by

$$\alpha_1(1) = \alpha_1(2) = \beta_1(1) = \beta_1(2) = \frac{\mu_1^2\gamma_{21}}{\mu_2\gamma_{12}} .$$

The resulting network has, then, an invariant measure given by the fundamental theorem as a product of the invariant measures for each node. The invariant measure for node 1 will be summable if $\delta_1 < 1$ and $\delta_2 < 1$, that is if

$$\mu_1\gamma_{21} < \mu_2\gamma_{12} \quad \text{and} \quad \mu_1^2\gamma_{21} < \mu_2^2\gamma_{12} .$$

If in addition the invariant measures for nodes 2 and 3 are summable, then the product form is proportional to a stationary distribution for the Markov process, and in equilibrium the states of the three nodes are independent. Independence may thus be obtained in an irreducible network with no identifiable exogenous arrival streams. If the process is observed at, say, those instants when two class 1 items in node 1 are uniting to form a class 2 item and if the units so involved are left out of the description of the system, then the resulting Markov chain has a stationary distribution identical to that of the process.

6. An Insensitive Clustering Network

We begin this Section with another example of a clustering node. Let D be a countable set of unit types, and let

$$S_1 = \{\underline{n} = (n(d),\ d \in D) : \sum_{d\in D} n(d) < \infty\} .$$

Define the operator $T_{d_1d_2d_3} : S_1 \to S_1$ by

$$T_{d_1d_2d_3}(n(d),\ d \in D) = (n(d) - I[d \in \{d_1,d_2,d_3\}], d \in D)$$

and let the non-zero transition rates of the collection Q_1 be given by

$$q_1(\underline{n},\ T_{d_1d_2d_3}\underline{n}) = \alpha_1(d_1,d_2,d_3) \prod_{i=1}^{3} \left[\frac{\lambda(d_i)}{\rho(d_i)}\right] n(d_i) \tag{19}$$

$$q_1(T_{d_1d_2d_3}\underline{n},\ \underline{n}) = \alpha_1(d_1,d_2,d_3) . \tag{20}$$

An invariant measure for Q_1 is

$$\pi_1(\underline{n}) = \prod_{d\in D} \left[\frac{\rho(d)}{\lambda(d)}\right]^{n(d)} \frac{1}{n(d)!}$$

Set $C = D^3$ and identify transitions (19) and (20) as signalling respectively the departure or arrival of an item of class (d_1,d_2,d_3). The system (Q_1,π_1,C) is then quasi-reversible with $\alpha_1(d_1,d_2,d_3) = \beta_1(d_1,d_2,d_3)$.

Consider now a second node which operates as follows. Items of class $(d_1,d_2,d_3) \in C$ arrive in a Poisson stream of rate $\alpha_2(d_1,d_2,d_3)$. They pass independently through the node, an item labelled (d_1,d_2,d_3) on arrival taking a random period of time whose distribution is determined by (d_1,d_2,d_3) and which can be represented by a passage time in a countable state space Markov process. Let the mean of this random period be $m(d_1,d_2,d_3)$. Upon arrival at the node an item of class (d_1,d_2,d_3) is allocated a second label (d_1',d_2',d_3') with probability $P(d_1,d_1')P(d_2,d_2')P(d_3,d_3')$, where $P: D^2 \to [0,1]$ is a transition probability matrix, and on departure it leaves as an item of class (d_1',d_2',d_3'). Without difficulty (although not without tedium) it is possible to define formally the node (Q_2,π_2,C) corresponding to this description and to show that it is quasi-reversible with arrival and departure rates $\alpha_2(d_1,d_2,d_3)$, $\beta_2(d_1,d_2,d_3)$ where

$$\beta_2(d_1,d_2,d_3) = \sum_{d_1'} \sum_{d_2'} \sum_{d_3'} \alpha_2(d_1',d_2',d_3')P(d_1',d_1)P(d_2',d_2)P(d_3',d_3). \quad (21)$$

We now intend to link the nodes (Q_1,π_1,C) and (Q_2,π_2,C) together. To satisfy condition (8) we must ensure that

$$\alpha_1(d_1,d_2,d_3) = \beta_1(d_1,d_2,d_3) = \alpha_2(d_1,d_2,d_3) = \beta_2(d_1,d_2,d_3) \quad (22)$$

Suppose that the transition matrix P admits a positive invariant measure $(\rho(d),\ d \in D)$, and use the symbol $\leftrightsquigarrow$ to indicate the communication relation induced by P. From the equality (21) it follows that condition (22) is met when

$$\alpha_1(d_1,d_2,d_3) = \alpha_2(d_1,d_2,d_3) = \rho(d_1)\rho(d_2)\rho(d_3)f(d_1,d_2,d_3)$$

for any function $f: D^3 \to [0,\infty)$ satisfying

$$d_1 \leftrightsquigarrow d_1',\ d_2 \leftrightsquigarrow d_2',\ d_3 \leftrightsquigarrow d_3' \Longrightarrow f(d_1,d_2,d_3) = f(d_1',d_2',d_3') .$$

The resulting network will then have invariant measure given by the fundamental theorem.

As an application of the above discussion, consider a communication network as illustrated in Figure 2. A collection of centres are connected by channels. Two centres may be in communication via a joining channel, in which case the triple so formed must be disjoint

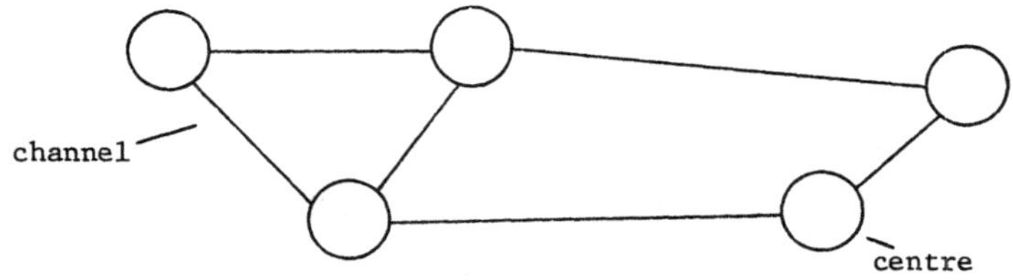

Figure 2. Communication Network

from all other such triples. If a centre or channel is not a member of a triple call it idle. To relate this communication network to the preceding discussion identify centres and channels as the basic units, so that each centre or channel is labelled with an element $d \in D$. The idle units then correspond to the occupants of node 1, and a linked triple corresponds to an item in node 2. Each time a centre or channel takes part in linked triple its state d changes in accordance with the transition probabilities P, but independently of the states of the units to which it is linked. To model the fact that centres and channels retain their geographical identify assume the transition matrix P has a number of closed communicating classes, one for each centre and one for each channel. To indicate which triples are geographically feasible let $f(d_1,d_2,d_3) = 1$ when the triple (d_1,d_2,d_3) identifies two centres and a channel which physically joins them, and let $f(d_1,d_2,d_3) = 0$ otherwise.

The requirements that there be exactly one unit associated with each of the closed communicating classes determined by P and that linked triples be geographically feasible identifies a closed communicating class S* for the overall network Q. If $\Sigma\rho(d) < \infty$ then the invariant measure for Q will be summable over S*, and so its restriction to S* will be the unique stationary distribution. Various consequences follow from the form of this distribution. For example, the equilibrium probability that $((d_1^i,d_2^i,d_3^i),\ i = 1,2,\ldots,I)$ gives the list of linked triples (with their current states) and that $(d_1^o, d_2^o, \ldots, d_{N-3I}^o)$ gives the list of idle units (with their current states) is

$$B \left(\prod_{j=1}^{N-3I} \frac{\rho(d_j^o)}{\lambda(d_j^o)} \right) \prod_{i=1}^{I} \rho(d_1^i)\rho(d_2^i)\rho(d_3^i)m(d_1^i, d_2^i, d_3^i) \tag{23}$$

when B is a normalizing constant, obtainable by summation. Observe the influence on this probability of $\lambda(d)$, the propensity of a node in state d to link, and $m(d_1, d_2, d_3)$, the mean link time of a triple (d_1, d_2, d_3). Various possibilities for the link time distributions are available. For example suppose that for each $d \in D$ we have a distribution F_d. If X_d is a random variable with distribution F_d the link time of the triple (d_1, d_2, d_3) could be distributed as, say,

$$\min(X_{d_1}, X_{d_2}, X_{d_3}) \tag{24}$$

or

$$X_{d_1} + X_{d_2} + X_{d_3} . \tag{25}$$

The technical restriction to passage times prevents the choice $X_{d_1} X_{d_2} X_{d_3}$, but observe that when we can write $m(d_1, d_2, d_3) = m(d_1)m(d_2)m(d_3)$ the product form (23) separates further.

Now focus attention on a single unit. The sequence of states taken by the unit is easily described, forming a Markov chain with transition matrix P. However the sequence of link times associated with the unit has a much more complicated structure depending not only on the unit's own state but also on the states of the units to which it happens to be linked. The resulting pattern of dependence in the sequences of link times associated with the various units is markedly more complex than occurs, for example, in the dependent sequences of service requirements associated with the various customers in a closed network of symmetric queues.

We shall now attempt to formulate the model of this Section within the framework provided by Matthes, Koenig, Nawrotski, Jansen and Schassberger ([11], [12], [15], [18], [19]). Using the terminology of [19], units can be identified as the elements of a generalized semi-Markov scheme provided link times of distinct triples have a common distribution, and the partial balance conditions are then found to be satisfied. However the scheme is not disconnected, since more than one element can be activated at the same time. If link times are generated from the forms (24) or (25) units can again be identified as elements, but the resulting formulation violates conditions imposed by the framework of a generalized semi-Markov scheme. Of course the correct formulation arises when we identify the set of elements with the set of possible linked triples - the items of the network formulation.

Various generalizations of the model of this Section can be carried through without disturbing its tractability: for example units can link to form larger clusters, and the linear factor n(d) in transition rate (19) can be generalized to reflect, perhaps, duplicate channels responding passively to link demands from centres. The guiding principle in the exploration of such generalizations is that they must leave node 1 quasi-reversible.

7. References

[1] Barbour, A. D. (1976) Networks of queues and the method of stages. Adv. in Appl. Probab. 8, 584-591.

[2] Baskett, F., Chandy, K. M., Muntz, R. R., and Palacios, F. G. (1975) Open, closed and mixed networks of queues with different classes of customers. J. Assoc. Comput. Mach., 22, 248-260.

[3] Beutler, F. J., Melamed, B., and Zeigler, B. P. (1977) Equilibrium properties of arbitrarily interconnected queueing networks. In P. R. Krishnaiah (Ed.), Multivariate Analysis IV, North-Holland, Amsterdam. pp. 351-370.

[4] Chandy, K. M., Herzog, U. and Woo, L. (1975) Parametric analysis of queueing networks. IBM J. Res. Develop. 19, 36-42.

[5] Chandy, K. M., Howard, J. H., and Towsley, D. F. (1977) Product form and local balance in queueing networks. J. Assoc. Comput. Mach., 24, 250-263.

[6] Kelly, F. P. (1975) Networks of queues with customers of different types. J. Appl. Probab., 12, 542-554.

[7] Kelly, F. P. (1976) Networks of queues. Adv. in Appl. Probab., 8, 416-432.

[8] Kelly, F. P. (1979) Reversibility and Stochastic Networks, Wiley, New York.

[9] Kendall, D. G. and Reuter, G. E. H. (1957) The calculation of the ergodic projection for Markov chains and processes with a countable infinity of states. Acta Math., 97, 103-143.

[10] Kingman, J. F. C. (1969) Markov population processes. J. Appl. Probab., 6, 1-18.

[11] König, D., and Jansen, U. (1974) Stochastic processes and properties of invariance for queueing systems with speeds and temporary interruptions. Transactions of the Seventh Prague Conference on Information Theory, Statistical Decision Functions, and Random Processes, and of the 1974 European Meeting of Statisticians, Czechoslovak Academy of Sciences. pp. 335-343.

[12] König, D., Matthes, K., and Nawrotzki, K. (1967) Verallgemeinerungen der Erlangschen und Engsetschen Formeln (Eine Methode in der Bedienungstheorie), Akademie-Verlag, Berlin.

[13] Lam, S. S. (1977) Queueing Networks with population size constraints. IBM J. Res. Develop., 21, 370-378.

[14] Lavenberg, S. S. and Reiser, M. (1980) Stationary state probabilities at arrival instants for closed queueing networks with multiple types of customers. J. Appl. Probab., 17, 1048-1061.

[15] Matthes, K. (1962) Zur Theorie der Bedienungsprozesse. Transactions of the Third Prague Conference on Information Theory, Statistical Decision Functions, and Random Processes, Czechoslovak Academy of Sciences. pp. 513-528.

[16] Melamed, B. (1979) On Poisson traffic processes in discrete-state Markovian systems with application to queueing theory. Adv. in Appl. Probab., 11, 218-239.

[17] Muntz, R. R. (1972) Poisson Departure Processes and Queueing Networks, IBM Research Report RC4145. A shortened version appeared in Proceedings of the Seventh Annual Conference on Information Science and Systems, Princeton, 1973, pp. 435-440.

[18] Schassberger, R. (1977) Insensitivity of steady-state distributions of generalized semi-Markov processes, Part I. Ann. Probab., 5, 87-99.

[19] Schassberger, R. (1978) Insensitivity of steady-state distributions of generalized semi-Markov processes, Part II. Ann. Probab., 6, 85-93.

[20] Schassberger, R. (1978) The insensitivity of stationary probabilities in networks of queues. Adv. in Appl. Probab., 10, 906-912.

[21] Sevcik, K. C. and Mitrani, I. (1981) The distribution of queueing network states at input and output instants. J. Assoc. Comput. Mach., 28, 358-371.

[22] Towsley, D. (1980) Queueing network models with state-dependent routing. J. Assoc. Comput. Mach., 27, 323-337.

[23] Walrand, J. and Varaiya, P. (1980) Interconnections of Markov chains and quasi-reversible queueing networks. Stochastic Process. Appl., 10, 209-219.

[24] Whittle, P. (1967) Nonlinear migration processes. Bull. Inst. Internat. Statist., 42, 642-647.

[25] Whittle, P. (1972) Statistics and critical points of polymerization processes. Supplement Adv. in Appl. Probab., 199-220.

Statistical Laboratory, University of Cambridge, 16 Mill Lane, Cambridge CB2 1SB, ENGLAND

Discussant's Report on
"Networks of Quasi-reversible Nodes,"
by F. P. Kelly

This elegant presentation contains some important results. The proof of the product form theorem given in Section 2 is a nice illustration of the technique which consists in guessing Q' to verify some invariant measure. The idea of considering invariant measures instead of invariant probability measures pays off in Section 4.

It is probably useful to complement the algebraic aspects of the theory emphasized in that presentation with some comments on the probabilistic interpretation of the concepts and results.

Notice that, in the stationary case, equations (6) [resp. (7)] say that the rates of the arrival processes [resp. the reversed departure processes] at time t are independent of x_t. Hence the equivalence with the conclusions (i), (ii).

The product form theorem relates independence properties: quasi-reversibility and product form. To explain why the calculations of Section 2 of [23] go through, I would like to sketch a probabilistic argument which isolates the role of quasi-reversibility and hopefully contributes to the intuitive understanding of those results.

Consider J nodes which are quasi-reversible under an invariant distribution π corresponding to a Poisson arrival process A_t of rate λ (all the counting processes are vector valued and indexed by $C\times\{1,\ldots,J\}$, where C is the set of classes). (See figure, with $d = 0$ for the time being.) The exogeneous arrival process E_t is Poisson with rate γ; after leaving the nodes the customers are possibly fed back by an independent routing. Assume that λ is a possible vector of average rates for the flows through the nodes. The claim is that π is invariant for the states of the nodes in that network.

This is the argument. For $d > 0$, introduce a pure delay d in the links between nodes (see figure). Denote by x_t the resulting state process for the nodes, by A_t [resp. D_t] the total number of arrivals [resp. departures] at the nodes. Let F_t be the output of the delay box. Assume that x_0 has distribution π and that $F_{(0,d]} = \{F_t,\ 0 < t \leq d\}$ (the contents of the delay box at $t = 0$) is Poisson with rate $\lambda - \gamma$ and independent of x_0 and of $(E_t,\ t \geq 0)$. Then $A_{(0,d]} = F_{(0,d]} + E_{(0,d]}$ is Poisson with rate λ and independent of x_0. Thus (x_t, A_t, D_t) will behave for t in (0,d] as if the nodes were in isolation. By quasi-reversibility, it follows that $D_{(0,d]}$ is Poisson and independent of x_d. By independence of the routing, the same is true of $F_{(d,2d]}$. Also, x_t has distribution π for t in (0,d]. By induction, this proves that x_t has distribution π for all t. By letting d go to zero, one can then show that π must be an invariant distribution for the original network. (This is easy if the original network is a regular Markov chain.)

Notice also that the argument for $d > 0$ shows that the exit process $B_{(0,t]}$ is Poisson and independent of x_t and of $F_{(t,t+d]}$. This leads to the quasi-reversibility of the network. It is also clear that A_t is not Poisson in general: $A_{(t,t+d]}$ and $A_{(t+d,t+2d]}$ are generally not independent. The same argument shows that an invariant distribution for the open network remains invariant for the associated closed network.

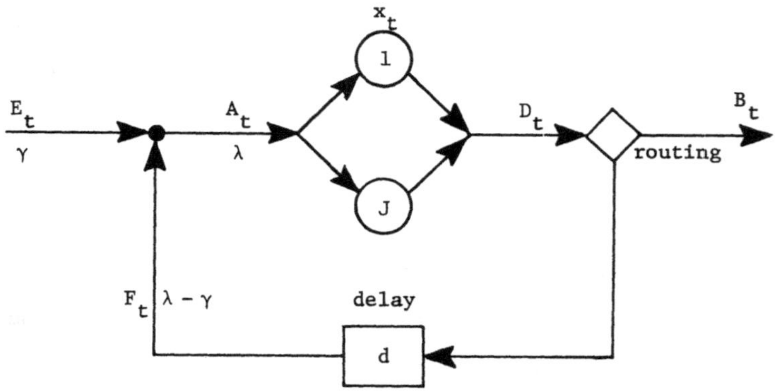

Discussant: Dr. J. Walrand, Cornell University, 1434 Hanshaw Road, Ithaca, New York 14850.

THE c - SERVER QUEUE WITH CONSTANT SERVICE TIMES AND A VERSATILE MARKOVIAN ARRIVAL PROCESS

Marcel F. Neuts

Abstract

Classical results of C. D. Crommelin on the c-server queue with constant service times and Poisson input are extended to the case of the versatile Markovian arrival process, introduced by the author.

The purely probabilistic analysis of a related problem in Markov chains leads to an algorithm for the evaluation of the stationary distributions of the queue length and waiting time at an arbitrary epoch.

As an illustration, the algorithmic steps are discussed in detail for the case where the arrivals form a Markov-modulated Poisson process.

Key Words

Theory of queues, computational probability, multi-server queues, constant service times, nonlinear matrix equations.

1. Introduction

Among the few tractable exact results for multi-server queues with non-exponential holding times, the derivations of the stationary distributions of the queue length and the virtual waiting time for the stable M/D/c queue, due to C. D. Crommelin [3,4], stand out by their clarity and simplicity.

As a tribute to this worker of the dawn of queueing theory, it is fitting to note that his papers reflect a genuine concern for the algorithmic implementation of results. At the time of their publication, the small numerical examples included in the papers must surely have required much tedious effort.

With the availability of the computer, much more extensive work on the M/D/c queue was carried out and presented in tabular form, among others by P. Kühn [8]. Direct numerical implementation of the classical analytic methods, based on transforms, presents serious difficulties for all but small values of c, the number of servers. It is also feasible to proceed by truncation of infinite stochastic matrices, as was apparently done in [8], but that approach does not fully utilize the rich structure of the Markov chain which appears in the M/D/c model. A method which does utilize such structure is proposed in Neuts [15]. For computations limited to the M/D/c case, the results in [15] are sufficient.

The present paper has a twofold purpose. Firstly, we show that the approach of Crommelin is valid for a wide class of Markovian arrival processes, which includes the Poisson process as its simplest, most tractable case. This class further includes particular cases such as Erlang (or phase) arrivals, the Markov-modulated Poisson process and many others, as discussed in [16]. Secondly, we shall give a detailed description of the specific algorithmic steps involved in implementing our results. For the sake of clarity, this description will be given for the Markov-modulated Poisson arrival process. For larger values of c, a substantial amount of computation is required, but by careful planning it may be carried out in a highly efficacious manner.

Both the description of the arrival process and the analysis of the block-partitioned Markov chain, defined in the sequel, are lengthy and require much notation. In order to keep the length of this paper within reasonable bounds, we shall not repeat theoretical arguments which we have presented elsewhere, but specific references to those will be given. For the notations and definitions of the arrival process, we refer to [16]. All notations introduced there will be used with the same significance here.

Crommelin's argument for the distribution of the queue length of the M/D/c queue goes as follows. Let c, λ, and a denote the number of servers, the Poisson arrival rate and the constant duration of the service times. The probability that there are $i \geq 0$, customers in the queue at time t is denoted by $\tilde{x}_i(t)$. The vector $\underline{\tilde{x}}(t)$ has the components $\tilde{x}_i(t)$, $i \geq 0$. Let $\tau \geq 0$, be a fixed time. Any customers <u>in service</u> at time τ and <u>only those</u> leave the system during the time interval $(\tau,\tau+a]$.

By an elementary counting argument, this leads to the equation

$$\underline{\tilde{x}}(\tau+a) = \underline{\tilde{x}}(\tau)\, \tilde{P}_1, \tag{1}$$

where $\tilde{P}_1$ is the stochastic matrix

$$\tilde{P}_1 = \begin{array}{c|cccccc|} 0 & a_0 & a_1 & a_2 & a_3 & a_4 & \cdots \\ 1 & a_0 & a_1 & a_2 & a_3 & a_4 & \cdots \\ 2 & a_0 & a_1 & a_2 & a_3 & a_4 & \cdots \\ \vdots & \vdots & \vdots & \vdots & \vdots & \vdots & \\ c-1 & a_0 & a_1 & a_2 & a_3 & a_4 & \cdots \\ c & a_0 & a_1 & a_2 & a_3 & a_4 & \cdots \\ c+1 & 0 & a_0 & a_1 & a_2 & a_3 & \cdots \\ c+2 & 0 & 0 & a_0 & a_1 & a_2 & \cdots \\ c+3 & 0 & 0 & 0 & a_0 & a_1 & \cdots \\ c+4 & 0 & 0 & 0 & 0 & a_0 & \cdots \\ \vdots & \vdots & \vdots & \vdots & \vdots & \vdots & \end{array} . \tag{2}$$

The quantities a_ν, $\nu \geq 0$, are the Poisson probabilities

$$a_\nu = e^{-\lambda a} \frac{(\lambda a)^\nu}{\nu!}, \quad \text{for } \nu \geq 0.$$

From (1), it follows that $\tilde{\underline{x}}(\tau+ka) = \tilde{\underline{x}}(\tau)\, \tilde{P}_1^k$, for $k \geq 0$. By letting k tend to infinity, we see that the stationary queue length density at an arbitrary time - if it exists - is given by the invariant probability vector $\underline{x}$ of the irreducible stochastic matrix $\tilde{P}_1$. It is well-known that $\tilde{P}_1$ has a unique, positive, invariant probability vector if and only if $\lambda a < c$. This is, of course, a particular case of the equilibrium condition for the GI/G/c queue.

The numerical solution of the system $\underline{x}\, \tilde{P}_1 = \underline{x}$, $\underline{x}\, \underline{e} = 1$, is discussed in [15] and references to earlier discussions and other applications are also given there.

2. The c-Server Queue with Constant Service Times and a Versatile Markovian Arrival Process

We now consider a service unit with c parallel servers with constant service times of length a and with arrivals according to the versatile Markovian point process (VMPP), which we introduced in [16].[1]

The matrices $P(\nu,t)$, $\nu \geq 0$, $t \geq 0$, introduced in [16], play an essential role here. For any specific description of the VMPP, they may be computed by the numerical integration of the differential equations (5) of [16]. In the simple case of Poisson arrivals, the $P(\nu,t)$ are scalars, given by the Poisson probabilities with parameter λt.

The states j, $1 \leq j \leq m$, of the irreducible Markov process with generator Q^*, defined in (3) of [16], will be called the <u>arrival phases</u>.

We shall first study the joint stationary density $\{x(i,j),\ i \geq 0,\ 1 \leq j \leq m\}$ of the queue length and the arrival phase at an arbitrary time. Let $\underline{x}$ denote the infinite vector with its components $x(i,j)$ written in lexicographic order.

We consider the stochastic matrix $\tilde{P}_i$, defined by

$$\tilde{P}_2 = \begin{vmatrix} P(0,a) & P(1,a) & P(2,a) & P(3,a) & \cdots \\ P(0,a) & P(1,a) & P(2,a) & P(3,a) & \cdots \\ \vdots & \vdots & \vdots & \vdots & \\ P(0,a) & P(1,a) & P(2,a) & P(3,a) & \cdots \\ P(0,a) & P(1,a) & P(2,a) & P(3,a) & \cdots \\ 0 & P(0,a) & P(1,a) & P(2,a) & \cdots \\ 0 & 0 & P(0,a) & P(1,a) & \cdots \\ 0 & 0 & 0 & P(0,a) & \cdots \\ \vdots & \vdots & \vdots & \vdots & \end{vmatrix} . \qquad (3)$$

The first $c + 1$ rows of $m \times m$ blocks $P(0,a), P(1,a), \ldots$ in $\tilde{P}_2$ are identical. The stair case pattern, obtained by shifting the rows one block to the right starts from the $(c + 1)$st row of blocks.

In order to cast the matrix $\tilde{P}_2$ into a form for which any theoretical results are known, we further partition the matrix into square blocks of order cm in the natural manner. We so obtain

$$\tilde{P}_2 = \begin{array}{c|ccccc|} \underline{0} & B_0 & B_1 & B_2 & B_3 & B_4 \cdots \\ \underline{1} & A_0 & A_1 & A_2 & A_3 & A_4 \cdots \\ \underline{2} & 0 & A_0 & A_1 & A_2 & A_3 \cdots \\ \underline{3} & 0 & 0 & A_0 & A_1 & A_2 \cdots \\ \underline{4} & 0 & 0 & 0 & A_0 & A_1 \cdots \\ \underline{5} & 0 & 0 & 0 & 0 & A_0 \cdots \\ & \vdots & \vdots & \vdots & \vdots & \vdots \end{array} . \qquad (4)$$

In the form (4), $\tilde{P}_2$ is readily seen to be a block-partitioned stochastic matrix of the M/G/1 type. General results on and other examples of Markov chains of the M/G/1 type were discussed in [9-13, 15, 19]. In Ramaswami [19], the VMPP was used as the input to a single server unit with general service times. A number of additional properties of the VMPP were established there. In particular and for use in the sequel, we note that in all non-trivial cases of the VMPP the matrices $P(\nu,a)$

are positive for $\nu \geq 1$ and $a > 0$. The matrix $P(0,a)$ has no zero columns. This readily implies that the matrices A_ν, $\nu \geq 1$, are positive and that A_0 does not have zero columns. It is further clear that $\tilde{P}_2$ is an irreducible stochastic matrix.

The set of states $\underline{k} = \{(kc+k',j),\ 0 \leq k' < c,\ 1 \leq j \leq m\}$ is called the level k. As discussed in the cited references, the analysis of Markov chains of M/G/1 type proceeds by considering the distributions of the first passage times between various pairs of levels of interest.

For use in the sequel, we now establish some elementary properties of the coefficient matrices A_ν, $\nu \geq 0$.

Lemma 1. The matrix $A = \sum_{\nu=0}^{\infty} A_\nu$, is a positive, stochastic, block-circulant matrix of order cm. Its invariant probability vector $\underline{\pi}^*$ is given by

$$\underline{\pi}^* = c^{-1}[\underline{\pi},\underline{\pi},\ldots,\underline{\pi}], \tag{5}$$

where $\underline{\pi}$ is the stationary probability vector of the generator Q^* - see Formula (3) of [16].

The column vector $\underline{\beta}^* = \sum_{\nu=1}^{\infty} \nu A_\nu \underline{e}$, is finite and the inner product $\rho = \underline{\pi}^* \underline{\beta}^*$, is given by

$$\rho = c^{-1} \mu^* a, \tag{6}$$

where $\mu^* = \underline{\pi} R'(1) \underline{e}$, is the stationary arrival rate of the VMPP as defined following Formula [12] of [16].

Proof. The structure of A is clear from its definition. The vector $\underline{\pi}^* A$ is of the form $[\underline{u},\underline{u},\ldots,\underline{u}]$, where $\underline{u} = c^{-1} \underline{\pi} \sum_{\nu=0}^{\infty} P(\nu,a)$. However, if follows from (7) in [16] that $\sum_{\nu=0}^{\infty} P(\nu,a) = \exp(Q^*a)$, so that $\underline{u} = c^{-1} \underline{\pi}$.

In order to prove the remaining statements, we partition the column vector $\underline{\beta}^*$ into c m-vectors $[\underline{\beta}_0,\underline{\beta}_1,\ldots,\underline{\beta}_{c-1}]'$. Elementary manipulations then show that for $0 \leq j \leq c-1$,

$$\underline{\beta}_j = \sum_{h=0}^{c-1} \sum_{k=1}^{\infty} k\, P(kc+h,a)\underline{e} + \sum_{h=1}^{c-1} \sum_{k=0}^{\infty} P(kc+h,a)\underline{e}\ .$$

Upon summation over j, we obtain

$$\sum_{j=0}^{c-1} \underline{\beta}_j = \sum_{\nu=1}^{\infty} \nu\, P(\nu,a)\underline{e} = \underline{\mu}(a)\ ,$$

by Formula (12) of [16]. Since $\underline{\pi}\,\underline{\mu}(a) = \mu^* a$, (6) follows.

Remark. The blocks of dimensions m x m which appear in the partition of the block circulant matrix A and the vectors $\underline{\beta}_j$, $0 \leq j \leq c-1$, may be computed by solving systems of linear differential equations. The details of that computation are given in the appendix.

Theorem 1. The queue under consideration is stable if and only if $\rho < 1$. In the stable queue, the joint stationary density $\{x(i,j),\ i \geq 0,\ 1 \leq j \leq m\}$ of the number of customers i and the arrival phase j at an arbitrary time is given by the components of the invariant probability vector $\underline{x}$ of $\tilde{P}_2$.

Proof. The proof is entirely analogous to that of Crommelin for the M/D/c queue, which was outlined in Section 1. Crucial to that argument is the fact that any customers in service at time τ and only those leave the system during the time interval $(\tau,\tau+a)$. By also keeping account of the arrival phases at the times τ and $\tau+a$, the argument is easily formalized.

The stationary density $\{x(i,j)\}$ - if it exists - is seen to be the invariant probability vector of $\tilde{P}_2$. The general results on Markov chains of the M/G/1 type imply that $\tilde{P}_2$ has an invariant probability vector if and only if $\underline{\pi}^*\ \underline{\beta}^* = \rho < 1$.

3. Determination of the Vector $\underline{x}$

We henceforth assume that the queue is stable. In this section, known properties of Markov chains of the M/G/1 type will be used to show how the vector $\underline{x}$ may be determined.

The minimal nonnegative solution G to the matrix equation

$$G = \sum_{\nu=0}^{\infty} A_{\nu} G^{\nu} , \tag{7}$$

is stochastic. Since A_0 has no zero columns and A_{ν} is positive for $\nu \geq 1$, the matrix G is positive. It is further known [10], that G is the unique solution to (7) in the set of substochastic matrices of order cm.

In [10] and [11], we discussed the properties and the significance[2] of the minimal nonnegative solution G*(z) to the matrix equation

$$G^*(z) = z \sum_{\nu=0}^{\infty} A_{\nu}[G^*(z)]^{\nu}, \quad 0 \leq z \leq 1. \tag{8}$$

It is clear that G = G*(1). The vector $\underline{\phi}$, defined by

$$\underline{\phi} = \left[\frac{d}{dz} G^*(z)\right]_{z=1-} \underline{e} , \tag{9}$$

plays an important role in what follows. As shown in [11], it is explicitly given in terms of G by

$$\underline{\phi} = (I-G+\tilde{G}) \; [I-A+\tilde{G}-\Delta(\underline{\beta}^*)\tilde{G}]^{-1} \; \underline{e} , \tag{10}$$

where $\tilde{G} = \underline{e} \cdot \underline{g}$, and $\underline{g}$ is the invariant probability vector of the matrix G. $\Delta(\underline{\beta}^*)$ is a diagonal matrix of order cm with the components of $\underline{\beta}^*$ as its diagonal elements.

Moreover

$$\underline{g} \; \underline{\phi} = (1-\rho)^{-1} = c(c-\mu^* a)^{-1} . \tag{11}$$

This equality provides us with an important accuracy check in numerical computations.

It is now convenient to partition several of the vectors and matrices in an appropriate way. The row vector $\underline{x}$ is partitioned into vectors $[\underline{x}_0, \underline{x}_1, \underline{x}_2, \ldots]$ of dimension cm. The column vector $\underline{\phi}$ is partitioned into c m-vectors as $[\underline{\phi}_0, \underline{\phi}_1, \ldots, \underline{\phi}_{c-1}]'$. The stochastic matrix G is partitioned into c^2 blocks of dimensions m x m according to

$$G = \begin{vmatrix} G(0,0) & G(0,1) & \cdots & G(0,c-1) \\ G(1,0) & G(1,1) & \cdots & G(1,c-1) \\ \vdots & \vdots & & \vdots \\ G(c-1,0) & G(c-1,1) & \cdots & G(c-1,c-1) \end{vmatrix} . \tag{12}$$

As the c rows of blocks in the matrices B_ν, $\nu \geq 0$, are identical and agree with the first row of blocks in the corresponding matrices A_ν, $\nu \geq 0$, it readily follows from (7) that

$$\sum_{\nu=0}^{\infty} B_\nu G^\nu = \begin{vmatrix} G(0,0) & G(0,1) & \cdots & G(0,c-1) \\ G(0,0) & G(0,1) & \cdots & G(0,c-1) \\ \vdots & \vdots & & \vdots \\ G(0,0) & G(0,1) & \cdots & G(0,c-1) \end{vmatrix} . \tag{13}$$

The matrix $\hat{G}$, defined by

$$\hat{G} = \sum_{r=0}^{c-1} G(0,r), \tag{14}$$

is a positive stochastic matrix of order m. We denote its invariant probability vector by $\underline{\hat{g}}$.

Theorem 2. The vector $\underline{x}_0$ is given by

$$\underline{x}_0 = (\underline{\hat{g}}\, \underline{\phi}_0)^{-1} [\underline{\hat{g}}\, G(0,0), \underline{\hat{g}}\, G(0,1), \ldots, \underline{\hat{g}}\, G(0,c-1)]. \tag{15}$$

The stationary probability that a customer arriving at time t does not have to wait is given by

$$\underline{x}_0\, \underline{e} = (\underline{\hat{g}}\, \underline{\phi}_0)^{-1}. \tag{16}$$

The stationary probability θ_0^* that the system is empty at time t is given by

$$\theta_0^* = (\underline{\hat{g}}\, \underline{\phi}_0)^{-1}\, \underline{\hat{g}}\, G(0,0)\, \underline{e} . \tag{17}$$

Proof. By an argument based on Markov renewal theory [9, 12, 13], we have earlier shown that $\underline{x}_0$ is given by

$$\underline{x}_0 = (\underline{\kappa}\, \underline{\kappa}^*)^{-1}\, \underline{\kappa} , \tag{18}$$

where $\underline{\kappa}$ is the invariant probability vector of the stochastic matrix $\sum_{\nu=0}^{\infty} B_\nu G^\nu$ and $\underline{\kappa}^*$ is the vector of means

$$\underline{\kappa}^* = \{\frac{d}{dz} \sum_{\nu=0}^{\infty} z\, B_\nu\, [G^*(z)]^\nu\}_{z=1}\, \underline{e}\ . \tag{19}$$

It is obvious from (13) that the row vector $\underline{\kappa}$ is given by

$$\underline{\kappa} = [\hat{\underline{g}}\, G(0,0),\ \hat{\underline{g}}\, G(0,1),\ldots,\hat{\underline{g}}\, G(0,c-1)]. \tag{20}$$

The vector $\underline{\kappa}^*$ is given by $\underline{e} + \sum_{\nu=1}^{\infty} B_\nu \sum_{r=0}^{\nu-1} G^r\, \underline{\phi}$, by virtue of (19). Since the rows of blocks in the matrices B_ν are all equal to the first row of the corresponding matrices A_ν, $\nu \geq 0$, we see that the vector $\underline{\kappa}^*$ is of the form $[\underline{u}^*,\underline{u}^*,\ldots,\underline{u}^*]'$, where the components of the m-vector $\underline{u}^*$ agree with the first m components of the vector $\underline{e} + \sum_{\nu=1}^{\infty} A_\nu \sum_{r=0}^{\nu-1} G^r\, \underline{\phi} = \underline{\phi}$. The vector $\underline{u}^*$ is therefore equal to $\underline{\phi}_0$ and we also trivially have that

$$\underline{\kappa}\, \underline{\kappa}^* = \hat{\underline{g}}\, \underline{\phi}_0.$$

This proves the first statement. The others are immediate.

Remarks. a. The vector $\underline{x}_1$ may also be explicitly related to the matrix G, but the resulting expressions are considerably more complicated. The particular form of the matrices B_ν, $\nu \geq 0$, again induces some simplifications, but these are not as striking as those for the vector $\underline{x}_0$.

By particularizing general formulas proved in [9], which are based on an argument first presented in [13], we obtain that

$$\underline{x}_1 = (\hat{\underline{\kappa}}\, \hat{\underline{\kappa}}^*)^{-1}\, \hat{\underline{\kappa}}\ , \tag{21}$$

where $\hat{\underline{\kappa}}$ is the invariant probability vector of the positive stochastic matrix

$$K = A_0(I-B_0)^{-1} \sum_{\nu=1}^{\infty} B_\nu\, G^{\nu-1} + \sum_{\nu=1}^{\infty} A_\nu\, G^{\nu-1}\ , \tag{22}$$

and the column vector $\hat{\underline{\kappa}}^*$ is given by

$$\underline{\hat{k}}^* = \underline{e} + A_0(I-B_0)^{-1}\,\underline{e}$$
$$+ \left\{A_0(I-B_0)^{-1}\left[\sum_{\nu=1}^{\infty} B_\nu - \sum_{\nu=1}^{\infty} B_\nu G^{\nu-1} + \sum_{\nu=2}^{\infty} (\nu-1)\, B_\nu \tilde{G}\right]\right. \tag{23}$$
$$\left. + \sum_{\nu=1}^{\infty} A_\nu - \sum_{\nu=1}^{\infty} A_\nu G^{\nu-1} + \sum_{\nu=2}^{\infty} (\nu-1)\, A_\nu \tilde{G}\right\} (I-G+\tilde{G})^{-1}\, \underline{\phi} \quad .$$

We note that it suffices to evaluate the matrices $\sum_{\nu=1}^{\infty} A_\nu = A - A_0$, $\sum_{\nu=1}^{\infty} A_\nu G^{\nu-1}$, and $\sum_{\nu=2}^{\infty} (\nu-1) A_\nu \tilde{G} = [\underline{\beta}^* + A_0\underline{e} - \underline{e}] \cdot \underline{g}$. The matrices $\sum_{\nu=1}^{\infty} B_\nu$, $\sum_{\nu=1}^{\infty} B_\nu G^{\nu-1}$, and $\sum_{\nu=2}^{\infty} (\nu-1) B_\nu \tilde{G}$, are then simply obtained by taking the first row of m x m blocks in the corresponding matrices with the coefficients A_ν and replicating that row c times.

The inverse $(I-B_0)^{-1}$ is given by

$$(I-B_0)^{-1} = \begin{vmatrix} I+CP(0,a) & CP(1,a) & \cdots & CP(c-1,a) \\ CP(0,a) & I+CP(1,a) & \cdots & CP(c-1,a) \\ \vdots & \vdots & & \vdots \\ CP(0,a) & CP(1,a) & \cdots & I+CP(c-1,a) \end{vmatrix} ,$$

where the matrix C, of order m, is given by

$$C = \left[I - \sum_{r=0}^{c-1} P(r,a)\right]^{-1} .$$

We see that only the inversion of a matrix of order m is required and that storage of the inverse $(I-B_0)^{-1}$ may be avoided.

Notwithstanding the apparent complexity of the preceding formulas, they have been successfully implemented in a large number of numerical examples for related models. There is merit in computing $\underline{x}_1$ directly from the preceding formulas, as the equation $\underline{x}_0 = \underline{x}_0\, B_0 + \underline{x}_1\, A_0$, provides yet another internal accuracy check on the computations.

Explicit formulas for the vectors $\underline{x}_i$, $i \geq 2$, may in principle also be derived by first passage arguments, but the resulting expressions are too complex to be algorithmically useful. The computation of these vectors may be performed by an iterative scheme, which we shall discuss in Section 7.

b. One of the clear advantages of the matrix methods, used here, is that many of the intermediate steps of the algorithm may be given interesting probabilistic interpretations. In the present model, some care is needed as all interpretations refer to the discrete parameter Markov chain, which is obtained by considering the queue length and the arrival phase at time points $\tau + ka$, $k \geq 0$.

We shall limit ourselves to one example. Suppose that at some time $\tau \geq 0$, in the stationary version of the queue, there are $i' = ic + k'$, $i \geq 1$, $0 \leq k' \leq c-1$, customers present and the arrival phase is j', $1 \leq j' \leq m$.

The expected number of <u>time slots</u> of length a until for the first time, the embedded Markov chain reaches the set of states $\{(k,j),\ 0 \leq k < c-1,\ 1 \leq j \leq m\}$, which corresponds to the case where one or more of the servers are free, is given by the component with index $k'm + j'$ of the column vector $\sum_{\nu=0}^{i-1} G^{\nu} \underline{\phi} = (I - G^{i} + i\,\tilde{G})(I - G + \tilde{G})^{-1}\,\underline{\phi}$. The proof that this interpretation is valid is elementary and rests on the interpretation of the coefficient matrices in the power series for $G^*(z)$, introduced in (8). The reader may find similar interpretations, discussed in greater detail, in [15].

4. The Mean Queue Length

In discussing the stationary mean queue length, we exploit the particular form of the matrix $\tilde{P}$ of Formula (3) to a larger extent than in Section 3. In order to do so, we now partition the vector $\underline{x}$ into m-vectors $[\underline{y}_0, \underline{y}_1, \underline{y}_2, \ldots]$ and we introduce the row vector of generating functions

$$\underline{Y}(z) = \sum_{i=0}^{\infty} \underline{y}_i\, z^{i}, \quad \text{for } |z| \leq 1. \tag{24}$$

Entirely routine manipulations now lead to

$$\underline{Y}(z)\ [z^{c}I - \tilde{P}(z,a)] = \sum_{\nu=0}^{c-1} \underline{y}_{\nu}\ (z^{c} - z^{\nu})\ \tilde{P}(z,a), \tag{25}$$

where the matrix $\tilde{P}(z,a) = \sum_{\nu=0}^{\infty} P(\nu,a)\, z^{\nu} = \exp\,[R(z)a]$, is as given in Formula (7) of [16].

We note that the vectors $\underline{y}_{\nu}$, $0 \leq \nu \leq c-1$, on the right hand side of (25) are given by (15), since $\underline{x}_0 = [\underline{y}_0, \underline{y}_1, \ldots, \underline{y}_{c-1}]$.

If we set $z = 1$, in (25), we obtain

$$\underline{Y}(1)\,[I-\exp(Q^*a)] = \underline{0}\,. \tag{26}$$

Since furthermore $\underline{Y}(1)\,\underline{e} = 1$, we obtain that

$$\underline{Y}(1) = \underline{\pi}\,, \tag{27}$$

as is to be anticipated.

The evaluation of the vector $\underline{Y}'(1)$ and of the <u>stationary mean queue length</u> $L = \underline{Y}'(1)\,\underline{e}$, is considerably more involved. Although they are complicated, the resulting formulas are well suited for numerical computation.

<u>Theorem 3</u>. Provided the second moments of all the group size densities in the VMPP are finite, the quantity $L = \underline{Y}'(1)\,\underline{e}$, is given by

$$\begin{aligned} L = 1/2\,(c-\mu^*a)^{-1}\,\{&\eta''(1) - c(c-1) \\ &+ \sum_{\nu=0}^{c-1} [2(c-\nu)\,\mu^*a + c(c-1) - \nu(\nu-1)]\underline{y}_{\nu}\underline{e} \\ &+ 2\sum_{\nu=0}^{c-1} (c-\nu)\,\underline{y}_{\nu} \cdot \underline{v}'(1)\}, \end{aligned} \tag{28}$$

and the vector $\underline{Y}'(1)$ by

$$\underline{Y}'(1) = (L-c)\underline{\pi} + \underline{\pi}\,\tilde{P}'(1,a)\,Z + \sum_{\nu=0}^{c-1} (c-\nu)\,\underline{y}_{\nu}\,\exp(Q^*a)\,Z. \tag{29}$$

The auxiliary items Z, $\underline{v}'(1)$ and $\eta''(1)$, appearing in (28) and (29) are given by

$$Z = [I - \exp(Q^*a) + \Pi]^{-1},$$

$$\underline{v}'(1) = Z\,\tilde{P}'(1,a)\underline{e} - \mu^*a\,\underline{e} = Z\,\underline{\mu}(a) - \mu^*a\,\underline{e},$$

$$\eta''(1) = \underline{\pi}\,\underline{\mu}_2(a) - 2(\mu^*a)^2 + 2\,\underline{\pi}\,\tilde{P}'(1,a)\,Z\,\tilde{P}'(1,a)\underline{e}\,.$$

The inner product $\underline{\pi}\,\underline{\mu}_2(a)$ is given by Formula (27) of [16]. It is the second factorial moment of the number of arrivals during (0,a) in the stationary version of the VMPP.

Proof. The method of the proof that follows was discussed in greater detail in [12] or [13]. The same method may also be used to yield expressions for the moments of higher order. Only the most important steps are repeated here.

Differentiation with respect to z in (25) and setting z = 1, yields

$$\underline{Y}'(1)\ [I - \exp(Q^*a)] = \sum_{\nu=0}^{c-1} (c-\nu)\ \underline{y}_\nu \exp\,(Q^*a) + \underline{\pi}\ \tilde{P}'(1,a) - c\ \underline{\pi}\ ,$$

a system of linear equations with a coefficient matrix of rank m - 1. Adding $\underline{Y}'(1)\ \Pi = L\ \underline{\pi}$, to both sides of the preceding equation and noting that $I - \exp\,(Q^*a) + \Pi$ is nonsingular [7] and that $\underline{\pi}\ Z = \underline{\pi}$, readily leads to Equation (29).

The mean queue length L cannot be determined from (29). To that end, a separate argument is needed.

Proceeding as in [12], we consider the Perron-Frobenius eigenvalue $\eta(z)$ of the positive matrix $\tilde{P}(z,a)$, for $0 < z \leq 1$, as well as corresponding positive left and right eigenvectors $\underline{u}(z)$ and $\underline{v}(z)$, chosen so that their components are analytic in z for $0 < z < 1$ and continuous on [0,1]. The vectors $\underline{u}(z)$ and $\underline{v}(z)$ further satisfy the defining and normalizing equations

a. $\underline{u}(z)\ \tilde{P}(z,a) = \eta(z)\ \underline{u}(z)$,

b. $\tilde{P}(z,a)\ \underline{v}(z) = \eta(z)\ \underline{v}(z)$,

c. $\underline{u}(z)\ \underline{v}(z) = 1$, (30)

d. $\underline{u}(z)\ \underline{e} = 1$,

e. $\underline{u}(1) = \underline{\pi}$, $\underline{v}(1) = \underline{e}$.

Postmultiplication by $\underline{v}(z)$ in (25) and differentiating the resulting equation with respect to z leads to

$$\underline{Y}'(z)\ \underline{v}(z) + \underline{Y}(z)\ \underline{v}'(z) = [z^c-\eta(z)]^{-1} \Big\{ \eta'(z) \sum_{\nu=0}^{c-1} \underline{y}_\nu\ \underline{v}(z)(z^c-z^\nu)$$

$$+ \eta(z) \left[\sum_{\nu=0}^{c-1} \underline{y}_\nu\ \underline{v}'(z)(z^c-z^\nu) + \sum_{\nu=0}^{c-1} \underline{y}_\nu\ \underline{v}(z)(cz^{c-1}-\nu z^{\nu-1}) \right]$$

$$- [cz^{c-1} - \eta'(z)]\ \underline{Y}(z)\ \underline{v}(z) \Big\} . \tag{31}$$

Letting z tend to one and applying l'Hospital's rule, we obtain

$$L + \underline{\pi}\ \underline{v}'(1) =$$

$$[c-\eta'(1)]^{-1} \Big\{ \sum_{\nu=0}^{c-1} [c(c-1)-\nu(\nu-1) + 2(c-\nu)\eta'(1)]\ \underline{y}_\nu\ \underline{e}$$

$$+ 2 \sum_{\nu=0}^{c-1} (c-\nu)\ \underline{y}_\nu\ \underline{v}'(1) + \eta''(1) - c(c-1)$$

$$- [c-\eta'(1)][L + \underline{\pi}\ \underline{v}'(1)] \Big\} , \tag{32}$$

since $\eta(1) = 1$. Next, we obtain explicit expressions for $\eta'(1)$, $\eta''(1)$ and $\underline{v}'(1)$. Equation (30) yields $\underline{u}'(1)\underline{e} = 0$. Differentiating in (30c) and setting $z = 1$, leads to $\underline{\pi}\ \underline{v}'(1) = 0$.

Differentiating in (30b), we obtain

$$[\eta(z)I - \tilde{P}(z,a)]\ \underline{v}'(z) = [\tilde{P}'(z,a) - \eta'(z)I]\ \underline{v}(z). \tag{33}$$

Premultiplication by $\underline{u}(z)$ and setting $z = 1$, gives

$$\eta'(1) = \underline{\pi}\ \tilde{P}'(1,a)\ \underline{e} = \mu^*a, \tag{34}$$

by Formula (12) of [16].

Setting $z = 1$, in (33) and using the same device as in the proof of (29), leads to

$$\underline{v}'(1) = Z[\tilde{P}'(1,a)-\eta'(1)I]\ \underline{e} = Z\ \underline{\mu}(a) - \mu^*a\ \underline{e}. \tag{35}$$

Differentiating in (33) and setting $z = 1$, we obtain

$$[I - \exp(Q^*a)]\ \underline{v}''(1) + 2[\eta'(1)\ I - \tilde{P}'(1,a)]\ \underline{v}'(1)$$
$$+ [\eta''(1)\ I - \tilde{P}''(1,a)]\ \underline{e} = \underline{0},$$

and premultiplying that equation by $\underline{\pi}$ yields

$$\eta''(1) = \underline{\pi}\, \tilde{P}''(1,a)\, \underline{e} + 2\, \underline{\pi}\, \tilde{P}'(1,a)\, \underline{v}'(1)$$

$$= \underline{\pi}\, \underline{\mu}_2(a) - 2(\mu^* a)^2 + 2\, \underline{\pi}\, \tilde{P}'(1,a)\, Z\, \tilde{P}'(1,a)\, \underline{e}\,. \qquad (36)$$

The quantity $\underline{\pi}\, \underline{\mu}_2(a)$ is finite if and only if the second moments of all the group size densities in the VMPP are finite. Equation (28) is now obtained from (32) by routine substitutions and simplifications.

Remarks. a. The simplifications in the case of Poisson arrivals are dramatic. The items $\underline{\pi}$, Z and $\underline{e}$ are all equal to the scalar one and $\tilde{P}'(1,a) = \lambda a$, $\underline{v}'(1) = 0$, and $\eta''(1) = (\lambda a)^2$.

The expression for the mean queue length L reduces to

$$L = 1/2\ (c-\lambda a)^{-1} \left\{ (\lambda a)^2 - c(c-1) + \sum_{\nu=0}^{c-1} [2(c-\nu)\lambda a + c(c-1) - \nu(\nu-1)]\, y_\nu \right\}.$$

As was shown in Theorem 5 of [15], this may also be written as

$$L = \lambda a + 1/2\ (c-\lambda a)^{-1} \left[(\lambda a)^2 + \lambda a + c^2 \sum_{\nu=0}^{c-1} y_\nu - c^2 - \sum_{\nu=0}^{c-1} \nu^2\, y_\nu \right].$$

b. It should be noted that the complexity of the formulas, even for such items as the stationary mean queue length, is inherent in all models which lead to block-partitioned matrices of the M/G/1 type. The analytic simplifications in such particular cases of the VMPP as Erlang or interrupted Poisson arrivals are so minor that it is much more transparent to obtain the expressions for these cases from the general formulas, rather than by ad hoc derivations from first principles.

c. By belabored but analogous calculations, expressions for $\underline{Y}''(1)$ and for the second factorial moment of the queue length may be derived. We have not done so, but the interested reader may find a guide to such calculations in those reported in [9].

5. The Virtual Waiting Time Distribution

Crommelin's argument, that gives the stationary distribution of the virtual waiting time for the M/D/c queue, may also be adapted to give the corresponding distribution for the present model.

We first present some preliminary material. For τ with $0 \leq \tau < a$, let $h_{ij}(\tau)$, $i \geq 0$, $1 \leq j \leq m$, be the stationary probability that at time 0, the arrival phase is j and there are at most i customers in the system who will still be there at time τ. The m-vector $\underline{h}_i(\tau)$, $i \geq 0$, has the components $h_{ij}(\tau)$, $1 \leq j \leq m$.

Lemma 2. The vectors $\underline{h}_i(\tau)$ and $\underline{y}_i$, $i \geq 0$, are related by

$$\sum_{r=0}^{i} \underline{h}_r(\tau)\, P(i-r,\tau) = \sum_{r=0}^{i} \underline{y}_r, \quad \text{for } i \geq 0. \tag{37}$$

Proof. The queue at time τ is made up by those customers present at time 0, who remain in the system up to time τ and by those arriving during the interval $(0,\tau]$. Since $\tau < a$, no customers who enter in $(0,\tau]$ can leave the system before time τ. Since the arrival process in $(0,\tau]$ is independent of the number present at time 0, who survive, the equations (37) follow.

For each fixed τ, the system of equations (37) is an infinite system of linear equations with the block upper triangular coefficient matrix

$$\begin{vmatrix} P(0,\tau) & P(1,\tau) & P(2,\tau) & P(3,\tau) & \cdots \\ 0 & P(0,\tau) & P(1,\tau) & P(2,\tau) & \cdots \\ 0 & 0 & P(0,\tau) & P(1,\tau) & \cdots \\ \vdots & \vdots & \vdots & \vdots & \end{vmatrix}.$$

By Formula (7) of [16], the matrix $P(0,\tau)$ is given by $P(0,\tau) = \exp[R(0)\tau]$, for $\tau \geq 0$, and is nonsingular. The equations (37) therefore uniquely determine the vectors $\underline{h}_i(\tau)$, $i \geq 0$, for each τ in $[0,a)$.

For future use, we note that

$$\sum_{i=0}^{\infty} \underline{h}_i(\tau)\, z^i = (1-z)^{-1}\, \underline{Y}(z) \exp[-R(z)\tau], \quad \text{for } |z| < 1, \tag{38}$$

where $\underline{Y}(z)$ is given by Formula (25).

Let now t be any nonnegative number and set $t^* = [\frac{t}{a}]$, and $\tau(t) = t - at^*$. Let $W_j(t)$ be the probability that in the stationary queue, the arrival phase at time 0 is j and that the virtual waiting time at 0 does not exceed t. Customers are served in the order of arrival.

Theorem 4. The probability $W_j(t)$, $t \geq 0$, $1 \leq j \leq m$, is given by

$$W_j(t) = h_{ct^*+c-1,j}[\tau(t)]. \tag{39}$$

Proof. The virtual waiting time at time 0 will exceed t if and only if c or more of the customers in the system at time 0 will still be present at time t. In that case, there must be $c + ct^*$ or more survivors in the system at time $\tau(t)$, since exactly c customers depart during each of the intervals $[\tau(t) + (\nu-1)a, \tau(t) + \nu a)$, $1 \leq \nu \leq t^*$. In order that the waiting time at 0 does not exceed t, there can therefore be at most $ct^* + c - 1$ survivors present at time $\tau(t)$. This clearly yields (39).

We note that for $\tau(t) = 0$, we obtain

$$W_j(t^*a) = \sum_{r=0}^{ct^*+c-1} y_{rj}, \quad \text{for } t^* \geq 0, \tag{40}$$

since $P(k,0) = \delta_{k0}I$, for $k \geq 0$, and hence $\underline{h}_i(0) = \sum_{r=0}^{i} \underline{y}_r$, for $i \geq 0$.

For $t = 0$, we obtain $W_j(0+) = \sum_{r=0}^{c-1} y_{rj}$, for $1 \leq j \leq m$. This is in agreement with Theorem 2. Letting t^* tend to infinity in (40), we see that $W_j(\infty) = \pi_j$, for $1 \leq j \leq m$. This is again as anticipated.

We shall now evaluate the conditional mean waiting time $\bar{W}_j$, $1 \leq j \leq m$, given that the arrival phase at time 0 is j. It is clear that $\bar{W}_j = \pi_j^{-1} b_j$, where b_j is the j-th component, $1 \leq j \leq m$, of the row vector $\underline{b}$, given by

$$\underline{b} = \int_0^\infty [\underline{\pi}-\underline{W}(t)]\, dt = \int_0^a \sum_{\nu=0}^{\infty} [\underline{\pi}-\underline{h}_{\nu c+c-1}^{(\nu)}(u)]\, du. \tag{41}$$

The row vector $\underline{W}(t)$ has the components $W_j(t)$, given in (39). The second integral in (41) is obtained by partitioning the domain of integration into intervals of length a and by using (39).

The evaluation of the vector $\underline{b}$ is quite involved and requires several interesting analytic manipulations. We introduce the following additional notation. The quantities $\omega_k = \exp(2\pi i \frac{k}{c})$, $0 \le k \le c-1$, are the c-th roots of unity. The matrix Ω is the Vandermonde matrix with $\Omega_{kj} = \omega_k^j$, for $0 \le j, k \le c-1$.

The row vector of generating functions $\underline{h}^*(z)$ is defined by

$$\underline{h}^*(z) = \int_0^a \sum_{i=0}^{\infty} [\underline{\pi}-\underline{h}_i(u)]\, z^i \, du, \qquad \text{for } |z| < 1 \tag{42}$$

and the vectors $\underline{h}_j^*(z)$, $0 \le j \le c-1$, are defined by

$$\underline{h}_j^*(z) = \int_0^a \sum_{\nu=0}^{\infty} [\underline{\pi}-h_{\nu c+j}(u)]\, z^{\nu c+j} \, du, \qquad \text{for } |z| < 1 . \tag{43}$$

By $\tilde{H}^*(z)$ and $H^*(z)$, we denote the c x c matrices whose rows are the vectors $\underline{h}^*(\omega_j z)$ and $\underline{h}_j^*(z)$, $0 \le j \le c-1$, respectively.

Theorem 5. The vector $\underline{h}^*(z)$ is given by

$$\underline{h}^*(z) = (1-z)^{-1} \{\underline{\pi}-\underline{Y}(z)\, R^{-1}(z)\{I-\exp[-R(z)a]\}\}, \text{ for } |z| < 1 . \tag{44}$$

The vector $\underline{b}$ is the row of index c-1 (i.e., the last row) of the matrix $H^*(1-)$. This matrix is given by

$$H^*(1-) = \Omega^{-1}\, \tilde{H}^*(1-). \tag{45}$$

For $1 \le j \le c-1$, the rows $\underline{h}^*(\omega_j)$ of $\tilde{H}^*(1-)$ are obtained by the substitution $z = \omega_j$, into the right hand side of (44).

The first row $\underline{h}^*(1-)$ of $\tilde{H}^*(1-)$ is given by the formula

$$\begin{aligned}\underline{h}^*(1-) = {} & (aL - 1/2\, a^2\mu^*)\underline{\pi} + \underline{Y}'(1)\, [\exp(-Q^*a)-I](\tau^*\Pi-Q^*)^{-1} \\ & + \underline{\pi}\, R'(1)\, \{\tau^{*-1}\, \Pi a - a(\tau^*\Pi-Q^*)^{-1} \\ & + [\exp(-Q^*a)-I](\tau^*\Pi-Q^*)^{-2}\},\end{aligned} \tag{46}$$

provided the second moments of the group size densities are all finite.

Proof. Formula (44) follows from (38) and (42) by a routine integration. The existence of the inverse $R^{-1}(z)$ for $|z| < 1$ follows easily from the properties of generators - see e.g., [19].

From (41) and (43), it is clear that $\underline{b} = \underline{h}_{c-1}^*(1-)$, provided the limit as $z \to 1-$ exists.

We readily obtain that

$$\underline{h}^*(\omega_k z) = \sum_{j=0}^{c-1} \omega_k^j \, \underline{h}_j^*(z), \qquad \text{for } 0 \leq k \leq c-1, \qquad |z| < 1,$$

or equivalently

$$H^*(z) = \Omega^{-1}\tilde{H}^*(z), \qquad \text{for } |z| < 1. \tag{47}$$

The components of the vector $\underline{h}^*(z)$ are generating functions of sequences of <u>positive</u> numbers. These generating functions are analytic inside the unit disk. They can therefore not have singularities on the circle $|z| = 1$, except possibly at $z = 1$. The vectors $\underline{h}^*(\omega_k)$, $1 \leq k \leq c-1$, are hence well-defined and given by setting $z = \omega_k$ in (44).

For $z \to 1-$, a more delicate passage to the limit is needed. We write $\underline{h}^*(z)$ as

$$\underline{h}^*(z) = \int_0^a (1-z)^{-1} \{\underline{\pi} \exp[R(z)u] - \underline{Y}(z)\} \exp[-R(z)u] \, du. \tag{48}$$

Provided all the densities of the group sizes in the VMPP have finite second moments, the matrix $\exp[R(z)u]$ has the expansion

$$\exp[R(z)u] = \exp(Q^*u) + (z-1) \sum_{\nu=1}^{\infty} \frac{u^\nu}{\nu!} \sum_{r=0}^{\nu-1} Q^{*r} R'(1) Q^{*\nu-r-1}$$
$$+ 1/2\, (z-1)^2 M_2(u) + o(z-1)^2, \qquad \text{as } z \to 1-, \tag{49}$$

where the matrix $M_2(u)$ is a bounded positive matrix for $0 \leq u \leq a$. Since also $\underline{Y}(z) = \underline{\pi} + (z-1)\, \underline{Y}'(1) + o(z-1)$, as $z \to 1-$, the integrand in (48) converges uniformly as $z \to 1-$.

We therefore obtain that

$$\underline{h}^*(1-) = \int_0^a \left[\underline{Y}'(1) - \underline{\pi} R'(1) \int_0^u \exp\, (Q^*v)\, dv\right] \exp(-Q^*u) du$$
$$= \underline{Y}'(1) \int_0^a \exp(-Q^*u) du - \underline{\pi} R'(1) \int_0^a \int_0^u \exp(-Q^*v)\, dv\, du. \tag{50}$$

The integrals in (50) may be explicitly evaluated in the same manner as shown in [14]. We then obtain that[3]

$$\int_0^a \exp(-Q^*u)\,du = \Pi\, a + [\exp(-Q^*a)-I]\, (\tau^*\Pi-Q^*)^{-1} ,$$

and

$$\int_0^a \int_0^u \exp(-Q^*v)\, dv\, du = 1/2\, \Pi a^2 + \tau^{*-1}\, \Pi a - a(\tau^*\Pi-Q^*)^{-1}$$
$$+ [\exp(-Q^*a)-I]\, (\tau^*\Pi-Q^*)^{-2} .$$

Upon substitution into (50), we obtain (46).

Remarks. a. From (46), we readily obtain that

$$\underline{h}^*(1)\, \underline{e} = a\, L - 1/2\, a^2\mu^* = \int_0^a (L-\mu^*u)\, du,$$

so that the quantity $a^{-1}\, \underline{h}^*(1)\underline{e}$ may be viewed as the expected number of survivors at a point chosen at random in $(0,a)$.

b. The evaluation of the vector $\underline{b}$ by use of the results in Theorem 5, involves complex arithmetic and several operations which are numerically delicate. We need, in particular, the matrices $\exp[-R(\omega_k)a]$, which are the inverses of the matrices $\exp[R(\omega_k)a]$, $0 \leq k \leq c-1$, discussed in the appendix. In contrast to the other, highly stable algorithmic procedures proposed in this paper, the implementation of the algorithm for $\underline{b}$ requires a further feasibility study, which we hope to do in the future.

6. The Cases of One or Infinitely Many Servers

There are major simplifications in the analysis of the queue with constant service times and VMPP input, when there is a single server or when there are infinitely many.

We shall first discuss the case $c = \infty$. Exactly the same argument as was used in Section 2, now shows that the vector $\underline{x}$ is the invariant probability vector of the matrix $\tilde{P}_3$, whose rows of blocks are all identical and given by $[P(0,a), P(1,a),\ldots]$. We partition the vector $\underline{x}$ into m-vectors $[\underline{y}_0,\underline{y}_1,\underline{y}_2,\ldots]$. The following theorem is then immediately verified.

Theorem 6. The stationary joint density of the number of customers in service and the arrival phase at an arbitrary time in the infinite server case is given by

$$\underline{y}_i = \underline{\pi}\, P(i,a), \quad \text{for } i \geq 0.$$

Proof. By direct verification.

Remark. As was shown in [20], no tractable analytic expressions for the corresponding probabilities can be found when the service times have a general distribution. This was shown to be the case, whenever the arrivals form a PH-renewal process which is not a Poisson process. It is a fortiori so for the case of VMPP arrivals. As was shown by V. Ramaswami [18], useful computational approaches to the moments of the queue length remain feasible.

The single server case (c=1) is a particular instance of the model treated by V. Ramaswami [19]. For this case, it is possible to obtain the joint density of the queue length and the arrival phase, immediately following departures. We shall not do so here, but refer to [19] for the details. The results, given there, become particularly tractable in the case of constant service times. We see that setting c = 1, in the formulas given in this paper, we also obtain major simplifications. These are particularly striking in those requiring the consideration of residue classes modulo c, as in the case of the waiting time.

We shall only consider the virtual waiting time in some detail. From Formula (39), we see that

$$W_j(t) = h_{t*}[\tau(t)], \quad \text{for } t \geq 0 .$$

The vector $\underline{b}$ is now given by

$$\underline{b} = \int_0^a \sum_{\nu=0}^{\infty} [\underline{\pi}-\underline{h}_\nu(u)]\, du = \underline{h}^*(1-) \; ,$$

and is therefore explicitly given by Formula (46), in which $\underline{Y}'(1)$ is given by (29) with c = 1.

The unconditional mean waiting time at an arbitrary epoch is now given by

$$\bar{W} = \underline{b}\, \underline{e} = L\, a - \frac{1}{2} \rho\, a.$$

This highly intuitive relation between $\bar{W}$ and L may also be obtained from the conservation law proved by S. Brumelle [1] and further discussed in [5]. It should, however, be noted that L remains an involved function of the parameters of the model. Furthermore, as is clear from Formula (46), there is no simple relationship between the vector $\underline{Y}'(1)$ and the vector with components $\bar{W}_j = \pi_j^{-1} b_j$, $1 \leq j \leq m$. It appears to be most unlikely that the conditional mean virtual waiting time and the conditional mean queue length, given the arrival phase, can be related to each other by the type of argument, given in [1] and [5].

7. Markov-Modulated Arrivals - Algorithmic Organization

The Markov-modulated Poisson process is characterized by the matrix Q*, which is a finite, irreducible generator, and by the vector $\underline{\lambda}$ of arrival rates. $\underline{\lambda}$ is a nonnegative, nonzero vector. During any sojourn of the Markov process with generator Q* in the state j, arrivals to the queue occur according to a Poisson process of rate λ_j, $1 \leq j \leq m$.

The matrix R(z) is then given by

$$R(z) = Q^* + (z-1)\, \Delta(\underline{\lambda}) \; , \tag{51}$$

where $\Delta(\underline{\lambda}) = \text{diag}\,(\lambda_1, \ldots, \lambda_m)$.

The matrices $P(\nu,t)$, $\nu \geq 0$, $t \geq 0$, are obtained by numerical integration of the simple system of differential equations

$$P'(0,t) = P(0,t)\ [Q^*-\Delta(\underline{\lambda})], \qquad (52)$$

$$P'(\nu,t) = P(\nu,t)\ [Q^*-\Delta(\underline{\lambda})] + P(\nu-1,t)\ \Delta(\underline{\lambda}), \quad \text{for } \nu \geq 1,$$

with initial conditions $P(\nu,0) = \delta_{\nu 0}\ I$, for $\nu \geq 0$.

The following formulas are routinely obtained by particularizing relations, proved in [16].

a. $\tilde{P}(z,t) = \exp\ \{[Q^*+(z-1)\Delta(\underline{\lambda})]t\}, \quad t \geq 0,$

b. $R'(1) = \Delta(\underline{\lambda}), \quad R''(1) = 0,$

c. $\mu^* = \underline{\pi}\ \underline{\lambda}, \quad \mu_2^* = 0.$

In most cases, it will be desirable to study the queue for several values of the service time a. Let us denote these values by $a_1,\ldots,a_N$, where $0 < a_1 < a_2 <\cdots< a_N < c/\mu^*$. It is convenient to rescale the arrival rates $\lambda_1,\ldots,\lambda_m$, so that $\mu^* = 1$ or $\mu^* = c$. For the sake of this discussion, let us choose the first of these. We note that this rescaling will not affect the computed probabilities and moments of the queue length. In interpreting results related to the waiting time, we need to remember that the stationary mean interarrival time is now chosen as the unit of time.

We know the first two factorial moments $\underline{\pi}\ \underline{\mu}(c) = c$, and $\underline{\pi}\ \underline{\mu}_2(c)$, (given by Formula (27) of [16]) of the number of arrivals during the interval $(0,c]$ in the stationary version of the arrival process. These may be used to determine a (conservative) index $\hat{\nu}$ at which the system of differential equations (52) may be truncated. This upper index is used in evaluating the matrices $P(\nu,a_r)$ for all the desired values a_r, $1 \leq r \leq N$.

It is also possible to integrate the differential equations (52) by increasing the index at which the system is truncated <u>adaptively</u>

as t increases. This elementary procedure is described in detail in [7], Chapter 2, and performs very satisfactorily in actual numerical work.

The equations (52) may be integrated by a classical procedure, such as the Runge-Kutta method. We only store the matrices $P(\nu,a_r)$, $\nu \geq 0$, for the current value of a_r. When all computations for that value are completed, we continue by integrating the equations (52) over the interval $[a_r,a_{r+1}]$, using the matrices $P(\nu,a_r)$ as the initial conditions.

We shall refer to an algorithm which exploits items computed at an earlier stage to evaluate efficiently the corresponding items for the next stage as a <u>nested algorithm</u>. The computation of the steady-state features of the present queueing model can be organized in a highly nested algorithm. Doing so results in an efficient overall algorithm, in spite of the apparent complexity of the individual steps.

We also note in passing that there are many accuracy checks on the various algorithmic steps. As examples, we note that the computed sum $\sum_\nu P(\nu,t)$ should be close to the matrix $\exp(Q^*t)$, which is itself easily evaluated by solving a system of linear differential equations. That system is usually much smaller than the truncated system (52).

The computation of the matrices $P(\nu,a_r)$, $\nu \geq 0$, is the initial step of the global algorithm (<u>Step 1</u>). It is important to note the following. Most algorithmic steps from here on <u>are not significantly simplified</u> by our choice of the Markov-modulated Poisson process as the arrival process. This is a common feature of the matrix methods, which we have developed. Particular assumptions are useful in simplifying the set-up computations, but whenever the matrix G is <u>positive</u>, they do not significantly affect the subsequent steps.

For the computation of the matrix G (<u>Step 2</u>), we distinguish between the smallest value a_1 and the subsequent a_r, $2 \leq r \leq N$. For the value a_1, the corresponding matrix G may be computed by successive substitutions in the equation (7). The matrix A_0 serves as the first iterate. By partitioning the matrix G as in (12), and by obvious care in programming, the special structure of the matrices A_ν, $\nu \geq 0$, may be exploited. It is never necessary to store the large matrices A_ν, but only the $P(\nu,a)$, $\nu \geq 0$. There are various elementary devices, which may be used to accelerate convergence of the method of successive substitutions. These are discussed in [11]. In the present case, however, the value a_1 typically corresponds to a highly stable queue and the convergence of the successive iterates to G is very rapid.

The iterative computation of G (for a_1) is stopped when the maximum element-wise difference between two successive iterates is small, say less than 10^{-8}. We then also verify that the last iterate is close to a stochastic matrix. For the computed value of G, we take a stochastic matrix obtained by a linear extrapolation based on the last two iterates, as was discussed in [9].

The matrices G, corresponding to the parameter values $a_2,\ldots,a_N$, are also computed by successive substitutions, except that the matrix G computed for the preceding a_r is used as the starting solution. The successive iterates are now stochastic matrices and convergence is no longer monotone.

The iterative computation of the matrix G requires two arrays of dimension $(cm)^2$. For each a_r, once the matrix G has been evaluated, one of these arrays may be released for use in the subsequent computations. The (truncated) right hand side of (7) is computed by Horner's algorithm.

In the computation of the invariant probability vector $\underline{g}$ of G (Step 3) any one of a number of classical techniques may be used.

Particularly when cm is large, we would favor an iterative technique, such as the power method [2].

In evaluating the vector $\underline{\phi}$ (Step 4) from Formula (10), the matrix A and the vector $\underline{\beta}^*$ are needed. In the appendix, a direct method for the computation of the blocks $U_0(1),\ldots,U_{c-1}(1)$, which arise in the block circulant matrix A, and for the evaluation of $\underline{\beta}^*$, is proposed. Provided the matrices $P(\nu,a)$, $\nu \geq 0$, have been accurately computed, it is much easier to compute the items $U_r(1)$ and $\underline{\beta}^*$ directly from their definitions. The direct method may be used as a (somewhat laborious) accuracy check, if one wishes to do so.

As we notice that the vector $(I-G+\tilde{G})^{-1}\,\underline{\phi}$ arises in Formula (23), we solve the system of linear euqations

$$[I-A+\tilde{G}-\Delta(\underline{\beta}^*)\tilde{G}]\ \underline{v} = \underline{e}. \tag{53}$$

We save the vector $\underline{v}$ for use in (23) and compute $\underline{\phi}$ from $\underline{\phi} = (I-G+\tilde{G})\underline{v}$. This avoids the inversion of the matrix $I-G+\tilde{G}$.

At this stage, we perform the accuracy check (Step 5) suggested by Equation (11). Both of the inner products $\underline{g}\ \underline{\phi}$ and $\underline{g}\ \underline{v}$ should agree with the easily computed quantity $(1-\rho)^{-1}$.

The computations of the matrix $\hat{G}$ (Step 6), the vector $\hat{\underline{g}}$ and the vector $\underline{x}_0$ by Formulas (14) and (15) do not present any difficulties. We now have the ingredients required for the computation of the stationary mean queue length L and the vector $\underline{Y}'(1)$ (Step 7). We notice, by considering Formulas (28) and (29) and the definitions of the auxiliary items in Theorem 3, that there are some useful algorithmic shortcuts here also.

The vectors $\underline{\pi}\ \tilde{P}'(1,a)\ Z$, $\tilde{P}'(1,a)\ \underline{e}$, and $\sum_{\nu=0}^{c-1} (c-\nu)\ \underline{y}_\nu$ are needed in several of the formulas and are computed only once. The row vector

$$\underline{u} = \sum_{\nu=0}^{c-1} (c-\nu)\ \underline{y}_\nu \cdot \exp\,(Q^*a)\ Z,$$

is computed by solving the system of linear equations

$$\underline{u}[I-\exp(Q^*a)-\Pi] = \sum_{\nu=0}^{c-1} (c-\nu)\, \underline{y}_\nu \cdot \exp(Q^*a) ,$$

preferably by a well-designed library routine.

The obvious relation

$$\underline{u}\ \underline{e} = \sum_{\nu=0}^{c-1} (c-\nu)\, \underline{y}_\nu ,$$

may serve as an internal accuracy check.

When it is desirable to compute additional components of the vector $\underline{x}$, we propose to proceed as follows. The vector $\underline{x}_1$ is computed by the procedure described in the remarks following Theorem 2 (Step 8). As we have already pointed out, there are also various useful shortcuts in that computation.

The equation

$$\underline{x}_0\, B_0 + \underline{x}_1\, A_0 = \underline{x}_0 , \qquad (54)$$

serves as an accuracy check (Step 9). For the problem under consideration, the matrix A_0 is nonsingular. It is a block upper triangular matrix with the nonsingular matrix P(0,a) as its diagonal blocks. It is therefore possible to compute $\underline{x}_1$ directly from (54), but we prefer the approach based on a probabilistic argument because of the available accuracy check. Because of the rapid loss of significance, we definitely advise against recursive computation of the vectors $\underline{x}_i$, $i \geq 2$, even though the nonsingularity of A_0 guarantees that the recursive evaluation is analytically valid.

Let us assume that we have presently evaluated the vectors $\underline{x}_0$ and $\underline{x}_1$ of dimension cm, or equivalently the m-vectors $\underline{y}_i$, $0 \leq i \leq 2c-1$. The particular structure of the matrix $\tilde{P}_2$, displayed in (3), is now useful in the construction of a Gauss-Seidel iterative scheme, which we have found to be highly efficacious in many similar computations (Step 10).

The steady-state equations for $i \geq 2c$ are rewritten in the equivalent form

$$\underline{y}_i = \left\{ \sum_{\nu=0}^{c-1} \underline{y}_\nu P(i,a) + \sum_{\nu=c}^{2c-1} \underline{y}_\nu P(i-\nu+1,a) + \sum_{\nu=2c}^{i-1} \underline{y}_\nu P(i-\nu+1,a) + \underline{y}_{i+1} P(0,a) \right\} [I-P(1,a)]^{-1}.$$

The inverse $S = [I-P(1,a)]^{-1}$ is a positive matrix of order m, which is computed once for every value of a. Since the vectors $\underline{\zeta}_i = \sum_{\nu=0}^{c-1} \underline{y}_\nu P(i,a) + \sum_{\nu=c}^{2c-1} \underline{y}_\nu P(i-\nu+1,a)$, $i \geq 2c$, are known, they also are computed only once.

The successive iterates $\{\underline{y}_i(n),\ i \geq 2c\}$ are computed for $n \geq 1$, by the scheme

$$\underline{y}_i(n+1) = \left[\underline{\zeta}_i + \sum_{\nu=2c}^{i-1} \underline{y}\ (n+1)\ P(i-\nu+1,a) + \underline{y}_{i+1}(n)\ P(0,a)\right] S, \quad (55)$$

for $i \geq 2c$.

The starting vectors $\underline{y}_i(0)$ may be selected in several convenient ways. For the smallest value a_1 of a, we propose to set $\underline{y}_i(0) = \underline{\zeta}_i$, for $i \geq 2c$; for the subsequent values of a, the previously computed vectors $\underline{y}_i$ will generally be an excellent starting solution.

It is part of the algorithm to adapt the upper index i_{max} for which we compute the vectors $\underline{y}_i$, as we may easily compute additional $\underline{\zeta}_i$ and augment the upper index i_{max} if the probability mass $\sum_{i>i_{max}} \underline{y}_i\, \underline{e}$ exceeds a given ε.

The previously computed vectors $\underline{Y}(1) = \sum_{i=0}^{\infty} \underline{y}_i = \underline{\pi}$, and $\underline{Y}'(1) = \sum_{i=1}^{\infty} i\, \underline{y}_i$, provide us again with powerful internal accuracy checks on this part of the algorithm.

The computations related to the waiting time (<u>Step 11</u>) are somewhat more delicate and belabored. This was already noted by Crommelin for the simpler case of the M/D/c queue.

We should note however that Formula (40) provides us with the values of the mass-functions $W_j(\cdot)$, $1 \leq j \leq m$, at time points which are multiples of the service time a. The implementation of (40) requires only a trivial amount of additional computation and the information on the waiting time, that is so provided, may be adequate for many practical purposes.

If it is desirable to compute the mass-functions $W_j(\cdot)$, $1 \leq j \leq m$, at other points t, we need to solve the block upper triangular system (37) for various values of $\tau = t - [\frac{t}{a}]\, a$. Since $P(0,a)$ is nonsingular, we may clearly evaluate the $\underline{h}_i(\tau)$, $i \geq 0$, recursively, but it is easy to provide examples where loss of significance can cause rapid deterioration of the numerical accuracy of the $\underline{h}_i(\tau)$ for the larger indices i. It is probably a safer course of action to solve a truncated version of (37) as a large finite system by using a well-designed library routine.

The evaluation of the conditional mean waiting times $\bar{W}_j$, $1 \leq j \leq m$, and of the unconditional mean waiting time $\bar{W} = \underline{b}\, \underline{e}$, requires the computation of the vector $\underline{b}$ of Theorem 5. In addition to the vector $\underline{h}^*(1-)$, given by Formula (46), we also need to evaluate the vectors $\underline{h}^*(\omega_j)$, $1 \leq j \leq c-1$, which are given by setting $z = \omega_j$, in (44). The components of these vectors are complex.

As was already noted at the end of Section 5, the vectors $\underline{h}^*(\omega_j)$ involve the matrices $\exp[-R(\omega_k)a]$, $0 \leq k \leq c-1$. These matrices may be computed by solving the linear systems of differential equations, discussed in the appendix.

Although none of the required steps are difficult, the computation of the vector $\underline{b}$ requires the evaluation of a considerable number of auxiliary quantities. A future study of the details of the implementation of this algorithm is desirable.

8. Appendix

8.1 The Explicit Procedure for the Vector $\underline{\beta}^*$

The vector $\underline{\beta}^*$, introduced in Lemma 1, may be explicitly computed, but the algorithm is somewhat belabored. As we have discussed in Section 7, we prefer to evaluate $\underline{\beta}^*$ in the course of the computation of the matrices $P(\nu,a)$, $\nu \geq 0$. The following derivations are however of potential utility to other models and are therefore included here.

If we introduce the m x m matrices

$$U_r(z) = \sum_{\nu=0}^{\infty} P(\nu c+r,a)\, z^{\nu}, \quad \text{for } 0 \leq r \leq c-1, \tag{A1}$$

then the matrix $A^*(z) = \sum_{\nu=0}^{\infty} A_{\nu} z^{\nu}$, is given by

$$A^*(z) = \begin{vmatrix} U_0(z) & U_1(z) & U_2(z) & \cdots & U_{c-1}(z) \\ zU_{c-1}(z) & U_0(z) & U_1(z) & \cdots & U_{c-2}(z) \\ zU_{c-2}(z) & zU_{c-1}(z) & U_0(z) & \cdots & U_{c-3}(z) \\ \vdots & \vdots & \vdots & & \vdots \\ zU_1(z) & zU_2(z) & zU_3(z) & \cdots & U_0(z) \end{vmatrix} . \tag{A2}$$

After routine calculations, we see that the vectors $\underline{\beta}_{\nu}$, $0 \leq \nu \leq c-1$, obtained by partitioning the vector $\underline{\beta}^*$, are given by

$$\underline{\beta}_0 = \sum_{r=0}^{c-1} U_r'(1)\, \underline{e}, \tag{A3}$$

$$\underline{\beta}_{\nu} = \underline{\beta}_0 + \sum_{r=c-\nu}^{c-1} U_r(1)\, \underline{e}, \quad \text{for } 1 \leq \nu \leq c-1 .$$

By Formula (A3), it clearly suffices to show how the matrices $U_r(1)$ and $U_r'(1)$ may be computed.

Let us set

$$F_r(z) = \sum_{\nu=0}^{\infty} P(\nu c+r,a)\, z^{\nu c+r} = z^r\, U_r(z^c) , \tag{A4}$$

for $0 \leq r \leq c-1$. It is then clear that

$$U_r(1) = F_r(1) , \tag{A5}$$

$$U_r'(1) = c^{-1}[F_r'(1) - r\, F_r(1)], \quad \text{for } 0 \leq r \leq c-1 ,$$

so that it suffices to show how the matrices $F_r(1)$ and $F_r'(1)$, $0 \le r \le c-1$, may be computed.

From (A4), $F_r(z)$ is clearly the matrix power series obtained by summing only the terms with indices, congruent to r (mod c), of the matrix power series

$$\tilde{P}(z,a) = \sum_{\nu=0}^{\infty} P(\nu,a)\, z^{\nu} = \exp\{R(z)a\}. \tag{A6}$$

As in the discussion of the mean waiting time in Section 5, we obtain

$$\tilde{P}(\omega_k z,a) = \sum_{r=0}^{c-1} \omega_k^r\, F_r(z) , \quad \text{for } 0 \le k \le c-1 . \tag{A7}$$

The quantities $\omega_k = \exp(2\pi i\frac{k}{c})$, $0 \le k \le c-1$, are the c-th roots of unity. As in Section 5, we introduce the square matrix Ω with elements $\Omega_{kj} = \omega_k^j$, $0 \le j,k \le c-1$.

If we introduce the cm x m matrices $\hat{F}(z)$ and $\hat{P}(z)$, partitioned into blocks as

$$\hat{F}(z) = [F_0(z), F_1(z),\ldots,F_{c-1}(z)]',$$

$$\hat{P}(z) = [\tilde{P}(z,a), \tilde{P}(\omega_1 z,a),\ldots,\tilde{P}(\omega_{c-1}z,a)]',$$

then (A7) may be written as

$$\hat{P}(z) = (\Omega\otimes I)\, \hat{F}(z) . \tag{A8}$$

Since the Vandermonde matrix Ω is nonsingular, it follows that

$$\hat{F}(z) = (\Omega^{-1}\otimes I)\, \hat{P}(z) . \tag{A9}$$

We now immediately obtain that

$$\hat{F}(1) = (\Omega^{-1}\otimes I)\, \hat{P}(1) , \tag{A10}$$

$$\hat{F}'(1) = (\Omega^{-1}\otimes I)\, \hat{P}'(1) .$$

The matrices $\tilde{P}(\omega_k,a)$ and $\tilde{P}'(\omega_k,a)$, for $0 \le k \le c-1$, are most easily computed by numerical solution of systems of differential equations. In order to set these up, we need to distinguish more carefully between differentiation with respect to z and t in $\tilde{P}(z,t)$.

From Equation (6) in [16], we readily obtain that

$$\frac{\partial}{\partial t} \tilde{P}(\omega_k,t) = \tilde{P}(\omega_k,t)\ R(\omega_k)\ , \tag{A11}$$

and

$$\frac{\partial}{\partial t} M_1(\omega_k,t) = M_1(\omega_k,t)\ R(\omega_k) + \tilde{P}(\omega_k,t)\ R'(\omega_k)\ , \tag{A12}$$

for $0 \leq k \leq c-1$, where

$$M_1(\omega_k,t) = \left[\frac{\partial}{\partial z} \tilde{P}(z,t)\right]_{z=\omega_k} = \tilde{P}'(\omega_k,t)\ . \tag{A13}$$

The initial conditions are $\tilde{P}(\omega_k,0) = I$, and $M_1(\omega_k,0) = 0$, for $0 \leq k \leq c-1$.

It should be noted that the actual numerical procedure is simple. It involves the following steps.

<u>Step 1</u>. Compute Ω^{-1}.

<u>Step 2</u>. For successive values of $a = hj$, $j \geq 0$, compute the matrices $\tilde{P}(\omega_k,a)$ and $M_1(\omega_k,a)$, $0 \leq k \leq c-1$, by numerical integration of the differential equations (A11) and (A12).

<u>Step 3</u>. Evaluate the matrices $F_k(1)$ and $F_k'(1)$, $0 \leq k \leq c-1$, by substitution in (A10).

<u>Step 4</u>. Evaluate $U_k'(1)$, $0 \leq k \leq c-1$, from (A5).

<u>Step 5</u>. Evaluate $\underline{\beta}^*$ by using (A3).

9. References

[1] Brumelle, S.L., (1971). On the Relation between Customer and Time Averages in Queues, <u>J. Appl. Probab.</u>, <u>8</u>, 508-520.

[2] Burden, R., Faires, J.D. and Reynolds, A., (1978). <u>Numerical Analysis</u>, Prindle, Weber and Schmidt, Boston, MA.

[3] Crommelin, C.D., (1932). Delay Probability Formulae when the Holding Times are Constant, <u>Post Office Electrical Engineer's Journal</u>, <u>25</u>, 41-50.

[4] Crommelin, C.D., (1934). Delay Probability Formulae, <u>Post Office Electrical Engineer's Journal</u>, <u>26</u>, 266-274.

[5] Heyman, D.P. and Stidham, S. Jr., (1980). The Relation between Customer and Time Averages in Queues, Oper. Res., 28, 983-994.

[6] Hunter, J.J., (May 1980). "Generalized Inverses and their Application to Applied Probability Problems", Tech. Report VTR-8006, Dept. of Industrial Engineering and Operations Research, Virginia Polytechnic Institute and State University, Blacksburg, VA.

[7] Kemeny, J. and Snell, J.L., (1960). Finite Markov Chains, Van Nostrand Publ. Co., Princeton, NJ.

[8] Kühn, P., (1976). Tables on Delay Systems, Inst. of Switching and Data Techniques, University of Stuttgart, Stuttgart, Germany.

[9] Lucantoni, D.M. and Neuts, M.F., (May 1978). "Numerical Methods for a Class of Markov Chains, arising in the Theory of Queues", Tech. Report No. 78/10, University of Delaware.

[10] Neuts, M.F., (1974). "The Markov Renewal Branching Process", Proceedings of a Conference on Mathematical Methods in the Theory of Queues, Kalamazoo, MI, Springer-Verlag, 1-21.

[11] Neuts, M.F., (1976). Moment Formulas for the Markov Renewal Branching Process, Adv. in Appl. Probab., 8, 690-711.

[12] Neuts, M.F., (1977). Some Explicit Formulas for the Steady-State Behavior of the Queue with Semi-Markovian Service Times, Adv. in Appl. Probab., 9, 141-157.

[13] Neuts, M.F., (1977). The M/G/1 Queue with Several Types of Customers and Change-over Times, Adv. in Appl. Probab., 9, 604-644.

[14] Neuts, M.F., (1978). Renewal Processes of Phase Type, Naval Res. Logist. Quart., 25, 445-454.

[15] Neuts, M.F., (1979). Queues Solvable without Rouché's Theorem, Oper. Res., 27, 767-781.

[16] Neuts, M.F., (1979). A Versatile Markovian Point Process, J. Appl. Probab., 16, 764-779.

[17] Neuts, M.F., (1981). Matrix-Geometric Solutions in Stochastic Models - An Algorithmic Approach, The Johns Hopkins University Press, Baltimore, MD.

[18] Ramaswami, V., (October 1978). The N/G/∞ Queue, Tech. Report, Dept. of Math., Drexel University, Philadelphia, PA.

[19] Ramaswami, V., (1980). The N/G/1 Queue and its Detailed Analysis, Adv. in Appl. Probab., 12, 222-261.

[20] Ramaswami, V. and Neuts, M.F., (1980). Some Explicit Formulas and Computational Methods for Infinite Server Queues with Phase Type Arrivals, J. Appl. Probab., 17, 498-514.

10. Endnotes

[1]What follows is an informal description of the versatile Markovian point process to aid in a cursory reading. Prior perusal of [16] is essential to a careful reading. We consider an (m+1)-state Markov process with one absorbing state m+1 and m transient states (phases). At each absorption, the process is restarted by choosing a new initial state from among the transient states. The epochs of absorptions define a PH-renewal process [14, 17]. The phase process with resetting is an irreducible Markov process with m states and generator Q*. In order to define the VMPP, we consider three types of arrival epochs. These are (a) the epochs of absorption (with resetting), (b) the epochs at which a change of phase occurs without an absorption and resetting, (c) the events in Poisson processes going on during the sojourns in the various phases. The rates of these Poisson processes may depend on the phase during which they are active.

The arrivals at each of these types of arrival epochs may occur in groups and the group sizes are allowed to depend on the type of phase change (or sojourn phase in case c) that occurs.

With a number of straightforward conditional independence assumptions, one so obtains a point process that may be analyzed in terms of a matrix formalism that mimics the simple expressions for the Poisson process and remains computationally highly tractable. The VMPP has a huge number of particular cases that have commonly been treated by ad hoc analyses.

The matrices $P(\nu,t)$, $\nu \geq 0$, $t \geq 0$, have the following significance. The element $P_{jj'}(\nu,t)$ is the conditional probability that during the interval $(0,t)$, ν (individual) items arrive and that the Markov process Q* is in the phase j', given that the phase at time 0 was j.

[2]The nonlinear matrix equation (8) is a generalization of the familiar equation for the generating function of the probability density of the number of customers served during a busy period in the classical M/G/1 queue. It is obtained by considering the number of transitions required to reach (for the first time) a state - say (k,j') - in the level k, starting from a state - say (k+1,j) - in the level k+1. The (conditional) probability generating function of the length of that first passage time is the element $G^*_{jj'}(z)$. The equation (8) is obtained by applying the law of total probability and natural conditional independence properties. One simply enumerates all possible states that can be reached in one transition from the state (k+1,j) and considers all paths leading to the level k from these states.

Unlike the scalar case, the discussion of the various properties of $G^*(z)$ and the corresponding moment matrices is involved. It defies all attempts at an informal summary and we must refer the careful reader to [10] and [11] for the details.

[3]As was discussed in [16], the formulas in which a generalized inverse of Q^* appears, involve a constant τ^* which is arbitrary, subject to very mild restrictions. The dependence on the value of τ^* is in appearance only. By considering various generalized inverses of Q^*, we may obtain a variety of equivalent, but formally distinct expressions for the integrals of interest. The use of generalized inverses in stochastic models is well surveyed in [6]. It is shown there that, in particular, the matrix $\Pi + Q^*$ is nonsingular, so that we may set $\tau^* = -1$ throughout.

University of Delaware, Newark, Delaware 19711

This research was supported by the National Science Foundation under Grant No. ENG-7908351 and the Air Force Office of Scientific Research under Grant No. AFOSR-77-3236.

Discussant's Report on
"The c - server Queue with Constant Service Times and a
Versatile Markovian Arrival Process,"
by Marcel F. Neuts

Professor Neuts has presented still another elegant, interesting, and useful model which can be solved using methods of computational probability. This provides continuing evidence of the power and generality of these methods, in particular for models which have M/G/1 or G/M/1 paradigm structure. One of the beautiful features of this approach is the capability to take an analytically tractable model which has unrealistic or inappropriate modeling simplifications and modify it, obtaining a more general and more realistic model which can be solved efficiently using algorithmic methods. In his paper Professor Neuts has taken the M/D/c model of C. D. Crommelin and generalized it by replacing the Poisson arrival process with a versatile Markovian point process (also known as a "Neuts Process"); the resulting N/D/c model, of course, has much wider applicability and can be solved with an efficient algorithm. Another fruitful generalization of analytically-solvable models is often obtained by replacing exponential holding distributions with phase-type distributions. This potential for algorithmic solution of more realistic and/or general models is very appealing. This raises the question of whether it is possible to roughly characterize (perhaps with a "metatheorem") the analytical models which can be generalized and extended into algorithmically-tractible models.

The "curse of dimension" apparently arises in some of these general models; for example, with the M/Ph/c queueing system a separate dimension is needed for each the servers. The M/G/1 and G/M/1 paradigms are both oriented towards a two-dimensional structure: a one-dimensional array of one-dimensional blocks. It would be useful to have efficient algorithmic solution techniques for higher dimensional paradigms, for example, a two- or higher- dimensional Quasi Birth and Death process.

Finally, as this computational area grows there will be need for more sophisticated analysis of numerical error in the algorithms. What are the possibilities here? Will it ever be possible to state the final numerical results and include an error bound?

Author's Response to Discussant's Report

I thank Professor Miller for his very kind comments on my paper. I should like to address briefly the general issues raised by him.

The key to the models, which may be analyzed as Markov chains of the GI/M/1 or M/G/1 types, is the recognition that these have embedded Markov processes with particular structures that lend themselves to detailed probabilistic arguments. These structures are fairly general and may be recognized by examining the transition pattern of the queue or the block pattern of the transition probability matrix (or operator) [2].

This line of investigation has some similarity with the theory of partial differential equations, where certain pervasive types also lend themselves to particular and successful methods of solution. As there also, the curse of dimensionality is both very real and strongly limiting to specific numerical solutions. The structural features of the models may still be used to prove useful asymptotic and qualitative results as, for example, in Neuts and Takahashi [1]. It may also be anticipated that other useful structures will be recognized as stochastic models and more widely considered from an algorithmic vantage point.

Concern for the algorithmic aspects of stochastic models is of recent date and several of the relevant equations are of types that have not been extensively studied in numerical analysis. With much computational experience and the use of internal accuracy checks, one may program the algorithms so as to have great confidence in their numerical results. A rigorous error analysis of our matrix methods and of the solution of various related integral equations appears to be possible and mathematically challenging. We hope that it will be undertaken in the near future.

References

[1] Marcel F. Neuts and Yukio Takahashi (1980), "Asymptotic Behavior of the Stationary Distributions in the GI/PH/c Queue with Heterogeneous Servers," Tech. Report No. 57B, Applied Mathematics Institute, University of Delaware.

[2] Richard L. Tweedie (1980), "Operator-Geometric Stationary Distributions for Markov Chains, with Application to Queueing Models," Tech. Report No. 68B, Applied Mathematics Institute, University of Delaware.

Discussant: Dr. Douglas R. Miller, Dept. of Operations Research, School of Engineering, George Washington University, Washington, DC 20052.

SIMULATION OUTPUT ANALYSIS FOR GENERAL STATE SPACE MARKOV CHAINS

Peter W. Glynn

Donald L. Iglehart

1. Introduction

The statistical analysis of simulation output has been the primary focus of recent research in simulation methodology. Methods have been developed which permit the simulator to construct confidence intervals for steady-state characteristics of the system being simulated. The principal methods in current use are autoregressive modeling, batch means, regenerative, and replication. With the exception of the replication method, all methods are based on just one simulation run. These methods for constructing confidence intervals are all based on central limit theorems for the underlying stochastic processes being simulated. Thus all methods are only valid asymptotically for long simulation runs.

In this paper we discuss three new methods for analyzing simulation output. All three methods are aimed at analyzing the simulation output of general state space Markov chains. This class of processes encompasses the embedded jump chain generated by a generalized semi-Markov process (GSMP). GSMP's are important for simulation since they may be used to model a general discrete event simulation. GSMP's have been discussed in recent papers by FOSSETT (1979), HORDIJK and SCHASSBERGER (1981), and WHITT (1980).

The first method, called the extended regenerative method (ERM), is based on some recent work on general state space Markov chains by ATHREYA and NEY (1978) and NUMMELIN (1978). This method involves a construction which creates regeneration points for Markov chains which do not hit a single point infinitely often. While this idea is very attractive in principle, there are a number of practical considerations which limit its application. However, the method can be used to increase the rate of regeneration points when using the standard regenerative method. For more details on this method and some related results see GLYNN (1981).

The second method, called the random blocking method (RBM), is based on blocks of the process which begin when the process enters a given set in the state space. This method is reminiscent of the regenerative method except the blocks created here are not independent and identically distributed. Details on this second method can be found in a forthcoming paper by GLYNN (1981).

The last method is a variation of the method of autoregressive modeling. This method, the multivariate autoregressive method (MARM), fits a multivariate autoregressive model to the simulation output data. The model fitting is done automatically based on Akaike's AIC-criterion; see AKAIKE (1976) for a full discussion of these criterion. A forthcoming paper by JOW (1981) will develop this method and other related methods for simulation applications.

This paper is organized as follows. Section 2 is devoted to a discussion of GSMP's and their relation to simulation. The ERM is covered in Section 3 and RBM in Section 4. Section 5 is devoted to the MARM and Section 6 to an illustrative example.

2. Generalized Semi-Markov Processes

In a discrete-event simulation a finite number of events occurring at random times cause changes in the state of the system being simulated. The number of events active at any given time is a function of the state of the system. This type of simulation is well modeled by the generalized semi-Markov process (GSMP) which we now describe.

Let S be the finite (or countable) set of states which describes the GSMP at the successive transition epochs and G the finite number of events which can cause a transition. If $G = \{e_1, e_2, \ldots, e_m\}$, then we let $G(s) = \{e_1(s), \ldots, e_{n_s}(s)\}$ denote the subset of events active when the GSMP is in state s. For each event active in state s, associate a clock which records the time until that event would trigger a state change. If in state s, the clock (associated with an event in G(s)) with the minimal reading triggers the next state change when it runs down to zero. Let C(s) denote the possible clock readings when the GSMP is in state s:

$$C(s) = \{c \in \mathbb{R}_+^m : c_i > 0 \text{ iff } e_i \in G(s);\ c_i \neq c_j \text{ for } i \neq j\},$$

where $\mathbb{R}_+^m$ is the Cartesian product of m copies of $[0,\infty)$. Next define the space Σ by

$$\Sigma = \bigcup_{s \in S} (\{s\} \times C(s))$$

and the process $(\underset{\sim}{S}, \underset{\sim}{C}) = \{(S_n, C_n): n \geq 0\}$ which lives on Σ and represents the state values and clock readings at the successive transition epochs. The process $(\underset{\sim}{S}, \underset{\sim}{C})$ is a general state space Markov chain whose transition kernel will be defined in Section 3. Finally, the GSMP is a piece-wise constant process, $\{X_t: t \geq 0\}$, constructed from the embedded jump process $(\underset{\sim}{S}, \underset{\sim}{C})$ in the usual manner.

3. The Extended Regenerative Method for GSSMC's

We start by formalizing the notion of a general state space Markov chain (GSSMC). Let E be a complete, separable metric space with $\mathcal{E}$ its associated Borel field. A function $P: E \times \mathcal{E} \to [0,1]$ is called a probability transition kernel if:

i) $P(x,\cdot)$ is a probability on $(E,\mathcal{E})$ for each x in E,

ii) $P(\cdot,B)$ is $\mathcal{E}$-measurable for all $B \in \mathcal{E}$.

One should think of $P(x,B)$ as representing the one-step transition probability of the chain passing from x into B. The analogous n-step transition probabilities are then given through the Chapman-Kolmogorov equations, namely

$$P^{n+1}(x,B) = \int_E P^n(y,B)\, P(x,dy)$$

where, of course, $P^0(x,B) = \delta_x(B)$ ($\delta_x(B)$ is 1 or 0 depending on whether or not $x \in B$). Given a kernel P and an initial probability μ on $(E,\mathcal{E})$, one can construct a measure P on $(\Omega,\mathcal{F}) = (E \times E \times \ldots, \mathcal{E} \times \mathcal{E} \times \ldots)$ such that

$$P_\mu\{X_0 \in B_0, X_1 \in B_1,\ldots,X_n \in B_n\}$$
$$= \int_{B_0} \mu(dx_0) \int_{B_1} P(x_0, dx_1) \ldots \int_{B_n} P(x_{n-1}, dx_n)$$

where $X_i(\omega) = \omega_i$ and $(\omega_0,\omega_1,\ldots)$ is a typical element of Ω. The above construction yields a process $\{X_n: n \geq 1\}$ that is endowed with the Markov property

$$P_\mu\{X_{n+1} \in B | X_0,\ldots,X_n\} = P_\mu\{X_{n+1} \in B | X_n\}$$

and is referred to as the GSSMC associated with kernel P and initial distribution μ.

Example 3.1. Let $E = \{0,1,\ldots\}$ and put $P(i,B) = \sum_{j \in B} p_{ij}$, where P is a stochastic matrix on E. This gives the classical countable state Markov chain.

Example 3.2. Let $E = \{W_n: n \geq 0\}$ be the waiting time process of a GI/G/1.queue. Suppose that u and v are the interarrival and service r.v.'s respectively, and that v-u has a density f(x). Then $E = [0,\infty)$ and

$$P(x,B) = \int_B f(y-x)dy + \delta_0(B) \int_0^x f(y)dy.$$

Example 3.3. Consider a model that imitates the dynamics of a lake. Let S_i, Z_i and X_i represent the volume of water stored in the lake, the inflow of water, and outflow of water respectively, at time i. Then, the mass-balance equation

$$S_i = S_{i-1} + Z_i - X_i \tag{3.4}$$

holds. If one now assumes that output increases linearly with storage through the relation $X_i = \alpha S_i$ $(0 < \alpha < 1)$, then (3.4) takes the form

$$X_i = \rho X_{i-1} + \varepsilon_i$$

where $\rho = 1/(1+\alpha)$ and $\varepsilon_i = (\alpha Z_i)/(1+\alpha)$. Finally, the further assumption that $\{\varepsilon_i\}$ is i.i.d. with

$$P\{\varepsilon_i \in B\} = p\delta_0(B) + (1-p) \int_B f(y)dy$$

yields a Markov chain model for the outflows $\{X_n\}$, where $E = [0,\infty)$ and

$$P(x,B) = p\delta_{\rho x}(B) + (1-p) \int_B f(y-\rho x)dy .$$

Example 3.4. The embedded jump process $(\underset{\sim}{S},\underset{\sim}{C})$ introduced in Section 2 is a GSSMC on state space Σ and with kernel

$$P((s,c), B)$$

$$= p(s'; s,i^*) \prod_{i\in N_{s'}} \int_{B_i} f(y; s',i,s,i^*)dy \prod_{j\in 0_{s'}} \delta_{c_i^*}(B_j)$$

where B is that subset of Σ corresponding to the GSMP entering state s' with the i'th clock of s' set to a value in B_i.

The numbers $p(s'; s,i^*)$ govern state transitions of the GSMP and represent the probability of a jump to s' from s, given that clock i* initiated the jump. The rest of the kernel governs clock readings. Those clocks $i \in N_{s'}$ (new clocks) are set stochastically according to a density $f(y; s',i,s,i^*)$, whereas those clocks $j \in 0_{s'}$ (old clocks) are inherited from the previous state and so must be set deterministically at the previous value c_j^*.

Note that by setting $\tau_n = c_0(i^*(C_0)) + \cdots + c_{n-1}(i^*(C_{n-1}))$ for $n \geq 1$ and $\tau_0 = 0$ ($c_j(k)$ is the value of the k'th clock just after the j'th jump of X_t), we can retrieve $\{X_t: t \geq 0\}$, the GSMP, from $\{(S_n, C_n): n \geq 0\}$ via

$$X_t = \sum_{k=0}^{\infty} \delta_t([\tau_k, \tau_{k+1}))S_k.$$

Let us turn now to a recurrence condition for GSSMC's first formulated by ATHREYA and NEY (1978) and NUMMELIN (1978). We say that $\{X_n: n \geq 0\}$ is (A,B,λ,φ,k) recurrent if there exist A, B ε $\mathcal{E}$, a positive number λ, an integer k, and a probability φ on B, such that:

i) $P\{\sum_{n=0}^{\infty} \delta_{X_n}(A) = +\infty \mid X_0 = x\}$ for all x in E,

ii) $P^k(x,\Gamma) \geq \lambda\varphi(\Gamma)$ for each x in A and measurable subset Γ of B.

In our setting, this recurrence notion is in fact equivalent to one first proposed by HARRIS (1956).

To gain some appreciation for the significance of these conditions, we observe that in the $(A,B,\lambda,\varphi,1)$ case, we can decompose P over A as

$$P(x,\Gamma) = \lambda\varphi(\Gamma) + (1-\lambda)\ Q(x,\Gamma)$$

where $Q(x,\cdot)$ is a probability on $(E,\mathcal{E})$. The key idea of Athreya, Ney and Nummelin was to exploit this decomposition via the following embedding.

Let $E' = E \times \{0,1\}$ and $\mathcal{E}'$ be the associated product σ-field. We extend P to a kernel P' on $(E',\mathcal{E}')$ by setting

$$P'((x,\delta),\ \Gamma \times \{0\}) = \begin{cases} (1-\lambda)\ P(x,\Gamma), & x \notin A \\ (1-\lambda)\ Q(x,\Gamma), & x \in A \end{cases}$$

$$P'((x,\delta),\ \Gamma \times \{1\}) = \begin{cases} \lambda P(x,\Gamma), & x \notin A \\ \lambda\varphi(\Gamma \cap B), & x \in A \end{cases}$$

For a probability μ on $(E,\mathcal{E})$, define μ' on $(E',\mathcal{E}')$ by

$$\mu'(\Gamma \times \{i\}) = \delta_0(i)\ (1-\lambda)\ \mu(\Gamma) + \delta_1(i)\lambda\ \mu(\Gamma).$$

Then, it can be readily verified that the Markov chain $X_n' = (X_n,\delta_n)$ on $(E',\mathcal{E}')$ associated with P', μ' has the property that the coordinate process $\{X_n: n \geq 0\}$ is the original GSSMC on $(E,\mathcal{E})$ associated with P, μ.

The importance of this embedding is that it furnishes one with a sequence of regeneration times T_k defined in terms of X_n' via

$$T_1 = \inf\{n \geq 1: X_{n-1} \in A,\ \delta_n = 1\}\ ,$$

$$T_k = \inf\{n > T_{k-1}: X_{n-1} \in A,\ \delta_n = 1\}\ , \qquad k > 1.$$

The regenerative character of X_n' guarantees "nice" ergodic behavior for X_n. For example, if we assume that $E'(T_2-T_1) < +\infty$, then the classical regenerative theory shows that (e.g., SMITH (1955))

$$\frac{1}{n}\sum_{k=1}^{n} f(X_k) \to r = E_\pi f(X) \qquad P'_{\mu'},\ \text{a.s.}$$

where

$$\pi(B) = E'\Big(\sum_{j=T_1}^{T_2-1} \delta_{X_j}(B)\Big)/E'(T_2-T_1).$$

Estimation of r is a common goal of simulators. The above discussion suggests an "extended regenerative method" (ERM) for producing confidence intervals for r based on the T_k sequence.

1. Generate the sequence $\{(X_n,\delta_n): n \geq 0\}$.

2. Let $Y_k = \sum_{j=T_k}^{T_{k+1}} f(X_j)$, $\alpha_k = T_{k+1} - T_k$ and form

$$\hat{r}(n) = \Big(\sum_{k=1}^{n} Y_k\Big)/\Big(\sum_{k=1}^{n} \alpha_k\Big)$$

$$s_{11}(n) = \frac{1}{n-1}\sum_{k=1}^{n} Y_k^2 - \frac{1}{n(n-1)}\Big(\sum_{k=1}^{n} Y_k\Big)^2$$

$$s_{12}(n) = \frac{1}{n-1}\sum_{k=1}^{n} Y_k\ \alpha_k - \frac{1}{n(n-1)}\Big(\sum_{k=1}^{n} Y_k\Big)\Big(\sum_{k=1}^{n} \alpha_k\Big)$$

$$s_{22}(n) = \frac{1}{n-1} \sum_{k=1}^{n} \alpha_k^2 - \frac{1}{n(n-1)} \left(\sum_{k=1}^{n} \alpha_k\right)^2$$

$$s^2(n) = s_{11}(n) - 2\hat{r}(n)\, s_{12}(n) + \hat{r}^2(n)\, s_{22}(n).$$

3. To form a $100(1-\delta)\%$ confidence interval for r choose z_δ so that $\Phi(z_\delta) = 1 - \delta/2$, where Φ is the standard normal distribution function. Then

$$\hat{I} = \left[r(n) - \frac{z_\delta\, s(n)\, n^{1/2}}{\sum_{k=1}^{n} \alpha_k},\ r(n) + \frac{z_\delta\, s(n)\, n^{1/2}}{\sum_{k=1}^{n} \alpha_k} \right]$$

is the desired confidence interval.

The above steps can all be justified under the assumption that $0 < E(Y_2 - r\alpha_2)^2 < +\infty$. Incidentally, although Step 1 appears to necessitate the ability to generate deviates distributed according to $Q(x,\cdot)$, this difficulty can be avoided by using an acceptance-rejection technique ([8], p. 57).

Example 3.1 (continued). Suppose that $X = \{X_n : n \geq 0\}$ is an irreducible, recurrent Markov chain with transition matrix $P = \{p_{ij}\}$. Then, X is $(\{0\}, E, 1, p_{0\cdot}, 1)$ recurrent and $T_k = N_k+1$, where N_k is the k'th hitting time of 0. The ERM reduces here to the classical regenerative technique (modulo a shift to the right).

Example 3.2 (continued). Assuming that $Ev \leq Eu$, $\{W_n : n \geq 0\}$ is $(\{0\}, E, 1, P(0,\cdot), 1)$ recurrent, and again $T_k = N_k+1$, when N_k is the k'th hitting time of 0, reducing our estimation technique to the usual one (again, modulo a shift).

Example 3.3 (continued). Assuming that $E\varepsilon_i < +\infty$ and that f is positive and continuous over $[0, +\infty)$, $\{X_n : n \geq 0\}$ is $([0,b], E, \lambda, \varphi, 1)$ recurrent where

$$\psi(y) = (1-p) \min_{0 \leq x \leq b} f(y-\rho x)$$

$$\lambda = \int_0^\infty \psi(y)dy; \ \varphi(y) = \psi(y)/\lambda \ .$$

Here the T_k's form a subsequence of the hitting times of [0,b] -- this reflects the fact that X_n returns to no point infinitely often. The ERM is applicable here, whereas the classical regenerative method is not.

Example 3.4 (continued). We say that the GSMP has a single set if there exists a state s' with which is associated only one clock. Then, given that the GSMP hits {s'} infinitely often, $(\underset{\sim}{S},\underset{\sim}{C})$ is $(A, \Sigma, 1, P((s',c), \cdot), 1)$ recurrent where $A = \{(s,c) \in \Sigma: s = s'\}$. Here, $T_k = N_k+1$ where the N_k's are consecutive hitting times of A.

It should be remarked that in the context of a GSMP, a simulator is often interested in the continuous-time quantity

$$\tilde{r} = \lim_{t\to\infty} \int_0^t f(X_s)ds/t \ .$$

This can be estimated by the ERM, provided that the definitions of Y_k and α_k are modified to

$$\tilde{Y}_k = \sum_{j=T_k}^{T_{k+1}-1} f(S_j) \ c_j(i^*(C_j))$$

$$\tilde{\alpha}_k = \sum_{j=T_k}^{T_{k+1}-1} c_j(i^*(C_j)) \ .$$

The above discussion focussed on $(A, B, \lambda, \varphi, 1)$ GSSMC's. A fundamental difficulty arises in the $(A, B, \lambda, \varphi, k)$ case, a difficulty that can be partially circumvented via an embedding that endows the chain with an environment that is "loosely regenerative" in the sense of SMITH (1955). An estimation procedure can then be developed that bears close resemblance to the "extended regenerative method" outlined above (see [8]).

Before leaving this topic, it should be mentioned that a number of difficulties remain in terms of implementation of the ERM. The most fundamental problem is that the decomposition of P over A requires an explicit form for the kernel over that set. The "event-scheduling" approach normally used by simulators to generate sample paths does not require an explicit representation for the kernel, and so a simulator is left with the burdensome task of calculating such a form. Additional difficulties arise in determining appropriate λ, φ, and A, although here simple numerical techniques would be applicable ([8], p. 52).

4. Random Blocking Method for GSSMC's

Our major incentive in studying GSSMC's here has been in terms of application to GSMP's. It turns out that GSMP's possessing no single set cannot be (A, B, λ, φ, 1) recurrent ([8], p. 60), and hence the ERM discussed in the previous section does not apply. This factor together with the ERM-related difficulties already mentioned, motivates development of other methods.

Suppose that $\{X_n: n \geq 0\}$ is an (A, B, λ, φ, k) GSSMC with invariant probability π. This will in fact be the case for the embedded jump process $(\underset{\sim}{S},\underset{\sim}{C})$, corresponding to a GSMP $\{X_t: t \geq 0\}$, provided that (see [9])

i) the state space S is finite

ii) the "road map" ;(s', s, i*) satisfies a natural irreducibility condition (see [7], p. 16)

iii) the densities f(·; s', i, s, i*) are positive and continuous on $[0,\infty)$, with finite mean.

For such an (A, B, λ, φ, k) recurrent $\{X_n: n \geq 0\}$ one can show that if $E_\pi|f(X)| < +\infty$, then

$$\frac{1}{n+1}\sum_{k=0}^{n} f(X_k) \to r = E_\pi f(X) \qquad P_\mu \quad \text{a.s.}$$

for any μ on $(E,\mathcal{E})$. A simulator commonly wishes to obtain an estimate for r, together with an associated confidence interval.

Let $T_1, T_2, \ldots$ be the consecutive hitting times of the set A. The "random blocking method" (RBM) hinges on the observation, due essentially to OREY (1959), that

$$V_k = (X_{T_k+1}, X_{T_k+2}, \ldots, X_{T_{k+1}})$$

is a Doeblin recurrent Markov chain with a single ergodic set (see DOOB (1953) for definitions and results). Under the assumption that V_k is aperiodic (this will generally be the case for GSMP's), functions of V_k will enjoy a central limit theorem with a variance constant of the form

$$\sigma^2 = \sigma^2(Z_0) + 2 \sum_{k=1}^{\infty} \operatorname{cov}(Z_0, Z_k). \tag{4.1}$$

By truncating the infinite sum and estimating the finite number of remaining terms in the series by the standard sample moments, we obtain the RBM.

1. Choose a truncation number m (the number of covariance terms of (4.1) to be retained).

2. Put

$$Y_k = \sum_{j=Tk+1}^{T_{k+1}} f(X_j), \quad \alpha_k = T_{k+1} - T_k \quad \text{and form}$$

$$\hat{r}(n) = \left(\sum_{k=1}^{n} Y_k\right) / \left(\sum_{k=1}^{n} \alpha_k\right)$$

$$c_{o\ell}(n) = \frac{1}{n-\ell} \sum_{k=1}^{n-\ell} Y_k \, Y_{k+\ell} - \frac{\hat{r}(n)}{n-\ell} \sum_{k=1}^{n-\ell} \alpha_k \, Y_{k+1}$$

$$- \frac{\hat{r}(n)}{n-\ell} \sum_{k=1}^{n-\ell} Y_k \, \alpha_{k+\ell} + \frac{\hat{r}^2(n)}{n-\ell} \sum_{k=1}^{n-\ell} \alpha_k \, \alpha_{k+\ell}$$

for $\ell = 0, 1, \ldots, m$

$$s^2(n) = c_{oo}(n) + 2 \sum_{\ell=1}^{m} c_{o\ell}(n).$$

3. A $100(1-\delta)\%$ confidence interval for r is

$$\hat{I} = \left[\hat{r}(n) - \frac{z_\delta \; s(n) \; n^{1/2}}{\sum_{k=1}^{n} \alpha_k} \; , \; \hat{r}(n) + \frac{z_\delta \; s(n) \; n^{1/2}}{\sum_{k=1}^{n} \alpha_k} \right] \; .$$

Justification of the above steps is possible, provided that

$$\int_A E_x\{|\sum_{j=1}^{T_1} f(X_j)|^{2+\delta}\}\pi(dx) < +\infty,$$

for some $\delta > 0$ ([9]). For GSMP's, the RBM can be modified in the same way as the ERM so as to provide confidence intervals for continuous-time quantities of the form (3.5).

Observe that when $m = 0$, this technique reduces to the "approximate regenerative method" (ARM) of CRANE and IGLEHART (1975). Hence one can think of the RBM as a second-order refinement of the ARM. It should be noted, however, that the RBM involves the undersirable element of having to make an a priori judgment as to an appropriate value of the trunca tion index m. This is a problem to which we intend to devote more attention.

5. Multivariate Autoregressive Method

The last method we shall discuss is the multivariate autoregressive method (MARM). Let $\{X_n : n \geq 0\}$ be a vector-valued strictly stationary process with mean vector $\underset{\sim}{\mu} = E\{X_0\}$ and covariance function $\underset{\sim}{\Sigma}(h) = E\{[\underset{\sim}{X}_n - \underset{\sim}{\mu}]\ [\underset{\sim}{X}_{n+h} - \mu]'\}$. Then under some regularity conditions (see BILLINGSLEY (1968), Theorem 20.1) the following central limit theorem holds as $n \to \infty$:

$$\sqrt{n}\left[\frac{1}{n}\sum_{j=0}^{n-1} \underset{\sim}{X}_j - \underset{\sim}{\mu}\right] \Longrightarrow N(0,\underset{\sim}{\Sigma}) \; ,$$

where $\underset{\sim}{\Sigma} = \underset{\sim}{\Sigma}(0) + 2\sum_{h=1}^{\infty} \underset{\sim}{\Sigma}(h)$.

To apply this result to simulation output analysis we need a method to estimate the covariance matrix $\underset{\sim}{\Sigma}$.

Let $\underset{\sim}{f}(\lambda)$ be the matrix-valued spectral density function for the process $\{X_n: n \geq 0\}$, namely,

$$\underset{\sim}{f}(\lambda) = \sum_{h=-\infty}^{\infty} \underset{\sim}{\Sigma}(h)\, e^{i\lambda h}, \quad \lambda \text{ real.}$$

Observe that $\underset{\sim}{\Sigma} = \underset{\sim}{f}(0)$. The function $\underset{\sim}{f}$ can be approximated arbitrarily closely by the corresponding spectral density of a multivariate autoregressive (MAR) process fitted to the process $\{\underset{\sim}{X}_n: n \geq 0\}$. To fit a MAR process of order p we must find matrices $\{\underset{\sim}{A}_k: 0 \leq k \leq p\}$ so that

$$\sum_{k=0}^{p} \underset{\sim}{A}_k \underset{\sim}{X}_{n-k} = \underset{\sim}{Z}_n ,$$

where the $\underset{\sim}{Z}_n$'s are i.i.d. normal with mean $\underset{\sim}{0}$ and known convariance $\underset{\sim}{B}$. We select the order p of the MAR process by applying the AIC-criterion of AKAIKE (1976), (1978). Once the model is fitted, it is a simple computation to find the spectral density function of the MAR process. This function evaluated at zero then provides our estimate for $\underset{\sim}{\Sigma}$.

This method and related ones will be developed in detail in JOW (1981). In Section 6 data will be presented from an application of the method to the lake model presented in Section 3.

6. Application to the Lake Model

In this section we illustrate the application of the three methods to the lake model presented in Example 2. Recall the model is generated by the recursion

$$X_n = \rho X_{n-1} + \varepsilon_n, \quad n \geq 0,$$

where the ε_n's are i.i.d. and $\rho < 1$. It can be shown that $X_n \Longrightarrow X$ as $n \to \infty$ with $E\{X\} = E\{Z_1\} = (1-\rho)^{-1} E\{\varepsilon_1\}$. For this simulation we have taken the ε_n's to have the distribution $P\{\varepsilon_n \leq x\} = 1 - (1-\rho)e^{-x}$, $x \geq 0$, which results in X being exponential with parameter 1. The simulations were carried out to estimate $E\{X\}$. For the ERM the return set $A = [0,b]$, where $b = (-\log \rho)/(1-\rho)$. This value can be shown to maximize the expected number of regenerations. The value $\rho = 0.75$ was used which makes $b = 1.15$. For the RBM the return set $A = [.75, 1.25]$. The infinite sum of covariance terms in the RBM was truncated at m ($m = 0, 5, 10, 25$). For the MARM the vector of observations used was $(X_n, X_n^2, 1_{[0, \ln 2]}(X_n))$. A total of 10,000 observations were generated in each of 30 replications.

Table 6.1. Simulation Results for Lake Model

number of observations	value estimated	theoretical values	ERM	RBM m = 0	RBM m = 5	RBM m = 10	RBM m = 25	MARM
1000	E{X}	1.0	1.018+/- 0.023	1.022+/- 0.023	1.022+/- 0.023	1.022+/- 0.023	1.022+/- 0.023	1.018+/- 0.024
	σ^2	7.0	7.307+/- 0.656	7.057+/- 0.635	6.757+/- 0.797	6.674+/- 1.184	6.115+/- 1.897	7.230+/- 0.527
	coverage probability	.90	0.900+/- 0.090	0.900+/- 0.090	0.933+/- 0.075	0.900+/- 0.090	0.767+/- 0.127	0.900+/- 0.090
5000	E{X}	1.0	1.015+/- 0.009	1.016+/- 0.009	1.016+/- 0.009	1.016+/- 0.009	1.016+/- 0.009	1.015+/- 0.009
	σ^2	7.0	7.191+/- 0.297	7.037+/- 0.299	7.181+/- 0.487	7.335+/- 0.591	7.842+/- 1.002	7.107+/- 0.247
	coverage probability	0.9	0.933+/- 0.075	0.933+/- 0.075	0.933+/- 0.075	0.933+/- 0.075	0.900+/- 0.090	0.967+/- 0.055
10,000	E{X}	1.0	1.006+/- 0.007	1.007+/- 0.007	1.007+/- 0.007	1.007+/- 0.007	1.007+/- 0.007	1.007+/- 0.007
	σ^2	7.0	7.062+/- 0.210	6.919+/- 0.185	7.026+/- 0.313	6.989+/- 0.361	7.018+/- 0.643	7.001+/- 0.153
	coverage probability	0.9	0.933+/- 0.075	0.933+/- 0.075	0.933+/- 0.075	0.900+/- 0.090	0.867+/- 0.102	0.933+/- 0.076

7. References

[1] Akaike, H. (1976). Canonical Correlation Analysis of Time Series and the use of an Information Criterion. Systems Identification: Advances and Case Studies, R. K. Mehra and D. G. Lainioties, eds., Academic Press, New York.

[2] Akaike, H. and G. Kitigawa (1978). A Procedure for the Modeling of Non-stationary Time Series. Ann. Inst. Statist. Math., 30, B, 351-353.

[3] Athreya, K. B. and P. Ney (1978). A New Approach to the Limit Theory of Markov Chains. Trans. Amer. Math. Soc., 245. 493-501.

[4] Billingsley, P. (1968). Convergence of Probability Measures. John Wiley, New York.

[5] Crane, M. A. and D. L. Iglehart (1975). Simulating Stable Stochastic Systems, IV: Approximation Techniques. Management Sci., 21, 1215-1224.

[6] Doob, J. L. (1953). Stochastic Processes. John Wiley and Sons, New York.

[7] Fossett, L. D. (1979). Simulating Generalized Semi-Markov Processes. Ph.D. Dissertation, Department of Operations Research, Stanford University.

[8] Glynn, P. W. (1980). An Approach to Regenerative Simulation on a General State Space. Technical Report 53, Department of Operations Research, Stanford University.

[9] Glynn, P. W. (1981). Forthcoming Technical Report, Department of Operations Research, Stanford University.

[10] Harris, T. E. (1956). The Existence of Stationary Measures for Certain Markov Processes. Proc. Third Berkeley Symposium on Mathematical Statistics and Probability, Vol. 2, Jersey Neyman, ed., University of California Press, Berkeley.

[11] Hordijk, A. and R. Schassberger (1982). Weak Convergence for Generalized Semi-Markov Processes. To appear in Stochastic Process. Appl.

[12] Jow, L. (1981). Forthcoming Technical Report. Department of Operations Research, Stanford University.

[13] Nummelin, E. (1978). A Splitting Technique for Harris Recurrent Markov Chains. Z. Wahrsch. Verw. Gebiete., 43, 309-318.

[14] Orey, S. (1959). Recurrent Markov Chains. Pacific J. Math., 9, 805-827.

[15] Smith, W. L. (1955). Regenerative Stochastic Processes. Proc. Roy. Soc. London, Ser. A, 232, 6-31.

[16] Whitt, W. (1980). Continuity of Generalized Semi-Markov Processes. Math. Oper. Res., 5, 494-501.

Stanford University

This research was supported by NSF Grant MCS79-09139, ONR Contract N00014-76-C-0578 (NR 042-343) and a Natural Sciences and Engineering Research Council of Canada Postgraduate Scholarship.

MODELS AND PROBLEMS OF DYNAMIC MEMORY ALLOCATION

V. E. Beneš

Abstract

We construct and analyze some stochastic models for optimal allocation of memory in fragmented arenas. The arenas, as well as the item types they carry, are of various shapes: linear, cubic, pentagonal pie, S^2 (surface of sphere). Formulated as a Markov dynamic programming problem, the choice of optimal memory allocation closely resembles the choice of an optimal route for a telephone call. Much can be guessed about optimal operation from scrutiny of the partial ordering of the possible states reduced under symmetries; especially, in many specific examples, a natural relation B (read "better than") of preference can be defined among alternative allocations, and a topological fixed-point argument based on B given, to solve the allocation problem without resort to numerical solution of the Bellman equation.

1. Introduction

The efficient allocation of storage space or memory is a fundamental problem in many kinds of industrial process, data systems, inventory situations, and the like. It is especially important [1] for time-shared computer design and operation, and we use terminology from this area henceforth.

In such a system, several programs are being executed at the same time; each program is stored in a memory while it is being carried out;

when it is finished, the memory locations it occupied become available for other incoming programs. We picture the operating system as follows: programs of various lengths arrive in a more or less random manner; if there is room, an arriving program is allocated some empty space in memory adequate for its size; it lives there for some random execution time, and then is removed; a program whose size cannot be accommodated is delayed, passed to other equipment, or rejected and lost.

The items in a memory are either *relocatable*, or they are fixed in place until their final removal at the program's termination. In the former case it is often approximately true that any memory spaces that are blank can be viewed (in principle) as a single usable (not necessarily connected) piece that will accept a new program; this case is exemplified by the extreme use of linked lists. In the latter case the memory is to some extent *fragmented*, and it may not be able to accept a new program whose length is less than the total amount of blank memory, because the blanks occur in a disjoint non-contiguous way, and the memory operates under the constraint that a program must occupy a connected portion of memory. We shall be concerned with the optimal allocation of memory in the second, fragmented case.

Here there arises a natural allocation problem, because after the system has operated for some time, the memory will be a jumble of filled parts and holes, depending on what sizes of programs arrived when, the how long it took to execute them. Faced with a newly arrived program of announced size, the system must find a good place for it in the memory if it can. The feature that programs can have many different sizes is important and realistic, and it makes the analytical problems hard [2]. There is a discussion of these problems in D. E. Knuth's book [3].

We study this kind of allocation problem in the context of a simple probabilistic model, and conclude that for many examples it can be solved by graphical and analytical methods based on domination arguments and dynamic programming.

2. Outline of Our Method

To allay the "curse of dimensionality" laid on us by the large number of memory states, we replace states by their equivalence classes under symmetries. By making some simple assumptions about the ways programs arrive and leave, we model the operating memory as a controlled Markov process: the control decisions place programs of particular type into available spaces in memory. This method of modeling allows a dynamic programming formulation of the allocation problem. As in other problems of this kind, the best control policy either must be guessed (and justified,) or it must be obtained by laboriously solving the attendant Bellman equation numerically.

We choose the former method, proceeding thus: for various examples, an examination of the possible states of the memory often quickly suggests which ones are "good" and "bad"; here by a "good" state we mean one in which the resident programs are efficiently packed so as to leave room for a wide variety of arriving programs, and by a "bad" state we mean one which is fragmented into a checkerboard pattern which will not accept large programs. Figures 1-9 illustrate these heuristics for various memory structures. From this initial intuitive idea of good and bad states we develop or guess a *preference* relation B (Figures 1-9 again.) among states that are alternative ways of entering a program; '$\alpha B \beta$' is to be interpreted as 'α is better than β'; this relation must sometimes be extended to states that are *not* alternatives. The next step is to guess that the optimal policy is to go to a "most

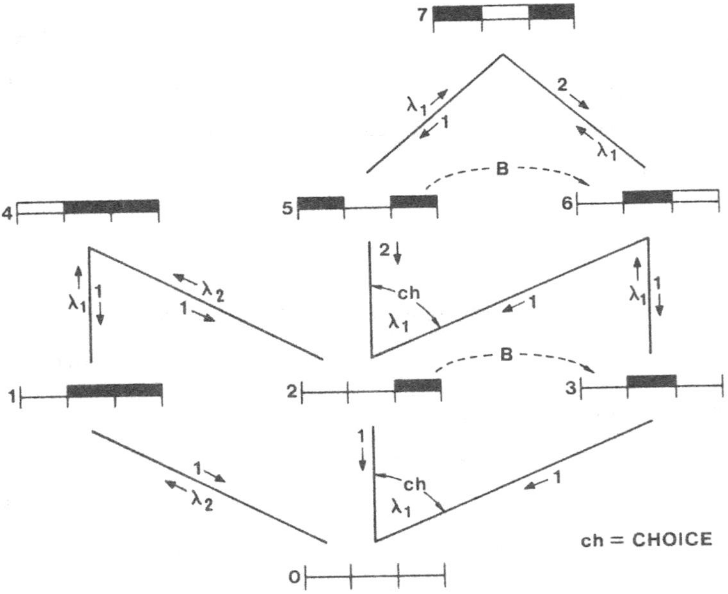

Figure 1. States of 3-Cell Linear Memory Reduced Under Symmetries, with Transition Rates

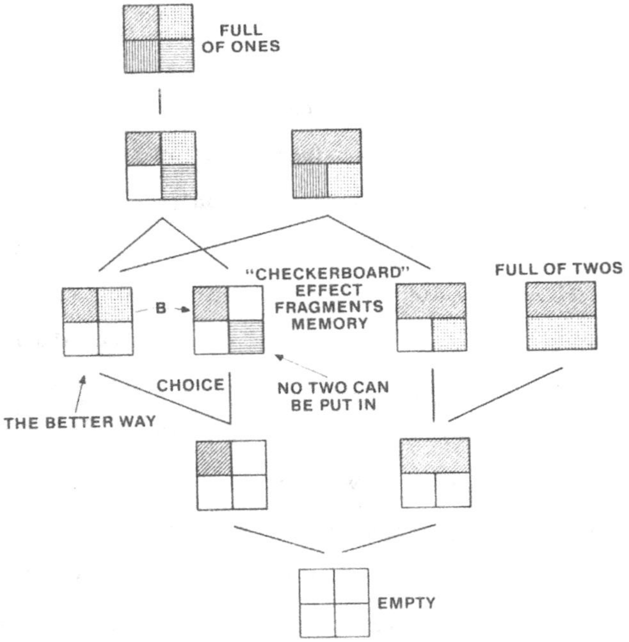

Figure 2. Square Arena Carrying "Ones" & "Twos"

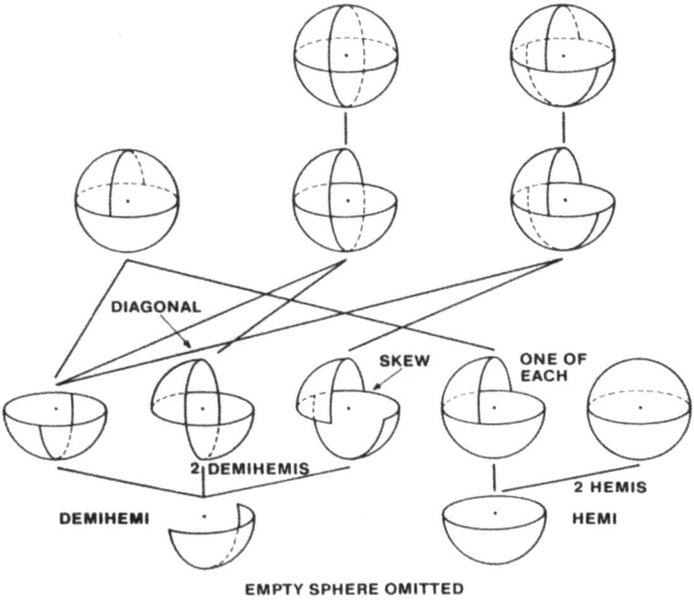

Figure 3. S_2-Arena Carrying Hemispheres and Demihemispheres

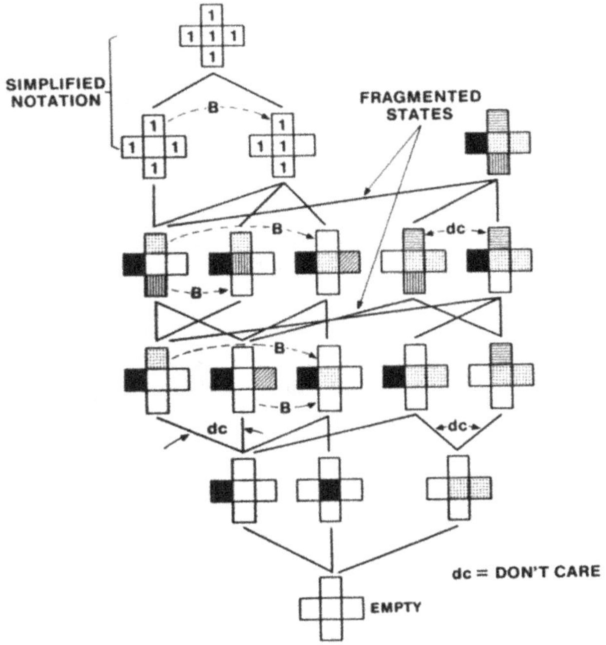

Figure 4. "Cross" Arena Carrying Ones & Twos

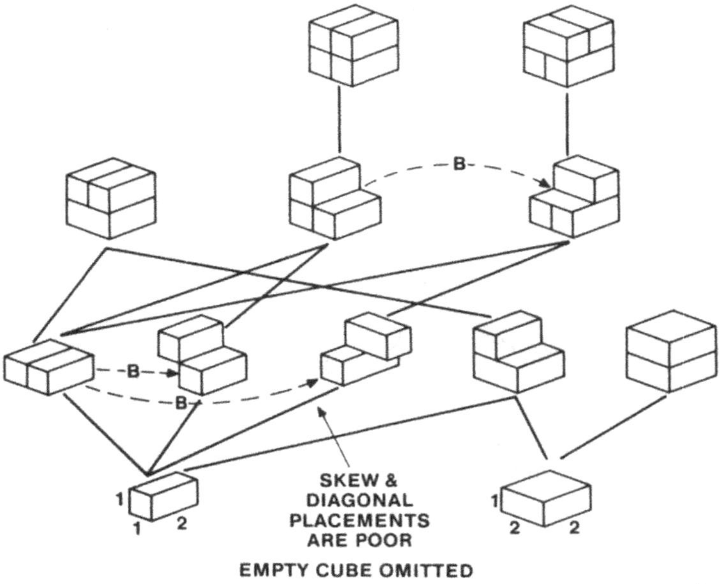

Figure 5. 2 × 2 × 2 Cubic Arena Carrying 1 × 1 × 2 & 1 × 2 × 2

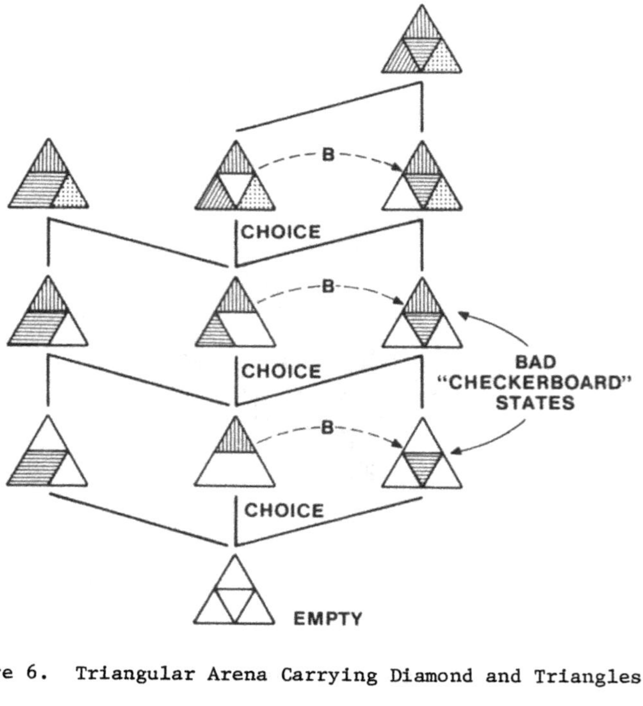

Figure 6. Triangular Arena Carrying Diamond and Triangles

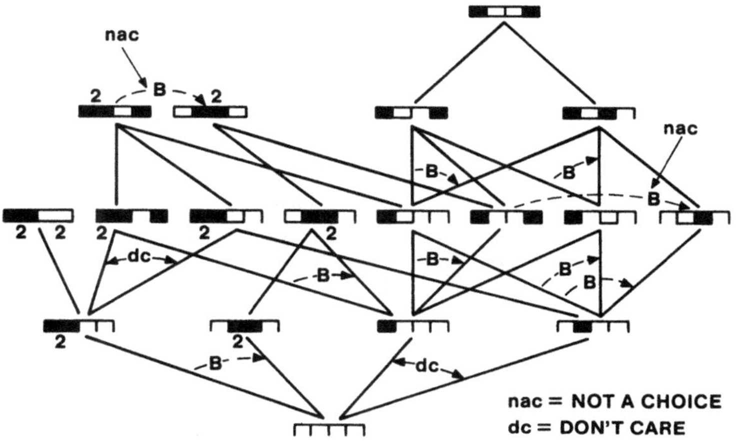

Figure 7. 4-Cell Linear Arena Carrying Ones & Twos

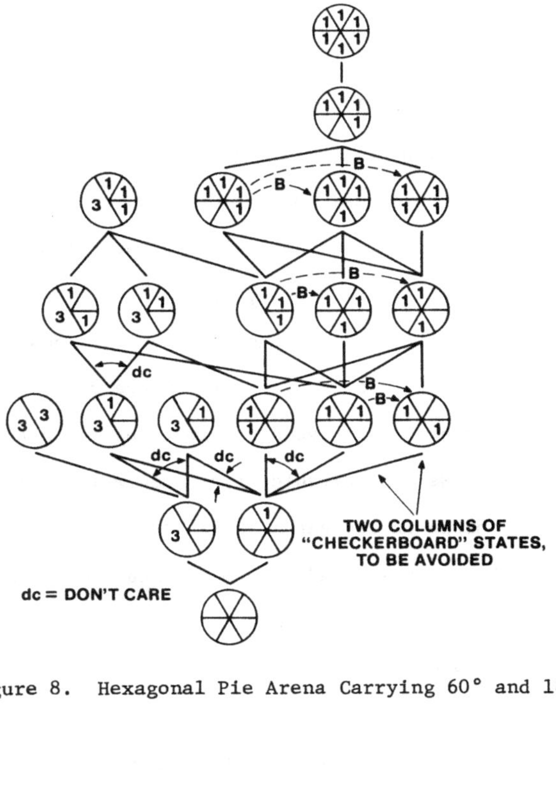

Figure 8. Hexagonal Pie Arena Carrying 60° and 180°

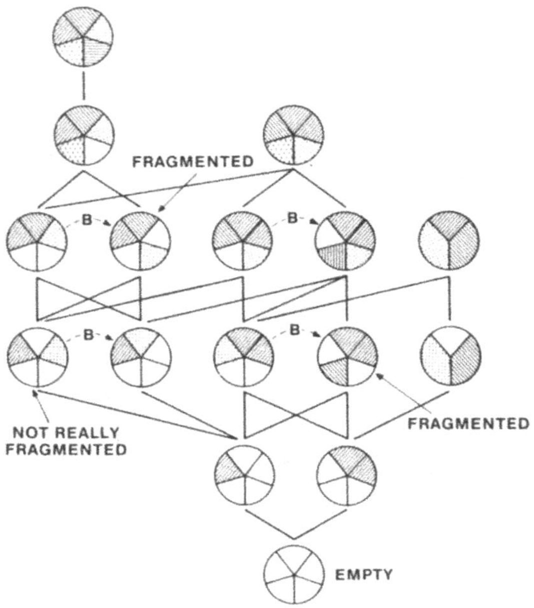

Figure 9. Pentagonal Pie Arena Carrying 72° and 144°

preferred" state among the alternatives.

The hard part, of course, is to show that the guess is right. A straightforward, though not easy, way to do this is by showing that the guess corresponds to or achieves the maxima in the Bellman equation. Mathematically, this amounts to proving that the Bellman equation has a solution in a certain simplex defined by inequalities based on the preference relation B. For "suitable" preference relations, this problem can be solved by successive approximations or by topological fixed point methods. Here a suitable preference will be one which is consistent or reasonable in the sense that, roughly, if a state (equivalence class) α is preferred to one β, then *neighbors* of α are preferred to various corresponding neighbors of β.

We give this consistency idea a precise formulation, and call it the *monotone property*, using it as follows: writing the Bellman equation in the form $v = Tv$ for a suitable nonlinear T, we note that if

B has the monotone property, then T maps the simplex

$$\{v \geq 0: \alpha B\beta \Rightarrow v_\alpha > v_\beta\}$$

corresponding to B *into itself*. It follows easily that T maps a compact convex subset K of the simplex into itself, and so various fixed point theorems guarantee that there is a fixed point of T in K, i.e., a solution of the Bellman equation satisfying the inequalities defining the simplex, i.e., a solution that justifies our guess as optimal. The verification that a proposed preference relation has the monotone property is a combinatorial and arithmetical process, and can often be carried out graphically by inspection of the state diagram.

It should be understood that the monotone property yields a very strong domination argument, and leads to optimal policies that are independent of the numerical values of the program arrival rates. The examples of Figures 1-9 have it, while that of Figure 10 seems to lack it. When it works, our method is truly an illustration of Bertrand Russell's quip that mathematics has all the advantages of theft over honest toil.

3. Summary

The basic structure of the set of possible states is described in Section 4. Reduction of this set under a symmetry group is achieved in Section 5; this reduction replaces states by their equivalence classes and greatly simplifies the treatment of examples. Packing rules and analogies to telephone routing appear in an informal discussion in Section 6. The Sections 7 and 8 are devoted to basic modeling assumptions, and to describing the process πx_t on the equivalence classes that models memory operation. Performance concepts are discussed in Section 9. Our method of justifying guessed allocations as optimal is

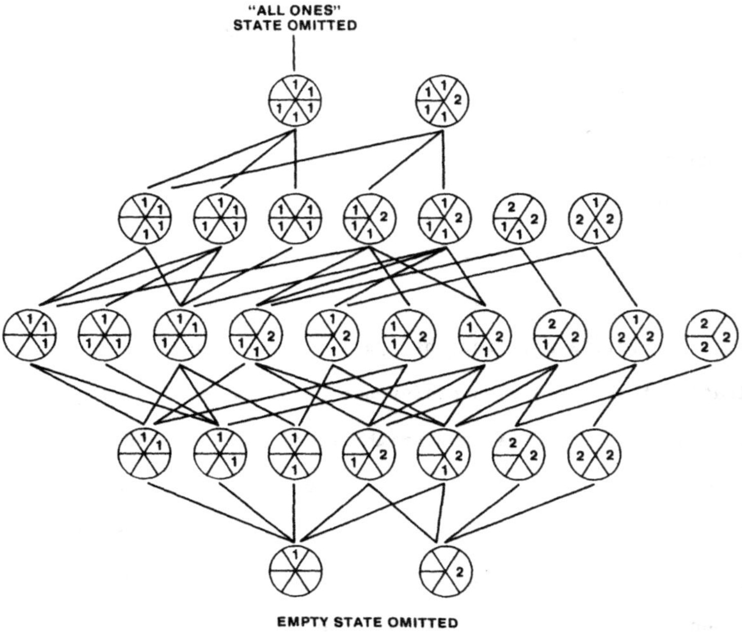

Figure 10. Hexagonal Pie, Apparently Lacks Monotone Property

introduced in Section 10 by a simple example, a small linear memory. This example is solved for various performance indices in Sections 11-13. The ideas behind these solutions are formulated in a general form in Sections 14-18.

4. States of the Memory

In order to attack the allocation problem it is very useful to have a mathematical description of the set S of possible states the memory assumes. The memories we have in mind, sometimes called *arenas*, always have a definite structure, size, and shape, and these, together with the types, (shapes, sizes, etc.) of the incoming programs, determine the possible states. Typically, the state is something like a labeled graph, and is not easily described by anything as simple as an n-tuple of integers. A suitable description of S arises quite naturally when we think of a state x as a choice of disjoint subsets or regions of the arena, of fitting size and shape, one for each item to be stored.

Not every such choice need represent a state; S may exclude some choices as wasteful, and others because of integral constraints. The choices are partially ordered by inclusion $\leq$, where $x \leq y$ means that x can be reached from y by removing zero or more subsets from y. It is thus reasonable that if y is a state, and $x \leq y$, then x is a state, i.e., that S should be closed under removal of items from memory. It can be seen from these properties that the set S of permitted states has the structure of a *semilattice*, that is, a partial ordering whose order relation is definable in terms of a binary operation $\cap$ that is idempotent, commutative, and associative, by the formula

$$x \leq y \iff x = x \cap y$$

Here for $x \cap y$ we can use usual set intersection: $x \cap y$ is exactly the state consisting of those items in memory and their respective locations there which are common to both x and y; of course, there may not be any, and then $x \cap y$ is the overall inf in the semilattice. The word 'cover' will be used in the usual way for partial orderings, to mean "lies immediately above."

An *assignment* is a specification of the sizes and shapes of the items to be placed in memory, without regard to where they might be put. An assignment is *realizable* if it can be put into the memory somehow; not every assignment is realizable. The set A of assignments forms a semilattice in the same way that the states do, and A is related to S thus: call two states x, y in S *equivalent* if they realize the same assignment, written $x \sim y$. The realizable assignments can then be identified with the equivalence classes of states under $\sim$, and there is a natural map γ: $S \to A$, the projection that takes each state x into the assignment $\gamma(x)$ it realizes, i.e., its equivalence class under $\sim$. The assignment concept is important because memory allocation choices are always made within an equivalence class under $\sim$.

With x and y states such that $x \geq y$, it is convenient to use $x - y$ to denote the state resulting from x by removing all the items in y; similarly for a and b assignments with $a \geq b$, $a - b$ is the assignment resulting from a by dropping all the items intended in b. It can now be seen that the map γ is a semilattice homomorphism of S into A, with the properties:

$$x \geq y \Longrightarrow \gamma(x) \geq \gamma(y)$$

$$x \geq y \Longrightarrow \gamma(x-y) = \gamma(x) - \gamma(y)$$

$$\gamma(x \cap y) \leq \gamma(x) \cap \gamma(y)$$

$$\gamma(x) = \emptyset \Longrightarrow x = 0 = \text{zero state, with nothing in memory}$$

We denote by A_x the set of states that are immediately above x in the partial ordering $\leq$ of S, and by B_x the set of those that are immediately below. A *unit assignment* is one consisting of a single program p, identified really only by its size or shape. A program p is blocked in a state x if x has no empty area large enough to accommodate p. It is convenient to write $\gamma(x) \cup p$ for the larger assignment consisting of $\gamma(x)$ and the program p together. Let also

$$A_{px} = A_x \cap \gamma^{-1}\{\gamma(x) \cup p\};$$

this is the subset of states that could result from x by entering p in memory, because $\gamma^{-1}\{\gamma(y)\}$ is the equivalence class of y under $\sim$. A_{px} is empty if and only if p is blocked in x. It can be seen that with F_x the set of programs that are not blocked in x, the family $\{A_{px}, p \in F_x\}$ forms the partition of A_z induced by the equivalence $\sim$. The notation $|x|$ indicates the number of programs stored in memory state x.

5. Reduction of States Under a Symmetry Group

It is obvious that the set of states can have astronomical cardinality, and also that many states are essentially equivalent (as far as performance is concerned) by virtue of being images of each other under various systematic symmetries. These facts suggest that for purposes of bookkeeping and analysis it is enough to consider equivalence classes of states as new macrostates, and thereby cut down the order of equations, the dimensions of vectors, etc. Such a program was originally suggested for telephone network applications [4] by S. P. Lloyd; we shall see that it is also valuable here.

The basic idea is to replace an individual state by its equivalence class under the symmetries to be considered, on the grounds that all the states in an equivalence class have the same stochastic behavior. The usually very large set of states is thereby reduced to more manageable size, without loss of essential information. The idea of "same stochastic behavior" can be described mathematically, and properly done it leads to a new Markov process on the equivalence states for a description of the operating system.

The equivalence classes under the symmetries form a partition of S, and we consider the class of all bijections g of S into itself which preserve this partition, i.e., map each equivalence class into itself. This class forms a group G under composition. The original equivalence relation can be recovered from G by the formula

$$x \equiv y \text{ iff there is a } g \in G \text{ such that } x = gy$$

and the projection map π that carries each x to its equivalence class can be defined as an orbit: $\pi x = \{gx: g \in G\}$. G acts transitively on each πx, and the πx are disjoint or identical, forming in fact the partition induced by $\equiv$.

The group G is now used to define "same stochastic behavior," as follows: Let x_t be a continuous parameter Markov process taking values on S, with a rate matrix $Q = (q_{xy})$. The group G and the equivalence $\equiv$ becomes relevant to x_t when the matrix Q is unaffected by the permutations of its rows and columns that correspond to $g \in G$. This relevance is recognized in the

(1) Definition: Q admits G iff $g \in G, x, y \in S \Longrightarrow$

$$q_{xy} = q_{(gx)(gy)}$$

If now $E = \{\pi x \colon x \in S\}$ is the set of equivalence classes then the projection $\pi : x \to \{gx \colon g \in G\} = \pi x$ defines a new "macroscopic" Markov process on E, with transition rates

$$q_{\alpha\beta} = \sum_{y \in \beta} q_{xy} \qquad x \in \alpha,\ \alpha \in E,\ \beta \in E.$$

The partial order $\leq$ on S naturally induces one on E according to $\alpha \leq \beta$ iff $\equiv \exists x, y, x \in \alpha,\ y \in \beta,\ x \leq y$. It is convenient to define, in analogy with A_x, B_x, A_{px}, and $|x|$, the notations

$$A_\alpha = \{\beta\colon \exists x, y\ x \in \alpha,\ y \in \beta,\ y \in A_x\}$$

$$B_\alpha = \{\beta\colon \exists x, y\ x \in \alpha,\ y \in \beta,\ y \in B_x\}$$

$$A_{p\alpha} = \{\beta\colon \beta \in A_\alpha \text{ and } \exists x \in \alpha\ \beta \cap A_{px} \neq \emptyset\}$$

$|\alpha| = |x|$ for $x \in \alpha$, (the same for all $x \in \alpha$).

6. Packing Rules

An algorithm or strategy for choosing memory locations for arriving programs can often substantially improve performance. Such gains can be realized in many systems in which the underlying combinatorial structure leads to several ways to taking an action, such as adding a program to memory, or routing a telephone call. Indeed, there are so many analogies between our allocation problem and call routing in telephony that we shall try systematically to exploit them. In telephone switching, methods for choosing routes for calls through the connecting network are called *packing rules*, in analogy with the problem of packing items of various sizes in a bin. This analogy is reasonably accurate as far as adding items is concerned; but both telephony and memory allocation involve the (usually) random removal of items to leave new "holes," a feature not present in the usual bin-packing problem. The point is that the items to be packed change irregularly, and one has to do the best he can from where he is, not just find the best packing for a given set of items. In this respect the bin-packing analogy breaks down.

For memory allocation, several simple rules have been advanced and used:

(1) *First fit*: search through the available space in some systematic way till you find a blank area which will take the arriving item. A telephonic analog is to search the central switches of a connecting network in a fixed order, and to choose the first feasible route.

(2) *Best fit*: search for the smallest (or in some other sense "best") blank space which will take the item. This rule is not a precise admonition, but depends on just what is meant by 'best fit'. It is therefore a whole class of rules each trying to express the intuitive idea that given the state of the memory and the item to go in, there is at least one best thing you could do to maximize throughput, minimize blocking, in short to optimize performance under whatever reasonable criterion you had chosen.

The meaning of "best" in a best fit rule is closely related to the horizon or domain over which one is optimizing. One might, for example, choose a location for a program purely on the basis of combinatorial properties of the available options: one leaves more room for later items, while the others are of a "checkerboard" type that excludes large items, so on this basis we choose the first; this is a reasonable local optimization which we hope is also globally optimal. Looking further, though, one might also want to look at the states that are *neighbors* of the available options, because we know that the system will have to move to one of these later on; so we try to formulate a notion of "best fit" suitable for an expanded, two-step situation. Carrying this line of thought to its logical extreme will result in defining "best fit" in terms of a complex dynamic programming methodology. This is in fact precisely what we shall do, but it will turn out that the notion of "best fit" so obtained is precisely the one suggested by some simple intuitive combinatorics.

The best fit idea has its telephonic analog. For many years it was a folk theorem among switching engineers that the best packing rule placed calls through the most heavily loaded part of the network that would accept them. Subsequently, examples of networks were found [5] for which such a rule is demonstrably optimal in the sense that it minimized the probability of blocking, as well as other criteria.

7. Modeling

Some basic probabilistic assumptions and the models they lead to are described next. We shall be interested primarily in "best fit" packing rules which are provably optimal; emphasis shall be on the possible combinatorial meanings of 'best fit' and on the probabilistic criteria of performance which are to be optimized. For these reasons we want to keep the probabilistic assumptions as simple as possible, and let the combinatorics speak for itself. Let us assume that

(i) Programs arrive at random, in a Poisson process of rate $\lambda > 0$

(ii) Types of programs, e.g., sizes and shapes, are i.i.d. variates with support a finite set σ, and independent of the Poisson arrival process. For $p \in \sigma$, λ_p is the rate of items of type p, and $\sum_{p\in\sigma} \lambda_p = \lambda$.

(iii) The lengths of time that programs stay in the memory arena are i.i.d. variates, independent of the arrivals and the types, with a negative exponential distribution of mean unity.

To complete the model for an operating memory we add an allocation algorithm r (packing rule) that tells where to put programs when there is a choice. This takes the form of a function

$$r: p,x,y \to r_{pxy} \qquad p \in F_x,\ y \in A_{px}$$

such that r_{pxy} is the fraction of times y results from x when item p comes along, and r is being used. The function r must satisfy the admissibility constraints $r \geq 0$ and

$$\sum_{y\in A_{px}} r_{pxy} = 1,$$

the last indicating that all of the arrival rate λ_p is spread over A_{px}.

Putting all the model assumptions together, we can say that each admissible r will give rise to a Markov process x_t taking values in S, with a generator $(q_{xy}) = Q$ given by

$$q_{xy} = \begin{cases} 1 & \text{if } y \in B_x \\ \lambda_p r_{pxy} & \text{if } y \in A_{px},\ p \in F_x \\ -|x| - \sum_{p\in F_x} \sum_{y\in A_{px}} \lambda_p r_{pxy} & \text{if } x = y \\ 0 & \text{otherwise} \end{cases}$$

The allocation problem consists in optimizing performance by judiciously choosing an admissible r.

(2) Remark: In speaking of r_{pxy} as the fraction of times p gives rise to y from x we are allowing a form of convexification. We could also let r depend on time or past history. Both these fillips are gratuitous, though; it turns out that the desired optima are achieved by time-independent r assuming only 0 and 1 as values.

8. The Macroscopic Process πx_t on E

It is intuitively clear that best performance is unaffected by knowing which state of an equivalence class α we start in; the transient effect is the same for all $x \in \alpha$. Similarly if $y \in A_{px}$ and $g \in G$ = natural symmetry group then going to y from x when p comes along is just as good (or just as bad) as going to gy from gx. Therefore we shall consider only allocation strategies r that are *consistent* in that they do the same thing with p in equivalent states, i.e.,

$$r_{pxy} = r_{p(gx)(gy)}, \qquad \text{for } g \in G.$$

In short, the matrix (r_{pxy}) is to admit G.

We can now write down the generator for the macroscopic process πx_t on E. Its transition rates are given by

$$q_{\alpha\beta} = \sum_{y\in\beta} q_{xy}, \qquad x \in \alpha,\ \beta \neq \alpha.$$

This quantity is zero unless β and α are neighbors, and if they are

$$q_{\alpha\beta} = \begin{cases} \sum_{y\in\beta} 1_{y\in B_x} & \text{if } \alpha \text{ covers } \beta \text{ and } x \in \alpha \\ \\ \lambda_p \sum_{y\in\beta} r_{pxy} & \text{if } \beta \text{ covers } \alpha,\ x \in \alpha,\ p \in F_x \end{cases}$$

It is now shown that each of these sums as defined has the same value for all $x \in \alpha$. For take any other $z \in \alpha$; there is a $g \in G$ such that $z = gx$, and so

$$\sum_{y\varepsilon\beta} r_{pzy} = \sum_{y\varepsilon\beta} r_{p(gx)y} = \sum_{y\varepsilon\beta} r_{p(gx)(gy)} = \sum_{y\varepsilon\beta} r_{pxy},$$

the third equality resulting from the fact that equivalence classes $\beta \varepsilon E$ are fixed by $g \varepsilon G$. Similarly for z, x both in α

$$\sum_{y\varepsilon\beta} 1_{y\varepsilon B_z} = \sum_{y\varepsilon\beta} 1_{y\varepsilon B_{gx}} = \sum_{y\varepsilon\beta} 1_{gy\varepsilon B_{gx}} = \sum_{y\varepsilon\beta} 1_{y\varepsilon B_x},$$

since $y \varepsilon B_x$ iff $gy \varepsilon B_{gx}$, so that $1_{y\varepsilon B_x}$ admits G. The numbers

$$h_{\alpha\beta} = \sum_{y\varepsilon\beta} 1_{y\varepsilon B_x}, \qquad x \varepsilon \alpha$$

= number of ways a departure could take $x \varepsilon \alpha$ into a state $y \varepsilon \beta$

represent the multiplicities inherent in the hangup or departure process, while those

$$r_{p\alpha\beta} = \sum_{y\varepsilon\beta} r_{pxy}, \qquad x \varepsilon \alpha$$

define an allocation strategy for the projected Markov process πx_t. Since $x, z \varepsilon \alpha \Longrightarrow F_x = F_z$, we use F_α to denote the set of program types which can be stored in a state of α. Recalling that $A_{p\alpha}$ is the set of equivalence classes reachable from an $x \varepsilon \alpha$ by storing item type p, we write the generator $Q = (q_{\alpha\beta})$ of the process πx_t as

$$q_{\alpha\beta} = \begin{cases} h_{\alpha\beta} & \text{if } \beta \varepsilon B_\alpha \\ \lambda_p r_{p\alpha\beta} & \text{if } \beta \varepsilon A_{p\alpha},\ p \varepsilon F_\alpha \\ -|\alpha| - \sum_{p\varepsilon F_\alpha} \sum_{\gamma\varepsilon A_{p\alpha}} \lambda_p r_{p\alpha\gamma} & \text{if } \alpha = \beta \\ 0 & \text{otherwise} \end{cases}$$

The allocation problem consists in optimizing performance by choosing the $r_{p\alpha\beta}$. Henceforth we replace x_t by πx_t and S by E. Variables $x, y, z, \cdots$ will now denote vectors of dimension $|E|$.

9. Performance

The operational quality of memory allocation systems such as those under discussion can be viewed in two ways. In one, the rejection of a new program because of blocking or overflow is seen as catastrophic: when it happens the system breaks down and all bets are off; the object of allocation is to put off the catastrophe on the average as far into the future as possible, so the appropriate index is the mean time to a blocked program, which should be suitably large. In the second, blocking is considered to be a much more minor inconvenience; blocked programs are as a matter of course shunted to other devices or delayed, and as in telephony, we are interested in the fraction of arrivals that cannot be accommodated, so the object of control is to achieve efficiency in the sense of minimizing this fraction. It will turn out in many cases that the same policy achieves both extrema.

Thus two criteria of memory system performance will be of principal interest here: the mean time until a program cannot be accepted, and the fraction of arrivals that cannot be accepted. The first is self-explanatory; as for the second, it is analogous to the probability of blocking in telephony, and will be so called. We remark that blocking is not an event in the sample space whose probability is defined by our model, and that it is an asymptotic ratio calculable in the model and best defined by describing how to measure it. We count the number $a(t)$ of programs arriving during $(0,t)$, and also the number $b(t)$ of blocked arrivals. The long-term fraction of arrivals that are blocked is then

$$\lim_{t\to\infty} \frac{b(t)}{a(t)}$$

called the *probability of blocking*. This definition is justified by showing that the limit exists and is constant with probability 1; indeed its value is

$$1 - \lambda^{-1} \sum_{\alpha \varepsilon E} p_\alpha \sum_{p \varepsilon F_\alpha} \lambda_p \tag{3}$$

where $\{p_\alpha, \alpha \varepsilon E\}$ are the stationary probabilities corresponding to the generator Q, solving $p'Q = 0$.

The analogy to blocking in networks is more than superficial. Blocking results in networks when terminals are idle but cannot be connected to each other because of unfavorable combinatorial conditions in the (state of the) network; and blocking results in memory allocation when the total blank area exceeds the size of the arriving program, but is *fragmented* into pieces no one of which is large enough. It should be mentioned that another kind of congestion also occurs in both kinds of system, namely *overflow*, caused less by poor combinatorics than by sheer overload. Thus a trunk group will overflow when all trunks are busy, and a memory will overflow when it is totally full, and in neither case is the problem combinatorial. The memories to be considered here will be finite and hence will overflow, and we shall count overflowing arrivals as blocked.

(4) Remark: Since the set of admissible r is convex and the constraint defining $\{p_\alpha, \alpha \varepsilon E\}$ is linear, it follows from (3) that the optimal allocation problem for least blocking is one of *linear programming*. We do not, however, attack it by this familiar method, because we have found that in the simple probabilistic setting postulated here, it becomes possible to prove that certain combinatorial hints (as to what is good allocation,) obtainable by looking at the partial ordering of states, are right on the mark, and yield optimal policies. These hints are described next.

10. Small Linear Memory

Our approach is best introduced by means of a very simple example. A linear memory, depicted by an interval, is to be used to store items which have length 1 or 2, and which cannot be divided up among unconnected blank areas in memory. We restrict attention to the simplest case, and assume that the size (length) of the memory is exactly 3 and that no stored item can start or stop at other than an integral position. These assumptions lead to a finite number of possible states. Identifying them modulo their natural group of symmetries leads to the reduced 8-state diagram shown in Figure 1. Numbers 0 to 7 name the states, and λ_1, λ_2 are the respective arrival rates of items of size 1, 2.

We now offer the following heuristics: states 2 and 3 are both ways of entering a single item of length 1; in 3 this item "hogs the center," precluding storage of an item of length 2, if it should come along; in 2, though, there is still room for a program of length 2. Also, states 5 and 6 are both ways of storing two programs of length 1; either can result from 2 if a program of length 1 arrives; as far as new programs are concerned, 5 and 6 are of equal value, both leading to 7; but the departure of the "outer" item in 6 would lead to 3, already discerned to be poor, while the departure of either item in 5 leads to 2. (Note how the fact that states are equivalence classes is at work here.) Our guess (henceforth called g) is then that 2 is preferable to 3, and 5 to 6; this guess g is depicted in Figure 1 by the preference relation B, read "is better than": the state at the start of the B-labeled arrow is better than the one at the point.

We have thus arrived at a natural or plausible allocation algorithm g. Since its heuristic justification above depended on looking ahead to what could happen after a particular allocation, it is really a "best fit" algorithm in a nonlocal sense, since 'best fit' refers not only to the state in question but also to its neighbors. The interesting questions are these: Is g indeed optimal? If so, how can it be *proved* to be optimal? They *can* be attacked and answered numberically (for particular parameter values) by the methods of dynamic programming; however, it turns out that for this example (and many others) some simple domination arguments quickly show that the algorithm g we guessed at, and analogs of g for other examples, are optimal for at least three criteria.

11. Maximum Expected Number of Steps Until a Program is Blocked

By a *step* let us mean either the arrival of a new program or the departure of one in memory, so that counting steps is like measuring time. Our first result is

(5) Proposition: For each $0 \leq i \leq 7$, g maximizes the expected number of steps until the first blocked program, given that the memory starts in state i.

This will follow from a stronger result:

(6) Proposition: For each $0 \leq i \leq 7$ and $n \geq 1$, g simultaneously maximizes all the quantities

$$\Pr\{\text{no program is blocked in } n \text{ steps} \mid \text{start in } i\}.$$

Proof. Put, for $0 \leq i \leq 7$

$$p_i(n) = \Pr\{\text{no blocked programs in } n \text{ steps} \mid \begin{matrix}\text{start at } i \text{ and an} \\ \text{optimal policy is used}\end{matrix}\}$$

The dynamic programming equations are

$$p_0(n+1) = \frac{\lambda_2}{\lambda_1+\lambda_2} p_1(n) + \frac{\lambda_1}{\lambda_1+\lambda_2} \max\{p_2(n), p_3(n)\}, \quad p_0(1) = 1$$

$$p_1(n+1) = \frac{1}{1+\lambda_1+\lambda_2} p_0(n) + \frac{\lambda_1}{1+\lambda_1+\lambda_2} p_4(n), \quad p_1(1) = \frac{1+\lambda_1}{1+\lambda_1+\lambda_2}$$

$$p_2(n+1) = \frac{1}{1+\lambda_1+\lambda_2} p_0(n) + \frac{\lambda_1}{1+\lambda_1+\lambda_2} \max\{p_5(n), p_6(n)\} + \frac{\lambda_2 p_4(n)}{1+\lambda_1+\lambda_2}, \; p_2(1) = 1$$

$$p_3(n+1) = \frac{p_0(n)}{1+\lambda_1+\lambda_2} + \frac{\lambda_1 p_6(n)}{1+\lambda_1+\lambda_2}, \; p_3(1) = \frac{1+\lambda_1}{1+\lambda_1+\lambda_2}$$

$$p_4(n+1) = \frac{p_1(n)+p_2(n)}{2+\lambda_1+\lambda_2}, \; p_4(1) = \frac{2}{2+\lambda_1+\lambda_2}$$

$$p_5(n+1) = \frac{2p_2(n)}{2+\lambda_1+\lambda_2} + \frac{\lambda_1 p_7(n)}{2+\lambda_1+\lambda_2}, \; p_5(1) = \frac{2+\lambda_1}{2+\lambda_1+\lambda_2}$$

$$p_6(n+1) = \frac{p_2(n)+p_3(n)}{2+\lambda_1+\lambda_2} + \frac{\lambda_1 p_7(n)}{2+\lambda_1+\lambda_2}, \; p_6(1) = \frac{2+\lambda_1}{2+\lambda_1+\lambda_2}$$

$$p_7(n+1) = \frac{2p_6(n)}{3+\lambda_1+\lambda_2} + \frac{p_5(n)}{3+\lambda_1+\lambda_2}, \; p_7(1) = \frac{3}{3+\lambda_1+\lambda_2}$$

Note that $p_2(1) > p_3(1)$, $p_5(1) = p_6(1)$. Assume $p_2(n) \geq p_3(n)$ and $p_5(n) \geq p_6(n)$, some $n \geq 1$. Then

$$p_2(n+1) - p_3(n+1) = \frac{\lambda_1}{1+\lambda_1+\lambda_2}[-p_6(n) + \max\{p_5(n), p_6(n)\}] + \frac{\lambda_2 p_4(n)}{1+\lambda_1+\lambda_2} \geq 0$$

and

$$p_5(n+1) - p_6(n+1) = \frac{2p_2(n)}{2+\lambda_1+\lambda_2} - \frac{p_2(n)+p_3(n)}{2+\lambda_1+\lambda_2} \geq 0 \, .$$

Thus $p_2 \geq p_3$ and $p_5 \geq p_6$. The maximal expectations in (5) are given by $e_i = \sum_{n=1}^{\infty} p_i(n)$, $i = 0,1,\ldots,7$, so (5) follows from (6).

12. Maximum Time to the First Blocked Program

We next show that if we measure time in the ordinary way with a clock, instead of counting state changes and blocks, then algorithm g maximizes the mean time until a program is blocked.

(7) Proposition: For each $0 \leq i \leq 7$, algorithm g meximizes

E{time until the first blocked program | start in i}

Thus g puts off the catastrophe as long as possible, on the average. For m_i = maximal mean time to a block, starting from i, the dynamic programming or Bellman equation is:

$$m_0 = \frac{1}{\lambda_1+\lambda_2} + \frac{\lambda_2}{\lambda_1+\lambda_2} m_1 + \frac{\lambda_1}{\lambda_1+\lambda_2} \max\{m_2,m_3\} \tag{8}$$

$$m_1 = \frac{1}{1+\lambda_1+\lambda_2} + \frac{m_0+\lambda_1 m_4}{1+\lambda_1+\lambda_2}$$

$$m_2 = \frac{1}{1+\lambda_1+\lambda_2} + \frac{\lambda_2 m_4 + {}_1\max\{m_5,m_6\}+m_0}{1+\lambda_1+\lambda_2}$$

$$m_3 = \frac{1}{1+\lambda_1+\lambda_2} + \frac{\lambda_1 m_6+m_0}{1+\lambda_1+\lambda_2}$$

$$m_4 = \frac{1}{2+\lambda_1+\lambda_2} + \frac{(m_1+m_2)}{2+\lambda_1+\lambda_2}$$

$$m_5 = \frac{1}{2+\lambda_1+\lambda_2} + \frac{2m_2+\lambda_1 m_7}{2+\lambda_1+\lambda_2}$$

$$m_6 = \frac{1}{2+\lambda_1+\lambda_2} + \frac{m_2+m_3+\lambda_1 m_7}{2+\lambda_1+\lambda_2}$$

$$m_7 = \frac{1}{3+\lambda_1+\lambda_2} + \frac{2m_6+m_5}{3+\lambda_1+\lambda_2}$$

We can write the Bellman equation as $x = Tx$ for $x \in R_+^8$ and T the nonlinear continuous map of R^8 defined by the right-hand side of (8). The following *wedge* inequality will be used repeatedly:

$$|(a \vee b) - (c \vee d)| \leq |(a - c)| \vee |(b - d)|, \quad a \vee b = \max\{a,b\} \tag{9}$$

(10) Lemma: T^2 is a contraction for the sup norm on R^8.

Proof. It is easily verified that there is a $u_0 \in (0,1)$ such that

$$|(Tx)_i - (Ty)_i| \leq u_0 ||x - y|| \quad \text{any } x,y \in R^8 \tag{11}$$

unless $i = 0$ or 2. Now by the wedge inequality (9)

$$(Tx)_0 - (Ty)_0 = \frac{\lambda_2(x_1 - y_1)}{\lambda_1 + \lambda_2} + \frac{\lambda_1}{\lambda_1 + \lambda_2}((x_2 \vee x_3) - (y_2 \vee y_3))$$

$$|(T^2x)_0 - (T^2y)_0| \leq \frac{\lambda_2}{\lambda_1 + \lambda_2} |(Tx)_1 - (Ty)_1|$$

$$+ \frac{\lambda_1}{\lambda_1 + \lambda_2} (|(Tx)_2 - (Ty)_2| \vee |(Tx)_3 - (Ty)_3|)$$

$$\leq \frac{\lambda_2 u_0 + \lambda_1}{\lambda_1 + \lambda_2} ||x - y||$$

by (11) and the fact that T does not increase distance:

$$||Tx - Ty|| \leq ||x - y||.$$

Similarly

$$(Tx)_2 - (Ty)_2 = \frac{\lambda_2(x_4 - y_4) + (x_0 - y_0) + \lambda_1(x_5 \vee y_5) - \lambda_1(x_6 \vee y_6)}{1 + \lambda_1 + \lambda_2}$$

$$|(T^2x)_2 - (T^2y)_2| \leq \frac{(1 + \lambda_2)u_0 ||x - y|| + \lambda_1 ||x - y||}{1 + \lambda_1 + \lambda_2}$$

The contraction constant for T^2 can be chosen to be

$$u = \max\left\{u_0,\ \frac{\lambda_2 u_0+\lambda_1}{\lambda_1+\lambda_2}\ ,\ \frac{(1+\lambda_2)u_0+\lambda_1}{1+\lambda_1+\lambda_2}\right\} \varepsilon\ (0,1).$$

It follows from Lemma (10) that T^2 has exactly one fixed point. Since a fixed point of T is one of T^2, T has at most one fixed point, so a solution of the Bellman equation (8) must be unique. However if x is a fixed point of T^2 then $Tx = T(T^2x) = T^2(Tx)$, so that Tx is also; but then x = Tx by uniqueness. So any fixed point of T^2 is also one of T, and there is exactly one of either. Consider next the action of T on the simplex

$$\{x \ \varepsilon\ R_+^8 : x_2 \geq x_3,\ x_5 \geq x_6\}$$

It is easily verified from the equations that T maps the simplex into itself. Moreover if x is in the simplex, then the maxima in the definition of T can be identified, and so Tx can be written as w + Ux, where w is a fixed positive vector, the forcing or inhomogeneous term of T, and U is a substochastic matrix. A simplification of the argument of Lemma (10) shows that U^2 is a contraction, with the same constant u as T. Thus for x in the simplex

$$\begin{aligned} T^2x &= w + U(w+Ux) \\ &= w + Uw + U^2x \end{aligned}$$

so if $\mu = \dfrac{\|w + Uw\|}{1 - u}$, then T^2 maps the compact convex set

$$K = \{x \ \varepsilon\ R_+^8 : \|x\| \leq \mu,\ x_2 \geq x_3,\ x_5 \geq x_6\}$$

into itself. By an extension of Brouwer's theorem, there is a fixed point of T^2 in K which must be m by uniqueness. Thus

$$m_2 \geq m_3 \quad \text{and} \quad m_5 \geq m_6,$$

which means that algorithm g achieves equality in the dynamic programming equations, and so is optimal.

(12) Remark: A similar argument yields a direct proof of Proposition (5), using the equations:

$$(\lambda_1+\lambda_2)e_0 = \lambda_2 e_1 + \lambda_1 \max\{e_2,e_3\}$$

$$(1+\lambda_1+\lambda_2)e_1 = e_0 + \lambda_1 e_4 + \lambda_2$$

$$(1+\lambda_1+\lambda_2)e_2 = \lambda_2 e_4 + \lambda_1 \max\{e_5,e_6\}$$

$$(1+\lambda_1+\lambda_2)e_3 = e_0 + \lambda_1 e_6 + \lambda_2$$

$$(2+\lambda_1+\lambda_2)e_4 = e_1 + e_2 + \lambda_1 + \lambda_2$$

$$(2+\lambda_1+\lambda_2)e_5 = 2e_2 + \lambda_1 e_7 + \lambda_2$$

$$(2+\lambda_1+\lambda_2)e_6 = e_2 + e_3 + \lambda_1 e_7 + \lambda_2$$

$$(3+\lambda_1+\lambda_2)e_7 = 2e_6 + e_5 + \lambda_1 + \lambda_2 .$$

13. Maximum Probability of Not Having Blocked Yet

Our final result for the simple linear memory is the continuous-time analog of Prop. (6).

(13) Proposition: Algorithm g maximizes the quantities

$$\Pr\{\text{no program is blocked in } (0,t) \mid \text{start at } i\},$$

simultaneously for all $t > 0$, $0 \leq i \leq 7$.

Proof: Let $p_i(t)$, $0 \leq i \leq 7$, be the chance that no arriving program is blocked in $(0,t)$ if the memory starts at i and an optimal policy is used. Thus $p_i(t)$ is the (t,i)-pointwise maximum

$$p_i(t) = \max\{\Pr \text{ no program is blocked in } (0,t) \mid x_0 = i\}$$

The dynamic programming equations are now the following ordinary DEs:

$$\frac{dp_0}{dt} = -(\lambda_1+\lambda_2)p_0 + \lambda_2 p_1 + \lambda_1 \max\{p_2,p_3\} \tag{14}$$

$$\frac{dp_1}{dt} = -(1+\lambda_1+\lambda_2)p_1 + \lambda_1 p_4 + p_0$$

$$\frac{dp_2}{dt} = -(1+\lambda_1+\lambda_2)p_2 + \lambda_1 \max\{p_5,p_6\} + \lambda_2 p_4 + p_0$$

$$\frac{dp_3}{dt} = -(1+\lambda_1+\lambda_2)p_3 + p_0 + \lambda_1 p_6$$

$$\frac{dp_4}{dt} = -(2+\lambda_1+\lambda_2)p_4 + p_1 + p_2$$

$$\frac{dp_5}{dt} = -(2+\lambda_1+\lambda_2)p_5 + 2p_2 + \lambda_1 p_7$$

$$\frac{dp_6}{dt} = -(2+\lambda_1+\lambda_2)p_6 + p_2 + p_3 + \lambda_1 p_7$$

$$\frac{dp_7}{dt} = -(3+\lambda_1+\lambda_2)p_7 + 2p_6 + p_5$$

with the initial condition $p_i(0)$, $i = 0,1,\ldots,7$.

The system (14) has a natural integrating factor in the "diagonal" terms on its right-hand side, so we define the integral transformation T as follows: Let

$$q_i = \lambda_1+\lambda_2 + \text{number of items stored in state } i$$

$$= \text{coeff. of "diagonal" term in RHS (14)},$$

and let $g\colon R^8 \to R^8$ be the maximization nonlinearity implicit in RHS (14). Put

$$(Tf)_i(t) = e^{-q_i t} + \int_0^t e^{-q_i(t-s)} g(f(s))_i ds.$$

It is easily seen that T describes the effect of using the "diagonal" integrating factors and the initial values in (14), and that (14) is equivalent to $Tp = p$, so that a fixed point of T is a solution of (14) and vice versa.

With $t_0 > 0$ fixed, consider the set K of functions $f:[0,t_0] \to R_+^8$ with the properties: $0 \leq f_i \leq 1$, $f_i \in \mathrm{Lip}(2a)$, $f_2 \geq f_3$, $f_5 \geq f_6$, $f_i(0) = 1$, where $a = 3 + \lambda_1 + \lambda_2 = |\text{largest coeff. in RHS (14)}|$. It is easy to see that K is convex, and compact in the uniform topology. We claim that T is continuous and maps K into itself. Continuity follows by a familiar argument from the Lipschitz nature of the right-hand side of (14). To show $TK \subseteq K$, take $f \in K$ and write Tf in the form

$$(Tf)_i(t) = e^{-q_i t} + \int_0^t e^{-q_i(t-s)} \sum_{\substack{j\in E\\ j\neq i}} q_{ij} f_j(s)\,ds,$$

with $q_{ij} \geq 0$, $E = \{0,1,\ldots,7\}$. This is possible, using a fixed matrix Q, because $f \in K$, so the maxima occurring in T can be identified. Noting now that $\sum_{\substack{j\in E\\ j\neq i}} q_{ij} \leq q_i$ and using $0 \leq f_j \leq 1$, we find that $0 \leq (Tf)_i \leq 1$. The final requisite inequalities

$$(Tf)_2 \geq (Tf)_3, \quad (Tf)_5 \geq (Tf)_6$$

follow at once from $f_2 \geq f_3$ and $f_5 \geq f_6$ and the definition of T. The Lip property of Tf is apparent, so $TK \subseteq K$.

The uniform topology is locally convex, and we conclude from Tychonov's fixed point theorem [6] that there is an element $f \in K$ such that $Tf = f$. Since the solution p of the DEs (14) is unique, we find $p = f$ and $p \in K$. Hence $p_2 \geq p_3$ and $p_5 \geq p_6$, pointwise in time. This means that algorithm g achieves equality in the dynamic programming equations, and so is optimal for every $t \leq t_0$. But t_0 is arbitrary.

(15) Remark: In the equations the property $p_2 \geq p_3$, $p_5 \geq p_6$ could also be proved by successive approximations $p^{(n)}$ of Picard type satisfying $p_2^{(n)} \geq p_3^{(n)}$, $p_5^{(n)} \geq p_6^{(n)}$.

14. Generalization: Maximizing the Expected Time to a Blocked Program

In this section we abstract, from the simple examples considered so far, a general setup in which we can derive optimal allocation rules for a wide class of memory structures. Optimal allocations will be guessed at and justified from the nonlinear Bellman equation by the same kind of topological fixed point method used for the linear memory. We retain the integral constraints and other assumptions that gave rise to finite S and E. The dynamic programming equations for maximizing the mean time m_α to a blocked program, given that the memory starts in α, are these:

$$m_\alpha = \frac{1}{|\alpha|+\lambda} + \sum_{\beta\varepsilon B_\alpha} h_{\alpha\beta} \frac{m_\beta}{|\alpha|+\lambda} + \frac{1}{|\alpha|+\lambda} \sum_{p\varepsilon F_\alpha} \lambda_p \max_{\beta\varepsilon A_{p\alpha}} m_\beta$$

$$= (Tm)_\alpha$$

for a suitable nonlinear T defined by the right-hand side of (16). It will turn out, Lemma (22) infra, that some power of T is a contraction, so there is a unique solution of x = Tx.

Now suppose that we could stare at the (reduced) state diagram (E, $\leq$) long and hard, and come up with an intuitive relation B (read "better than") of preference among states that represent alternative ways of putting in programs. To justify our guess that B is the key to optimal allocation we would look for a suitable connection between B and the map T that embodies the Bellman equation. The examples considered earlier suggest asking whether T preserves the ordering of components according to B, i.e., looking at the *simplex* of vectors $x\colon E \to R_+$ such that

$$\alpha B\beta \Longrightarrow x_\alpha \geq x_\beta.$$

Of course, T will not preserve just any such simplex; clearly a special relation must exist between B and T, similar to that used in (6) and (13) for the linear memory. Such a relation, based on a similar idea [5] used in routing telephone traffic, is described as follows:

(17) Definition: A preference relation B defined on E is said to have the *monotone property* iff $\alpha B \beta$ implies that

(17.1) There is a relation $D = D(\alpha, \beta)$ on E such that

(i) $D \subseteq (B_\alpha \times B_\beta) \cap B$

(ii) Each $\alpha \in B_\beta$ occurs in the range of D $h_{\beta\gamma}$ times

(iii) Each $\delta \in B_\alpha$ occurs in the domain of D $h_{\alpha\delta}$ times

(17.2) $F_\beta \subseteq F_\alpha$

(17.3) $\gamma \in A_{p\beta} \Longrightarrow \exists \delta \in A_{p\alpha} \ \delta B \gamma$

It is understood that B holds only between states that are equivalent in the sense of storing the same programs, with repetitions allowed.

The purpose of the monotone property is to express exactly the following intuitive but partly vague idea: the preference B is such that if $\alpha B \beta$ then (i) any program not blocked in β can also be accommodated in α so as to lead to a state that is preferable to that reached from β, and (ii) states reachable from β by a program departure are matched by preferable states similarly reachable from α. In other words, if xBy, then the *neighbors* of α hold B to suitable corresponding *neighbors* of β. The map T, derived by following a Kolmogorov generator by a maximization, is intimately linked to the same concept of "neighbor," and thence springs the relevance of the monotone property, expressed specifically in the next result:

(18) Lemma: If B has the monotone property, and $x\colon E \to R^+$ is a vector in the simplex defined by $\alpha B \beta \Longrightarrow x_\alpha \geq x_\beta$, then $\alpha B \beta$ implies

(i) $(Hx)_\alpha \geq (Hx)_\beta$, and

(ii) $\max_{\gamma \varepsilon A_{p\alpha}} x_\gamma \geq \max_{\delta \varepsilon A_{p\beta}} x_\delta$.

The monotone property expresses a strong kind of consistency or continuity of B, and it will be used to prove strong domination results of the form: if $\alpha B \beta$, then the best you can do (in performance) starting from α is no worse than anything you can achieve with a start in β.

Proof of (18): Let D be the relation in (17.1): Since $D \subseteq B$ and x is in the simplex,

$$(Hx)_\alpha = \sum_{\gamma \varepsilon B_\alpha} h_{\alpha\gamma} x_\gamma = \sum_{(\gamma,\delta)\varepsilon D} x_\gamma \geq \sum_{(\gamma,\delta)\varepsilon D} x_\delta = \sum_{\delta \varepsilon B_\beta} h_{\beta\delta} x_\delta = (Hx)_\beta$$

Also, (ii) of (18) follows from (17.3).

(19) Remark: Needless to say, not every memory structure has defined on it a preference B satisfying the monotone property. But those that do are common enough to form a significant subclass for which the allocation problem is easily solved. Figures 1-9 show the reduced state diagram for a number of examples, two of which are actually equivalent. In each case the relation B entered in the figure has the monotone property. In Figure 10 the monotone property apparently is not present. The examples show repeatedly how our evaluation of the fragmented or "checkerboard" states can be graphically parlayed into the preference relation B which is the key to proving optimality. The 'don't care' entries mean that the choice is immaterial, or that B holds in both directions.

(20) Lemma: With $k = \max_{\alpha \varepsilon E} |\alpha|$, there is a $u \varepsilon (0,1)$ such that for all $\alpha \varepsilon E$, and $||z|| = \max_{\alpha \varepsilon E} |z_\alpha|$

$$\left|(T^{k-|\alpha|}x)_\alpha - (T^{k-|\alpha|}y)_\alpha\right| \leq u||x-y||$$

Proof: By downward induction on $\leq$. Suppose first that α is maximal in the partial ordering $\leq$ induced on E. Then α is completely blocking: no programs can be entered in α. Thus

$$(Tx)_\alpha - (Ty)_\alpha = \sum_{\beta\varepsilon B_\alpha} h_{\alpha\beta} \frac{x_\alpha - y_\alpha}{|\alpha|+\lambda}, \text{ with } \sum_{B\varepsilon B_\alpha} h_{\alpha\beta} = |\alpha|$$

$$|(Tx)_\alpha - (Ty)_\alpha| \leq \frac{|\alpha|}{|\alpha|+\lambda} ||x-y|| = u_\alpha ||x-y||$$

Since the probabilistic part of T is substochastic, T never increases distance. That is, we have

$$||Tx - Ty|| \leq ||x-y||$$

Hence putting $T^{k-|\alpha|}x$ and $T^{k-|\alpha|}y$ for x and y respectively and iterating we find

$$|(T^{k-|\alpha|}x)_\alpha - (T^{k-|\alpha|}y)_\alpha| \leq \frac{|\alpha|}{|\alpha|+\lambda} ||x-y||.$$

If now $A_\alpha \neq \emptyset$, so that $|\alpha| < k$, we can form the induction hypothesis that there is a number $u_{|\alpha|+1} \varepsilon (0,1)$ such that $\forall_\gamma$ with $|\gamma| = |\alpha| + 1$

$$|(T^{k-|\beta|}x)_\gamma - (T^{k-|\beta|}y)_\gamma| \leq u_{|\alpha|+1}||x-y||. \tag{21}$$

The uniformity of $u_{|\alpha|+1}$ over the γ's follows from the finiteness of E. Then

$$(T^{k-|\alpha|}x)_\alpha - (T^{k-|\alpha|}y)_\alpha$$

$$= \sum_{\beta\varepsilon B_\alpha} \frac{h_\alpha{}^\beta}{|\alpha|+\lambda} \left[(T^{k-|\alpha|-1}x)_\beta - (T^{k-|\alpha|-1}y)_\beta\right]$$

$$+ \sum_{p\varepsilon F_\alpha} \frac{\lambda_p}{|\alpha|+\lambda} \left\{\max_{\gamma\varepsilon A_p\alpha} (T^{k-|\gamma|}x)_\gamma - \max_{\gamma\varepsilon A_{px}} (T^{k-|\gamma|}y)_\gamma\right\}$$

Using (21) and the wedge inequality (9), we can conclude that

$$|(T^{k-|\alpha|}x)_\alpha - (T^{k-|\alpha|}y)_\alpha| \le \left(|\alpha|+u_{|\alpha|+1} \sum_{p\varepsilon F_\alpha} \lambda_p\right) \frac{||x-y||}{|\alpha|+\lambda}$$

Now pick $u = \max_{0\le\ell\le k} u_\ell$, where $u_k = 0$ and for $\ell < k$,

$$u_\ell = \max_{|\alpha|=\ell} \left(\ell + u_{\ell+1} \sum_{p\varepsilon F_\alpha} \lambda_p\right) \frac{1}{\ell+\lambda} \; \varepsilon \; (0,1).$$

(22) Lemma: For $k = \max_{\alpha\varepsilon E} |\alpha|$, T^k is a contraction with constant u.

Proof: (20)

(23) Remark: Defining the vector $z: E \to R_+$ by $z_\alpha = (|\alpha|+\lambda)^{-1}$ = mean time in α till an arrival or a departure, and the map T_1 by $T_1x = Tx - z$, the dynamic programming equation becomes

$$x = z + T_1x.$$

The arguments for (20) show that T_1^k is also a contraction with constant u. T_1x is a subconvex combination of various components of x with the property

$$T_1(x+y) \le T_1x + T_1y \tag{24}$$

(25) Lemma: With u the contraction constant for T^k and T_1^k, there is an *a priori* bound

$$a = (1-u)^{-1} \sum_{k=0}^{k-1} ||T^iz||$$

such that all solutions $x \ge 0$ of $x = Tx$ (or $x = z + T_1x$) satisfy $||x|| \le a$.

Proof: Starting with $x = z + T_1x$, $x \ge 0$, use (24) k times along with positivity to conclude

$$||x|| \le \sum_{i=0}^{k-1} ||T_1^iz|| + ||T_1x||$$

$$\le \sum_{i=0}^{k-1} ||T_1^iz|| + u||x|| \le a.$$

(26) Remark: T maps the ball $\{\|x\| \le a\}$ into itself. For if $\|x\| < a$ then $\|Tx\| = \|z + T_1 x\|$, and so

$$\|Tx\| \le \sum_{i=0}^{k-1} \|T_1^i z\| + \|T_1^k x\|$$

$$\le (1-u)a + ua \le a$$

(27) Theorem: If B has the monotone property, then there is a unique solution m of the dynamic programming equation x = Tx in the simplex

$$K = \{y \in R_+^{|E|} : \|y\| \le a,\ \alpha B \beta \Longrightarrow y_\alpha \ge y_\beta\}.$$

Proof: By Lemma (18) and Remark (26), T maps K into itself continuously. K is compact and convex, so by an extension of Brouwer's fixed point theorem there is a fixed point x of T in K. Since T^k is a contraction, there is exactly one fixed point anywhere.

(28) Remark: If in addition to having the monotone property, the preference B is such that every $A_{p\alpha}$ contains a state (equivalence class) β which bears B to ("is better than") every other state in $A_{p\alpha}$, then the choice β is optimal when storing p in state α.

15. Generalization: Maximizing the Probability of Not Having Blocked Yet

We continue with the general analog of the results in Section 13 for the small linear memory. Let $p(t)\colon E \to [0,1]$ be vectors interpreted as

$$p_\alpha(t) = \Pr\left\{\begin{matrix}\text{no program was blocked}\\ \text{during } [0,t]\end{matrix}\ \middle|\ \begin{matrix}\text{start at } \alpha \text{ and an optimal}\\ \text{policy is used}\end{matrix}\right\}$$

The (dynamic programming) equations for p(t) are the DEs: $p(0) = 1$, with

$$\dot{p}_\alpha = -q_\alpha p_\alpha + \sum_{\beta \varepsilon B_\alpha} h_{\alpha\beta} p_\beta + \sum_{p \varepsilon F_\alpha} \lambda_p \max_{\beta \varepsilon A_{p\alpha}} p_\beta, \quad q_\alpha = q_{\alpha\alpha} = |\alpha| + \lambda$$

or their integral analogues (more convenient here):

$$p_\alpha(t) = e^{-q_\alpha t} + \int_0^t e^{-q_\alpha(t-s)}\left\{\sum_{\beta\varepsilon B_\alpha} h_{\alpha\beta}p_\beta(s) + \sum_{p\varepsilon F_\alpha} \lambda_p \max_{\beta\varepsilon A_{p\alpha}} p_\beta(s)\right\}ds.$$

$$= (Tp)_\alpha(t), \quad \text{to define } T.$$

As before, we shall single out a special simplex based on the preference relation B to use in a fixed point argument; here, though, the simplex will be in a Banach space. We choose any $t_0 > 0$,

$$K = \{f\colon [0,t_0] \to R_+^{|E|},\ 0 \le f_\alpha \le 1,\ f_\alpha \in \mathrm{Lip}(2q_\alpha),\ f(0) \equiv 1,$$
$$\text{and } \alpha B\beta \Longrightarrow f_\alpha \ge f_\beta\}$$

We shall assume that B has the monotone property, and that each A_{px} has a B-maximal element. Then for $f \in K$ the maxima occurring in T can be identified, and there is a constant matrix $Q = (q_{\alpha\beta})$ such that for $f \in K$

$$(Tf)_\alpha(t) = e^{-q_\alpha t} + \int_0^t e^{-q_\alpha(t-s)} \sum_{\beta\neq\alpha} q_{\alpha\beta} f_\beta(s)ds$$

Since $\sum_{\beta\neq\alpha} q_{\alpha\beta} \le q_\alpha$, it follows easily that $0 \le (Tf)_\alpha \le 1$, and that $|\frac{d}{dt}(Tf)_\alpha| \le 2q_\alpha$. It also follows, from Lemma (18) and the monotone property, that

$$\alpha B\beta \Longrightarrow (Tf)_\alpha \ge (Tf)_\beta, \text{ pointwise in } [0,t_0].$$

It is easy to verify that T is continuous in the uniform topology of $C([0,t_0] \to R^{|E|})$, and that K is convex. By the Ascoli-Arzela theorem, K is compact. Thus T maps K continuously into itself, and since $C([0,t] \to R^{|E|})$ is locally convex, it follows from Tychonov's fixed point theorem [6] that there is a fixed point p of T in K. Uniqueness of p follows from standard arguments for DEs with Lipschitz right-hand sides. We have thus proved this result:

(29) Theorem: Let B have the monotone property, and let there be a

B-maximal element in each $A_{p\alpha}$. Let r* be a policy that chooses a B-maximal element in each A_{px}. Then the unique solution p of the dynamic programming equation has the simplex property

$$\alpha B\beta \Longrightarrow p_\alpha(t) \geq p_\beta(t), \qquad 0 \leq t \leq t_0$$

Under r* the probability of not having blocked by t, starting from α, is exactly $p_\alpha(t)$, and r* is an optimal allocation rule, and simultaneously maximizes all the quantities

$$\Pr\{\text{no blocked program in } [0,t] \mid \text{start in } \alpha\}, \quad (\alpha \in E,\ 0 \leq t \leq t_0).$$

Since t_0 was arbitrary the results extend to any $t \geq 0$.

16. Least Probability of Blocking

We now take up the problem of choosing an allocation rule so as to minimize the chance of blocking. Our principal result is that typically, except for some minor hedges and caveats, the same rule that maximizes the mean time to a blocked program will also yield the least blocking probability. This is a natural result, but hardly obvious; in both cases the immediate effect of the rule is to visit the "bad" (blocking" states as little as possible, because it is only in them that the risk of untoward events, catastrophic for the mean time criterion, and just poor for the "chance of blocking" criterion, is really felt.

However, the theory we have been able to put together for probability of blocking is less complete and more complicated than that in Section 14-15 for the time to the first rejected program. This is because the dynamic programming equations for optimizing an average rate are less transparent than those for mean time to block: the mean length of time stayed in a state is involved in ways that make the simple "simplex" approach used earlier inapplicable. We have found a way around this difficulty by generalizing the admissible rules to allow the rejection of unblocked programs. This device yields the inequalities needed for a "simplex" type argument, but provides no way of telling when (if ever) an unblocked program should be rejected. Our results are therefore of the conditional form: if indeed no unblocked program should ever be rejected, then the following rule is best:... We believe that almost always in practice there is no point in rejecting unblocked programs, although pathological examples show that this can occur, as in telephone routing [5]. Thus to find the best rule modulo such refusals is essentially a solution of the problem; this is no longer true if the different kinds of program are weighted in importance.

We henceforth assume that the admissible rules r are extended to allow, at any time, the random summary rejection of an unblocked program. In the present problem the probability of blocking is a ratio of rates; the denominator is the constant λ because there is no finite-source effect. As a result, minimizing blocking reduces to maximizing the rate of accepted programs. That is, when a time-independent allocation rule r is adopted, the formula for probability of blocking can be written as

$$\text{blocking} = \frac{\text{equil. rate of blocked or rejected programs if r is used}}{\text{equil. rate of arriving programs}}$$

$$= 1 - \lambda^{-1}\ \text{(rate of accepted programs)},$$

since the arrival rate is λ, independently of the state and of r. The equilibrium rate of accepted programs is just

$$\theta_r = \sum_{\alpha\varepsilon E} p_\alpha^r \sum_{p\varepsilon F_\alpha} \lambda_p r(p,\alpha)$$

where

$$r(p,\alpha) = \begin{cases} 0 \text{ if r rejects p in } \alpha \\ 1 \text{ if r accepts p in } \alpha \end{cases}$$

and $\{p_\alpha^r,\ \alpha \in E\}$ is the vector of equilibrium probabilities under r. The acceptance rate is of course also the departure rate

$$\Sigma p_\alpha^r |\alpha|,$$

here coincident with the mean number of programs in memory. By the law of large numbers, both rates are equal to

$$\lim_{t\to\infty} t^{-1} E_x^r\{\text{number of programs accepted in } (0,t)\}$$

where E_x^r denotes expectation with respect to the measure induced by starting the memory in state α and running it according to r. It is a small step of generalization to suggest looking at the problem of maximizing

$$\liminf_{t\to\infty} t^{-1} E_\alpha^r\{\text{number of programs accepted in } (0,t)\}$$

over all r, not just stationary ones. Evidently here

$$\sup_r \liminf_{t\to\infty} \le \lim_{t\to\infty} \sup_r$$

and our procedure will be to find, for "suitable" memory systems, a stationary r which achieves

$$\sup_r E_\alpha^r\{\text{number of programs accepted in } (0,t)\} \tag{30}$$

for each (t,α), proving that the inequality above is an equality, and achieving the maximum acceptance rate. The "suitable" memory systems will be those which enjoy the previous monotone property, and for which unblocked programs should never be rejected. The first step is to write dynamic programming equations for the suprema in (30).

17. Maximizing the Expected Number of Programs Accepted in (0,t)

With the abbreviations

$$s(\alpha) = \sum_{p\in F_\alpha} \lambda_p = \text{"success rate" in } \alpha$$

$$a_\alpha = |\alpha| + s(\alpha) = \text{change rate in } \alpha$$

and allowing now the rejection of unblocked programs, the usual infinitesimal arguments applied to the projection process πx_t lead to a Bellman equation for

$$E_\alpha(t) = \sup_r E_\alpha^r\{\# \text{ of programs accepted in } (0,t)\}.$$

This equation is, with $E(0) = 0$, and $x \vee y = \max\{x,y\}$,

$$\frac{d}{dt} E_\alpha = \sum_{\beta\varepsilon B_\alpha} h_{\alpha\beta}E_\beta + \sum_{p\varepsilon F_\alpha} \lambda_p(E_\alpha \vee 1 + \max_{\gamma\varepsilon A_{p\alpha}} E_\gamma) - a_\alpha E_\alpha. \tag{31}$$

We write the Bellman equation in the integrated form

$$E_\alpha(t) = \int_0^t e^{-a_\alpha(t-s)}\left\{\sum_{\beta\varepsilon B_\alpha} h_{\alpha\beta}E_\beta(s) + \sum_{p\varepsilon F_\alpha} \lambda_p(E(s) \vee 1 + \max_{\gamma\varepsilon A_{p\alpha}} E_\gamma(s))\right\}ds$$

$$= (TE)_\alpha(t)$$

to define a map T continuous in the sup norm topology of $C[0,t]^{|E|}$ for $t > 0$. The argument to be given will work for all t at once. So choose a time $t > 0$ and set

$$a = \max_{\alpha\varepsilon E} a_\alpha = \max_{\alpha\varepsilon E}(|\alpha|+s(\alpha)) \tag{32}$$

and let K be the set of functions $f: [0,t] \to R^{|E|}$ such that

(i) $0 \leq f_\alpha(s) \leq e^{as}$, $0 \leq s \leq t$, and $f(0) \equiv 0$

(ii) f_α is Lipschitz with modulus $\lambda + ae^{at}$

(iii) f is B-isotone, i.e., confined to the simplex defined by

$$\alpha B\beta \Longrightarrow f_\alpha(u) \geq f_\beta(u),\ 0 \leq u \leq t. \quad (\text{"simplex" property})$$

K is convex, and compact in the uniform topology; we show that when B has the monotone property, then T maps K into itself.

(33) Lemma: B has monotone property $\Longrightarrow TK \subseteq K$.

Proof: T is obviously nonnegative; for $f \in K$, since $\sum_{\beta \in B_\alpha} h_{\alpha\beta} + \sum_{p \in F_\alpha} \lambda_p = a_\alpha$, we find by (i) and (32) that

$$(Tf)_\alpha(u) \leq a_\alpha \int_0^u e^{-a_\alpha(u-s)} \left\{ \frac{s(\alpha)}{a_\alpha} + e^{as} \right\} ds$$

$$\leq 1 - e^{-a_\alpha u} + e^{(a-a_\alpha)u} \int_0^u a_\alpha e^{a_\alpha s} ds$$

$$\leq 1 - e^{-au} + e^{au}(1-e^{-au})$$

$$\leq e^{au},$$

so Tf has property (i). Tf is differentiable with $|\frac{d}{dt}Tf| \leq \lambda + ae^{at}$, $0 \leq u \leq t$, so Tf has (ii). To prove the simplex property (iii), we first note that by the monotone property, $\alpha B \beta$ and $p \in F_\beta$ imply for $f \in K$ that

$$\delta \in A_{p\beta} \Longrightarrow \exists \gamma \in A_{p\alpha} \ \gamma B \delta$$

$$\Longrightarrow \exists \gamma \in A_{p\alpha} f_\gamma \geq f_\delta$$

$$\max_{\gamma \in A_{p\alpha}} f_\gamma \geq \max_{\delta \in A_{p\alpha}} f_\delta, \text{ pointwise in time.}$$

Thus $\alpha B \beta$ implies, by (17.2), $F_\beta \subseteq F_\alpha$, so that also

$$\sum_{p \in F_\alpha} \lambda_p (f_\alpha \vee 1 + \max_{\gamma \in A_{p\alpha}} f_\gamma) \geq \sum_{p \in F_\beta} \lambda_p (f_\beta \vee 1 + \max_{\delta \in A_{p\beta}} f\ \delta)$$

The argument for (i) of Lemma (18) shows that $\alpha B \beta$ implies

$$(Hf)_\alpha \geq (Hf)_\beta$$

and so it follows that

$$(Tf)_\alpha \geq (Tf)_\beta,$$

completing the proof that $TK \subseteq K$.

We can now argue, as before, that the solution E(t) of the Bellman equation must lie in K, because the solution is unique (standard Lipschitz argument), and Tychonov's theorem guarantees the existence of a fixed point f = Tf in K, which then has to be E by uniqueness. It follows that if there is a B-maximum in each $A_{p\alpha}$, then B describes at least one optimal allocation policy, modulo refusals.

18. The Asymptotic Bellman Equation

Stochastic control problems in which an asymptotic or equilibrium rate is optimized usually lead to an equation involving an unknown constant θ (the optimal rate), as well as a function $e\colon S \to R^{|E|}$ that identifies the optimal policy in the same way that the gradient of the solution of the Hamilton-Jacobi equation does. The function e is no longer interpretable as a vector of probabilities or expectations, as were the quantities we used earlier. Still, it has a close relationship to such concepts, and we will study them now.

The Bellman equation for maximizing the rate of accepted programs, with refusals allowed, is

$$\theta = \sum_{\beta\varepsilon B_\alpha} h_{\alpha\beta} e_\beta + \sum_{p\varepsilon F_\alpha} \lambda_p (e_\alpha \vee 1 + \max_{\beta\varepsilon A_{p\alpha}} e_\beta) - a_\alpha e_\alpha, \tag{33}$$

the interpretation, the usual "verification" lemma, being that if one can find a θ > 0 and a vector $e\colon E \to R$ satisfying this equation, then there is a best stationary policy obtained by noting what states achieve the maxima in the equations. This policy is asymptotically as good as anything else (non-stationary, if you like) you could do.

Intuitively and formally (33) is easily connected to (31) by noting that e should give the second term in an expansion of the form

$$E_\alpha(t) = \theta t + e_\alpha + O(t^{-1}).$$

If we now substitute this form in (31), integrate from 0 to t, divide by t, and let $t \to \infty$ we find, using $E(0) = 0$

$$\theta = \lim_{t\to\infty} t^{-1} E_\alpha(t) = \lim_{t\to\infty} t^{-1} \int_0^t E_\alpha(s)ds$$

$$= \lim_{t\to\infty} t^{-1} \int_0^t \Big[\sum_{y\in B_x} h_{xy}(\theta s + e_y) + \sum_{p\in F_x} \lambda_p((\theta_s + e_x) \vee 1 + \max_{y\in A_{px}} (\theta s + e_y))$$

$$- a_x(\theta s + e_x) + 0(s^{-1}) \Big] ds$$

Since $a_\alpha = \sum_{\beta\in B_\alpha} h_{\alpha\beta} + \sum_{p\in F_\alpha} \lambda_p$, we find (33).

(34) Theorem: Let r be a stationary policy corresponding to a generator Q for πx_t and a stationary probability vector p_α, $\alpha \in E$. Let θ and $e: S \to R^{|E|}$ solve the asymptotic Bellman equation (33). Then the acceptance rate achieved by r is no larger than θ.

Proof: We describe r alternatively by a function $\emptyset$ such that $\emptyset(\alpha,p)$ is the state in $\{\alpha\} \cup A_{p\alpha}$ that r takes the system to if $p \in F_\alpha$ is to be put into memory when in state α. The rate achieved by r is then

$$\theta_r = \sum_{\alpha\in E} p_\alpha \sum_{p\in F_\alpha} 1_{\emptyset(\alpha,p)\neq\alpha}\, e_b$$

It follows from (33) that

$$\theta = \sum_{\beta\in B_\alpha} h_{\alpha\beta} e_\beta + \sum_{p\in F_\alpha} \lambda_p (e_\alpha \vee 1 + \max_{\beta\in A_{p\alpha}} e_\beta) - a_\alpha e_\alpha$$

$$\geq \sum_{p\in F_\alpha} \lambda_p 1_{\emptyset(\alpha,p)\neq\alpha} + (Qe)_a$$

Multiply by p_α and sum to find

$$\theta \geq \theta_r + e'Q'p;$$

the term $e'Q'p$ is zero, since the equilibrium condition for r is $Q'p = 0$ so $\theta \geq \theta_r$.

We now sketch how θ and $e: S \to R^{|E|}$ are to be found. First of all since θ is an asymptotic rate, we would expect it to have the form of an "average" $\theta = \sum_{\alpha\varepsilon E} p_\alpha s(\alpha)$, where $\{p_\alpha, \alpha \varepsilon E\}$ is the vector of stationary probabilities associated with the optimum policy. Assuming now that no program should ever be rejected, and that every $A_{p\alpha}$ has a B-maximum, it follows that there is a stationary policy, call it r*, which achieves the sup in (30) for each t, and is defined by choosing to go to a B-highest state in each $A_{p\alpha}$, whenever p is to be put into memory. Let such a B-maximum be denoted by $\emptyset(\alpha,p)$, i.e.,

$$\emptyset(\alpha,p)B\beta \text{ for any } \beta \varepsilon A_{p\alpha}$$

Then since r* achieves,

$$E_\alpha = s(\alpha) + \sum_{\beta\varepsilon B_\alpha} h_{\alpha\beta}E_\beta + \sum_{p\varepsilon F_\alpha} \lambda_p E_{\emptyset(\alpha,p)}$$

We claim that if $p^*_{\alpha\beta}(t)$ and p^*_α are respectively the transition probabilities and the equilibrium probabilities associated with the policy r*, then the Bellman equation (33) is solved by

$$\theta = \lambda \sum_{\alpha\varepsilon E} p^*_\alpha s(\alpha)$$

$$e_\alpha = \int_0^\infty \sum_{\beta\varepsilon E} [p^*_{\alpha\beta}(u)-p^*_\beta]ds \tag{35}$$

For it is easily seen that since r* achieves,

$$E_\alpha(t) = \int_0^t \sum_{\beta\varepsilon E} p^*_{\alpha\beta}(u)s(\beta)ds,$$

so that the backward Kolmogorov equation for $p^*_{\alpha\beta}(\cdot)$ implies

$$\sum_{\beta\varepsilon E} p^*_{\alpha\beta}(t)s(\beta) = \sum_{\beta\varepsilon B_\alpha} h_{\alpha\beta} \sum_{\gamma\varepsilon E} p^*_{\beta\gamma}(t)s(\gamma)$$

$$+ \sum_{p\varepsilon F_\alpha} \lambda_p \sum_{\gamma\varepsilon E} p^*_{\emptyset(\alpha,p)\gamma}(t)s(\gamma) - a_x \sum_{\gamma\varepsilon E} p^*_{\alpha\gamma}(t)s(\gamma).$$

Now we add and subtract $a_\alpha \sum_{B \varepsilon E} p^*_\beta s(\beta)$ on the right, and integrate from 0 to ∞; the integrals exist because the eigenvector p^* of Q associated with 0 is unique, and all the eigenvalues of Q are 0 or have real part strictly negative [7]. The left side is

$$\int_0^\infty \sum_{\beta \varepsilon E} p^*_{\alpha\beta}(t) s(\beta) dt = \sum_{\beta \varepsilon E} [p^*_{\alpha\beta}(\infty) - p^*_{\alpha\beta}(0)] s(\beta) = \theta - s(\alpha)$$

The right is

$$\sum_{\beta \varepsilon B_\alpha} h_{\alpha\beta} e_\beta + \sum_{p \varepsilon F_\alpha} \lambda_p e_{\emptyset(\alpha,p)} - a_\alpha e_\alpha$$

Since

$$E_\alpha(t) = \theta t + \int_0^t \sum_{\beta \varepsilon E} [p^*_{\alpha\beta}(u) - p^*_\beta] s(\beta)' = \theta t + e_\alpha + o(1)$$

as $t \uparrow \infty$ and $\alpha B \beta$ implies $E_\alpha(t) \geq E_\beta(t)$, we can let $t \to \infty$ to find that

$$\alpha B \beta \text{ implies } e_\alpha \geq e_\beta$$

By assumption, $1 + \max_{\beta \varepsilon A_{p\alpha}} E_\beta(t) \geq E_\alpha(t)$, so also

$$1 + \max_{\beta \varepsilon A_{p\alpha}} e_\beta \geq e_\alpha$$

Thus

$$e_{\emptyset(\alpha,p)} = e_\alpha \quad 1 + \max_{\beta \varepsilon A_{p\alpha}} e_\beta$$

and we have proved that θ and e as defined by (35) solve the asymptotic Bellman equation (33), and that e lies in the simplex $\alpha B \beta \Longrightarrow e_\alpha \geq e_\beta$. Note that e is defined by (33) only up to adding the same constant to all its components.

19. Acknowledgements

Problems of optimal memory allocation were emphasized by M. D. McIlroy as early as ten years ago. Some of them were posed to the author about a year ago by E. G. Coffman in conversation, whence this paper. It is a pleasure to thank them both for helping to formulate precise versions and encouraging this research.

20. References

[1] E. G. Coffman, Jr., Kimming So, M. Hofri, and A. C. Yao, "A Stochastic Model of Binpacking," Inf. & Control, 44 (1980), pp. 105-115, last paragraph.

[2] Ibid.

[3] D. E. Knuth, The Art of Computer Programming: Fundamental Algorithms (Vol. 1, 2nd Ed.), Addison-Wesley, Reading, Mass., 1973, p. 435 ff.

[4] V. E. Beneš, "Reduction of Network States Under Symmetries," Bell System Tech. J., 57, (1978), pp. 111-149.

[5] V. E. Beneš, "Programming and Control Problems Arising from Optimal Routing in Telephone Networks," Bell Systems Tech. J., 49, (1966), pp. 1373-1438.

[6] N. Dunford and J. T. Schwartz, Linear Operators: General Theory (Part I), Interscience, New York, 1958, p. 456.

[7] R. Bellman, Introduction to Matrix Analysis, McGraw-Hill, 1960, p. 294, exercise 8.

Bell Laboratories, Murray Hill, New Jersey 07974

PROBABILISTIC ANALYSIS OF ALGORITHMS

Jon Louis Bentley

Abstract

This paper is a brief introduction to the field of probabilistic analysis of algorithms; it is not a comprehensive survey. The first part of the paper examines three important probabilistic algorithms that together illustrate many of the important points of the field, and the second part then generalizes from those examples to provide a more systematic view.

1. Introduction

Most research in analysis of algorithms has studied the worst-case performance of various algorithms. Although such analyses have led to many interesting results and are important in some applications ("usually" isn't good enough in air traffic control, for example), the critical issue in many applications is the expected amount of resources used by an algorithm. With this motivation, a great deal of work has recently been devoted to designing and analyzing algorithms that are very efficient on the average.

This paper is a brief introduction to the field of probabilistic analysis of algorithms; it is not a comprehensive survey. In Section 2 we will survey three important probabilistic algorithms, and in Section 3 we will generalize from those examples to provide a more systematic view. Finally, conclusions are offered in Section 4.

2. The Algorithms

In this section we will investigate three probabilistic algorithms that together illustrate many of the important points of the field. In Subsections 2.1 and 2.2 we will investigate two algorithms that sort the elements of an array, and in Subsection 2.3 we will turn to a more subtle problem.

2.1 Quicksort

In this section we will consider the problem of efficiently sorting an array of values from an ordered set. Specifically, we are given an array X[1..N], and we must permute its elements so that

$$X[1] \leq X[2] \leq \cdots \leq X[N].$$

This problem is of tremendous practical importance, and it also serves to illustrate many important points in the field of algorithm design. For a thorough yet delightful study of the general problem of sorting, the reader is referred to Knuth [1973, Chapter 5].

The first sorting algorithm we will study is Quicksort; it is due to Hoare [1962] and is nicely described by Knuth [1973, Section 5.2.1] and Sedgewick [1975]. Quicksort is a classical example of a "divide-and'conquer" algorithm: it solves a problem by partitioning its input into two subproblems and then solving each recursively. An example of Quicksort at work is shown in Figure 1. Given the 8-element array shown in the upper part of the figure, it chooses the first element in the array (in this example 55) as a partitioning element. It then partitions the array around that element, placing 55 in X[4], everything less than 55 in X[1..3] (that is, to the left of 55), and everything greater than 55 in X[5..8] (that is, to the right of 55). The program then calls itself recursively to sort the two subarrays.

It is very easy to turn the above sketch into an almost complete program, which is shown below in pseudo-Pascal. It is a recursive procedure with the two integer parameters L and U, which denote the lower and upper bounds of the subarray to be sorted.

```
procedure Quicksort(L, U: integer);
  begin
  if U ≤ L then
     (*Quicksort was asked to sort an empty or one-element array;
      return at once without performing any action.*)
   else
    begin
    Partition around array element X[L] and assume that its final
      position is X[M];
    Quicksort(L,M - 1);
    Quicksort(M + 1,U)
    end
  end;
```

To sort the entire array X[1..N] one calls Quicksort(1,N). A simple inductive argument shows that the program correctly produces a sorted list, assuming that the partitioning routine is correct.

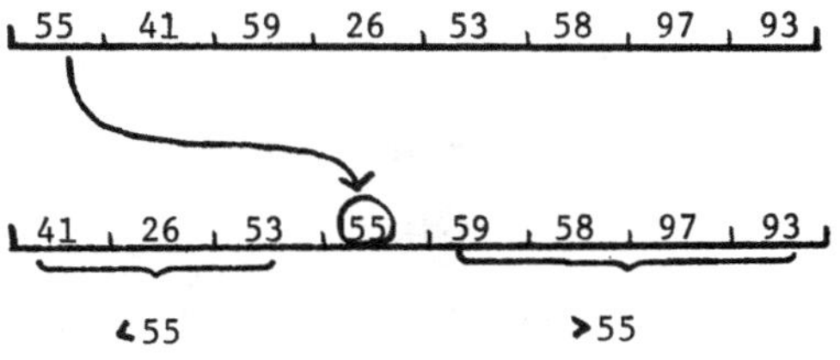

Figure 1. The Partitioning Phase of Quicksort

To analyze of the running time of Quicksort, we will let T(N) denote the number of comparisons made when N is the number of the elements in the array during a recursive call on the procedure (that is, when N = L - U + 1). If we assume that the ith largest element was chosen as the partitioning element, then we can recursively calculate the running time of the program as

$$T(N) = N - 1 + T(i-1) + T(N-i).$$

That is, the time required to sort N elements is equal to the time required to partition (which uses N - 1 comparisons) plus the time required to sort the left and right sets (which are respectively of sizes $i - 1$ and $N - i$). We also know the boundary conditions $T(0) = T(1) = 0$.

The above recurrence makes it very easy for us to analyze the running time of Quicksort. The best case occurs when we are always lucky enough to partition around the median element; that is, when i is approximately $N/2$. The previous recurrence simplifies to (approximately)

$$T(N) = 2T(N/2) + N - 1,$$

which establishes that the minimum number of comparisons ever made by Quicksort is $\sim N \lg N$. In the worst case, though, we might be unlucky enough always to choose the least element in the set as the partitioning element (that is, we always choose $i = 1$); this gives the recurrence

$$T(N) = T(N - 1) + N - 1,$$

which shows that the maximum number of comparisons made by Quicksort is proportional to $N^2/2$.

Although it is nice to know what happens in the best and worst cases, it is often more useful to know what happens on the average. To learn this, we must first describe the "average" input to the sorting algorithm; we usually just assume that all $N!$ possible permutations of input values are equally likely. If we assume this and ensure that the partitioning routine preserves the randomness inherent in the main file as it breaks the input into subfiles, it is easy to show that the average running time of Quicksort is

$$T(N) = N - 1 + (1/N) \sum_{1 \le i \le N} [T(i - 1) + T(N - i)].$$

(Note that the recurrence just assumes that each of the N values of i is chosen equiprobably.) This has solution $\sim 2NH_N$, or about $1.386\ N \lg N$.

Quicksort programs have very rapid average running times due to the combination of the small number of comparisons and an extremely efficient implementation; in fact, it is the fastest known sorting algorithm for large arrays. This might make us think that Quicksort is just the program to use in most applications, until we make two frightening observations.

1. In many applications, the input files either are already sorted or are close to being sorted.
2. The running time of Quicksort on a sorted file is proportional to $N^2/2$; that is, sorted files are a worst case for Quicksort.

We are now in an interesting state: we know that on random inputs our algorithm performs quite well, but that on common inputs it performs horribly. What can we do to escape this predicament? Hoare has given a remarkably elegant solution: we take whatever particular data the user might give us (be it random or not), and then scramble it up to make it appear random to our program!

There are two ways in which we might accomplish this. In the first method we make an initial pass over the data and "shuffle" it so that it is in random order (for information on shuffling arrays, see Knuth [1981, Section 3.4.2]). The second method is equivalent to shuffling but is often more convenient to implement: instead of always partitioning around the first element of the set, we choose an element randomly in X[L..U], and partition around that element. Both of these measures give an algorithm with an average number of comparisons of $\sim 1.386\, N \lg N$. Note that the average, now, is not quantified over all inputs, but rather says that for any fixed input, the average running time over all invocations of the algorithm is as claimed.

It is helpful to view the latter randomization method as sampling one element and then partitioning around it. This method is easily generalized to a sample of k elements, after which we partition around the median of the k. This technique is particularly effective when k is equal to three. (This modification has been analyzed in detail by Sedgwick [1975].)

This concludes our very brief investigation of Quicksort. Before studying the next sorting algorithm, we should list some of the general methods that we have seen in this particular example.

- Divide-and-conquer, a paradigm for constructing efficient algorithms.
- A distributional analysis of the expected running time of an algorithm (that of Quicksort under the assumption that all permutations are equally likely).
- A randomization algorithm that randomly shuffles its input to avoid high running time on common inputs.
- Sampling as a technique for increasing the speed of an algorithm.

2.2 Binsort

Quicksort achieves its very rapid sorting time by assuming only that the elements to be sorted came from a set with a comparison operation. In this subsection we shall study an asymptotically faster algorithm, called Binsort, that achieves greater speed by making additional assumptions. Initially, we will assume that the elements to be sorted are real numbers uniformly distributed on (0,1) (which we write as U(0,1)). As we progress further in this subsection we shall see that the above assumption can be relaxed considerably. The Binsort algorithm is quite old and well known (Knuth [1973, Section 5.2.1] attributes key ideas to Professor R. M. Karp); our investigation of Binsort will follow the development of Weide [1978], to whom are due many of the refinements and extensions that we shall see. The analysis of a related algorithm can be found in Devroye [1979].

To Binsort N reals from U(0,1) we will divide the segment (0,1) into N bins, place each real in its bin, and then sort within the bins. Specifically, the ith bin will contain all reals in [(i-1)/N,i/N) and will be represented by the array location B[i] (which is a pointer to a linked list of the reals contained in the bin). Our algorithm has the following structure.

1. Place reals into bins. Initialize an array B as defined above. Go through the input set, and place each real into the appropriate bin. Note that this step takes time proportional to N.
2. Sort within bins. Make a pass through the array B, and sort the elements in each bin using a standard worst-case sorting algorithm. It is easy to see that the expected number of records in each bin is unity and therefore it is easy to believe that the expected time required by this step is linear in N; a formal proof can be found in Weide [1978] and the other references cited above.
3. Concatenate the sorted lists. Go through the array B, and place the sorted list of records in each bin sequentially into the output list. This step takes time proportional to N.

Because Steps 1 and 3 both have linear running time (even in the worst case) and Step 2 has linear expected running time, the expected time of the entire algorithm is O(N). Its worst-case running time is proportional to O(N lg N).

It is easy to see why the algorithm performs so well: the bins divide the input into very small sets to be sorted. Unfortunately, it is also easy to see that if inputs drawn from certain nonuniform distributions are presented to this algorithm, then it will perform quite poorly. We will now try to remedy that behavior.

In the first extension of Binsort we will assume that the input reals are not uniformly distributed, but rather that they are drawn i.i.d. from some distribution with known C.D.F. F; for instance, we might know that we are sorting a set of reals drawn from a normal distribution with mean 0 and variance 1. We would like to achieve a system of bins of variable width, such as that shown in Figure 2a (with the P.D.F. of a truncated normal shown superimposed). Although this is quite contrary to the underlying structure of bins, we can achieve exactly the same effect by using the inverse C.D.F. F^{-1} to map each input onto a U(0,1) distribution. We then use the mapped value to determine the bin number; such a system is illustrated in Figure 2b.

Unfortunately, it is quite rare that a programmer is told to "write a routine to sort an array of N numbers, and, by the way, the inputs are known to come from such-and-such a distribution with an inverse C.D.F. of so-and-so"; the above approach might therefore seem useless. However, we can salvage much if (instead of assuming the C.D.F. as input) we sample a small set of points, calculate its Empirical C.D.F., and then use that E.C.D.F. to map from the input reals to (0,1). This scheme is illustrated in Figure 2c. Weide [1978] has shown that this scheme yields a linear expected-time sorting algorithm so long as the distribution satisfies several technical conditions. Furthermore, Weide's implementation of Binsort proved to be quite competitive with Quicksort for inputs of size at least five hundred.

Before we leave Binsort, we should briefly recap the general methods we have seen.

- Bins, a powerful algorithmic technique for localizing "near" objects in uniformly distributed data.
- Sampling to estimate the C.D.F. of an unknown distribution.

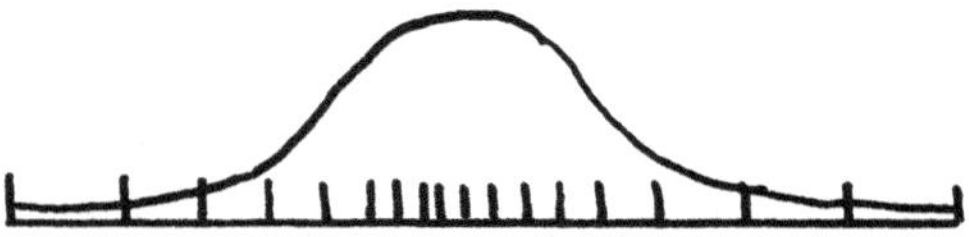

a.) Variable-width bins.

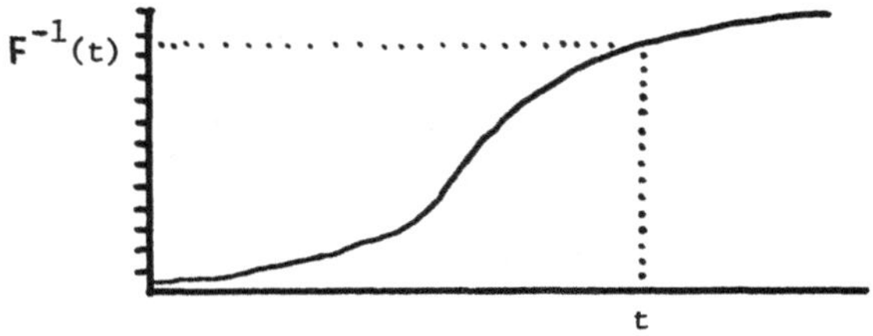

b.) Mapping by the C.D.F.

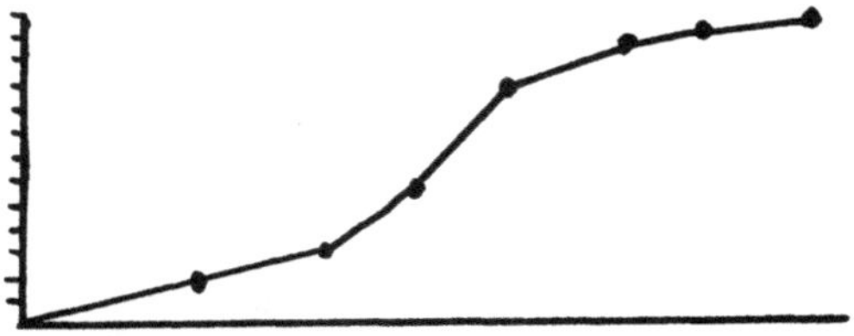

c.) Empirical C.D.F.

Figure 2. Modifications to Binsort

- A second distributional analysis of the expected running time of an algorithm (that of Binsort under the assumption that its inputs are i.i.d. from U(0,1)).
- A transformation that made arbitrary input distributions (from a certain class) appear as though they were from U(0,1).

2.3 Primality Testing

In this subsection we shall study the problem of determining whether a given input integer is prime or composite. One might at first think this is quite easy: to determine whether M is prime we merely divide it by all positive integers from 2 to $\lfloor M^{1/2} \rfloor$, and if none evenly divides M then it is prime. Unfortunately, if M is a 100-decimal digit, then this method will involve making approximately 10^{50} divisions of 100-digit numbers, which is remarkably expensive! As a matter of fact, no algorithms are currently known to test for primality that have running time only polynomial in the number of input bits of the integer. We will now study an algorithm due to Rabin [1976] that returns a probabilistic answer when asked if a certain integer is prime (we will define a "probabilistic answer" shortly). Before we do so, however, it will help us to consider a simpler problem.

In this simple problem we have two types of coins. Type I coins always come up heads when they are flipped, and Type II coins are fair (that is, they come up heads and tails equiprobably). The problem is, is it possible to determine a coin's type by flipping alone? There are two simple answers to this question.

- No: "It is clearly impossible to determine the coin's type by flipping alone. If it ever comes up tails, then I can accurately say that it is Type II. One the other hand, if I toss it one million times and it comes up heads each and every time, then it could be because it is a Type I coin or it could be that it is a Type II coin and I am just extremely unlucky."
- Yes: "It clearly is possible to determine the coin's type by flipping. I just flip it for some integer k times (say, k = 100). If it ever comes up tails, then I know that it is Type II; on the other hand, if it is heads all the time, then I'd be willing to bet that it is Type I."

Both answers are correct, because they answer different questions. The negative answer correctly asserts that it is in fact impossible to determine the coin's type by flipping alone, if we insist on absolute certainty. On the other hand, the positive answer gives us a way to get an answer that is usually correct. We must be very careful about the statement of the result given by the method; we do not claim that "the given coin is Type I with probability $1 - 2^{-k}$". That claim is ludicrous because the coin is either Type I or Type II (it is like claiming that nine is a square number with probability .7). Rather, the algorithm does claim that the probability of a Type II coin coming up heads k times in a row is 2^{-k}, and we can use that fact to draw what conclusions we may.

Rabin has made a brilliant observation that reduces the problem of primality testing to precisely the game we just studied. Specifically, he identified a predicate W (for witness) of two integer arguments that has the following remarkable properties.

Case I -- N is prime.
The predicate W(b,N) is false for $1 \leq b < N$.
Case II -- N is composite.
The predicate W(b,N) is true for at least half the integers b satisfying $1 \leq b < N$.

Details of the predicate W can be found in Knuth [1981, Section 4.5.4] or Rabin [1976] (a similar test is given by Solovay and Strassen [1977]).

Given the predicate W the primality testing problem is reduced to exactly the coin problem we saw previously. Certainly, we cannot use W to test exactly whether a given N is prime. On the other hand, if we compute W(b,N) for k random values of b satisfying $1 \leq b < N$, and W is false for all k values (say k = 100), then we should find this quite suspicious. Notice that such an outcome would not allow us to assert that N is prime, nor could we even say that it is prime with a certain probability (would you like to be quoted as saying that seven is prime with probability .8?), but we can say that the probability of this occurring if N is composite is only 2^{-k}.

The above probabilistic primality test is of great interest to number theorists, and furthermore has proved eminently practical in the public-key cryptosystem of Rivest, Shamir and Adleman [1978]. (Algorithmic aspects of that system have been described by Bentley [1979] and Knuth [1981, Section 4.5.4]). Furthermore, this problem has led us to the following algorithmic methods.

- Approximate answers, rather than exact answers to computational questions.
- A sophisticated use of randomized sampling.

3. A Systematic View

In the last section we studied probabilistic algorithms by investigating in detail three particular algorithms; in this section we shall take a more systematic view of the field. In Subsection 3.1 we shall examine some of the dimensions along which an individual problem may be viewed, and then in Subsection 3.2 we shall take a more distant view by examining various problem areas that have been studied. Both these vantages of probabilistic algorithms have much in common with the field of algorithm design in general; for an introduction to the general field, see Bentley [1979]. In Subsection 3.3 we shall study a topic specific to probabilistic methods: various techniques for constructing and analyzing probabilistic algorithms.

3.1 A Microscopic View

In this section we will briefly study the dimensions along which a particular algorthmic problem can be viewed. This discussion is quite short; the reader interested in more detail is referred to Bentley [1979].

The first choice one must make in approaching an algorithmic problem is the mathematical model of computation that will be used in the analysis of the problem. If we wish to model computation on most modern computers, we could choose a model such as an IBM System/360-370 or a Digital Equipment Corporation DPD-10, or a "polyunsaturated" computer such as Knuth's MIX machine, or a very simple model such as a single-register random-access machine (RAM). As the models become simpler, it is easier for us to analyze our problems, but our analyses are further removed from practice. We could also choose a more primitive model of computation, such as counting the number of some critical operation (such as comparisons or arithmetic operations), or parallel machines or even Turing Machines.

Once we have settled on the model, there are several independent choices we can make about the kind of algorithm we will develop and the kind of analysis we will perform on the algorithm.

- Efficiency vs. Efficacy -- An analysis of the efficiency of an algorithm tells how much of a certain resource it uses in achieving a solution; resources of common interest are time and space. An analysis of the efficacy of an algorithm tells how good its answer is (such as the relative error of a numerical algorithm or the probability of a composite number not being detected by Rabin's algorithm).
- Upper vs. Lower Bounds -- An upper bound is usually demonstrated by presenting a certain algorithm and analyzing its performance. A lower bound (which is usually much more difficult to achieve) says that all methods of solving a certain problem must require a certain amount of a given resource.
- Worst-case vs. Expected -- Will our analysis tell what happens in the very worst case, or are we more concerned with what happens on the average?
- Exact vs. Approximate Analysis -- How precisely shall we give our final answer? For a particular algorithm, we might be able

to give the increasingly refined answers of $O(N)$, $\sim 2N$, $2N -)(\lg N)$, and $2N - H_N$. The detail of analysis in a particular situation should be determined by how difficult and how valuable increasingly precise analyses are.

We have already seen examples of many of these dimensions. Our analysis of Quicksort is an approximate analysis giving an upper bound on the expected time efficiency of the algorithm; our sketch of Rabin's primality test gave an approximate analysis of the expected efficacy of his test.

An important point in the probabilistic analysis of algorithms is the source of randomness. There are two primary sources: either we can assume that the randomness is present in the input (that is, that the inputs are drawn randomly from some distribution), or we can induce randomness by having our algorithm take random steps during its execution.

3.2 A Macroscopic View

In this subsection we will take a more global view and investigate a number of problem areas in which probabilistic algorithms have been developed. For each area we will briefly describe the problem domain and then mention a few probabilistic algorithms that have been developed in the area. These lists are by no means complete; rather, the algorithms mentioned have been chosen because they illustrate important principles.

The first problem domain we will consider is that of algorithms that manipulate ordered sets. We have already seen two algorithms for sorting sets in rapid expected time; there are of course many others. Another important problem is searching: given a set of elements from a totally ordered set, and a new query element, is the query element a member of the stored set? The best searching algorithm for most applications it the constant expected-time algorithm of hashing. Yao and Yao [1977] showed that a method known as interpolation search also has a very fast expected running time. Another problem that arises in connection with ordered sets in selecting the ith largest element. Hoare [1962] describes a very rapid algorithm for this problem; that algorithm was later improved by Floyd and Rivest [1975], who used a more sophisticated sampling scheme to reduce its average running time. Details of these and many more algorithms on ordered sets can be found in Knuth [1973].

Much research has been devoted recently to finding graph algorithms with fast expected running time. Karp and Tarjan [1980] develop linear expected time algorithms for finding connected components, strong components, and blocks in random graphs. Lueker [1979] and Weide [1978] both present techniques for designing and analyzing greedy graph algorithms that are efficient and produce good answers when applied to random graphs.

Problems in geometry are especially appealing for probabilistic algorithms because of the wealth of probabilistic models that describe geometric phenomena. Bentley, Weide and Yao [1980] describe a host of linear expected-time algorithms for problems on point sets such as finding the nearest neighbor of each point in the set and constructing the Voronoi diagram of the point set. Unfortunately, their algorithms assume that the points are either uniformly distributed or come from distributions that satisfy pseudo-uniform properties. Bentley and Shamos [1978] gave a class of divide-and-conquer geometric algorithms that achieve linear expected time under much weaker stochastic assumptions. For instance, they showed that the convex hull of N points in the plane could be found in linear expected time for such distributions as points uniform on a polygon, uniform on a circle, and bivariate normal. They also gave a linear expected-time algorithm for linear programming in two variables. All of these algorithms exploit randomness in the input; Rabin [1976] gave a linear expected-time randomization algorithm that finds the closest pair in any point set (assuming nothing about the underlying distribution). Weide [1978] gives a simpler derivation of Rabin's result.

An especially important domain for probabilistic algorithms is that of algorithms for NP-complete problems, whose worst-case, exact-answer solutions are suspected to be computationally intractable. Karp [1977] has given an $O(N \lg N)$ time algorithm that almost surely produces a traveling salesman tour of N points uniform on a rectangle that is within any specified constant factor of optimal (the precise statement of this result is rather delicate). Lueker [1981] describes a heuristic for solving integer knapsack problems and shows that its performance is very close to optimal. We already mentioned the algorithms of Lueker [1979] and Weide [1978] that give approximate solutions to NP-hard graph theoretic problems.

This section has touched on only a few of the problem areas for which probabilistic algorithms have been developed; other areas include string algorithms, database problems, language theory, and number theory.

3.3 Techniques of Probabilistic Algorithms

In this subsection we will briefly study a number of techniques for constructing and analyzing probabilistic algorithms. For each technique, we will consider both the algorithmic technique and the mathematical methods used to analyze it, and then briefly survey some algorithms in which the technique has been used.

- Distributional analyses -- The backbone of most probabilistic algorithms is an analysis that shows that the algorithm performs to some given level under certain probabilistic assumptions. We saw sketches of such analyses for Quicksort and Binsort. Although it is nice if we can perform such an analysis rigorously using probability theory, we must often resort to simulations when our analytic techniques are not powerful enough.

- Transformations -- The idea underlying any transformation is to transform the problem to be solved into a problem we have already solved. For instance, we transformed the problem of Quicksorting a nonrandom file into the problem of Quicksorting a random file by shuffling the keys to be sorted. We used a more sophisticated transformation in Binsorting a nonuniform file by using an estimated C.D.F. to transform a given file to near-uniformity.
- Sampling -- Sampling uses a small subset of the input to infer cheaply and quickly useful information about the entire set. In Quicksort we saw how sampling can be used to estimate the median; in Binsort we use sampling to achieve a particularly Floyd and Rivest [1975] used sampling to avhieve a particularly efficient selection algorithm, and Weide [1978] presents a general discussion of sampling.
- Cells -- Binsort is an example of how cells can be used to identify locality in a uniform data set; we also saw how sampling can be used to transform a nonuniform set into a set that is apparently uniform. The idea of bins has been extended to higher-dimensional spaces by Bentley, Weide and Yao [1980].
- Divide-and-conquer -- Quicksort is a classic example of a divide-and-conquer algorithm: to solve a problem it partitions the inputs into two subsets and solves each recursively. The

convex hull and linear programming algorithms of Bentley and Shamos [1978] follow the same paradigm: to construct the convex hull of N points, we divide the points into two subsets, construct the hull of each recursively, and then merge the subhulls to form the hull of the entire set. We usually analyze the efficiency of divide-and-conquer algorithms by setting up and solving recurrence relations (just as we did for Quicksort). An important principle in the construction and analysis of divide-and-conquer algorithms is the preservation of independence -- we must show that if we assume that the input to the original problem is random in some sense, then partitioning into subproblems does not perturb that randomness. Such perturbations always complicate the analysis, and usually make the algorithm run slower.

- Greedy algorithms -- An algorithm that always takes a locally optimal step in hopes of producing a globally optimal answer is referred to as a greedy algorithm; for instance, certain greedy algorithms are known to produce the correct minimum spanning tree of a graph. There are many other problems for which the greedy method is known to produce an answer which is not necessarily optimal, but is expected to be quite close to optimal; for details on such algorithms, see Weide [1978].
- Oh-of-probability analysis -- Knuth [1980] has identified an important set of algorithms he refers to as "oh-of-probability" algorithms. The key idea in such algorithms is to identify typical inputs, and to show that the probability that any particular input is atypical is exceedingly small. One can then design an algorithm that is very efficient on typical inputs, and correct but slow on atypical inputs. If an

algorithm's running time on atypical inputs times the probability of an input being atypical is small enough, then the cost added by atypical inputs is very little. For an example of such an analysis, see the Voronoi diagram algorithm of Bentley, Weide and Yao [1980].

- Randomization -- This idea underlies many of the above techniques, but is important enough to merit mention as a separate technique. An algorithm using randomization "tosses coins" during its execution, and takes the next step as a function of the outcome of the toss. Quicksort used randomization to shuffle its input, and Rabin's [1976] primality algorithm used randomization to guess witnesses of compositeness.

A more detailed study of these and many other algorithmic techniques may be found in Weide [1978].

4. Conclusions

As stated in the introduction, the purpose of this paper is not to provide a comprehensive survey of the field of probabilistic analysis of algorithms, but rather to provide the reader with a brief introduction to a young field. To do this, we briefly studied three important algorithms in Section 2, and then took a more systematic view of the field in Section 3. We will now conclude the paper by summarizing exactly what the field of probabilistic analysis of algorithms has to offer to practitioners of computing.

The first thing that the field has to offer is a large and ever-increasing set of results in the form of efficient algorithms. We saw some details of three such algorithms in Section 2, and briefly surveyed a number of results in Subsection 3.2. More systematic collections of results can be found in the surveys of Janko [1981] and Lueker [1981]. In addition to being of fundamental theoretical importance, these algorithms have already been shown to be eminently practical.

Just as important as the results are the methods that the field supplies to practitioners of computing. In Subsection 3.3 we saw a set of particular algorithmic and analytic techniques that can be used to create and analyze probabilistic algorithms. A less tangible but equally useful set of ideas was given in Subsection 3.1: the parameters of a problem. Those allow us to identify and to speak precisely about the important aspects of the problem at hand; they give a perspective that often opens a new approach to the problem.

The probabilistic analysis of algorithms is an exciting young field with a host of open problems. Some were identified long ago and have withstood the repeated attacks of many investigators; others are of the very general form of "the best deterministic algorithm for this problem is such-and-such; is there a better probabilistic algorithm?". One of the most challenging open problems is to develop a methodology by which practitioners of computing will be able to employ routinely the results and methods of the field.

5. Acknowledgements

The careful comments of John McDermott and George Lueker are gratefully acknowledged. I would like to thank Dick Karp, Don Knuth, Bruce Weide and Andy Yao for sharing their excitement about the field as they taught me about the probabilistic analysis of algorithms.

6. References

[1] Bentley, J. L. [1979]. "An introduction to algorithm design," IEEE Comput. Magazine 12, February 1979, pp. 66-78.

[2] Bentley, J. L. and M. I. Shamos [1978]. "Divide and conquer for linear expected time", Inform. Process. Lett., 7, February 1978, pp. 87-91.

[3] Bentley, J. L., B. W. Weide and A. C. Yao [1980]. "Optimal expected-time algorithms for closest-point problems", ACM Trans. Math. Software, 6, December 1980, pp. 563-580.

[4] Devroye, L. [1979]. "Average time behavior of distributive sorting algorithms", Technical Report No. SOCS 79.4, March 1979.

[5] Floyd, R. W. and R. L. Rivest [1975]. "Expected time bounds for selection", Comm. ACM, 18, March 1975, pp. 165-172.

[6] Hoare, C. A. R. [1962]. "Quicksort", Comput. J., 5, April 1962, pp. 10-15.

[7] Janko, W. [1981]. "Bibliography of probabilistic algorithms", in preparation.

[8] Karp, R. M. [1977]. "Probabilistic analysis of partitioning algorithms for the traveling-salesman problem in the plane", Math. Oper. Res., 2, August 1977, pp. 209-224.

[9] Karp, R. M. and R. E. Tarjan [1980]. "Linear expected-time algorithms for connectivity problems", J. Algorithms, 1, December 1980, pp. 374-393.

[10] Knuth, D. E. [1973]. The Art of Computer Programming, volume 3: Sorting and Searching, Addison-Wesley, Reading, MA.

[11] Knuth, D. E. [1980]. Class lecture in course Computer Science 255, Stanford University, March 1980.

[12] Knuth, D. E. [1981]. The Art of Computer Programming, volume 2: Seminumerical Algorithms, Second Edition, Addison-Wesley, Reading, MA.

[13] Lueker, G. [1979]. Optimization problems on graphs with independent random edge weights, UC-Irvine Technical Report #131, Department of Information and Computer Science.

[14] Lueker, G. [1981]. "Algorithms with random inputs", to appear in Proceedings of Computer Science and Statistics: Thirteenth Annual Symposium on the Interface, Pittsburgh, PA, March 1981.

[15] Rabin, M. O. [1976]. "Probabilistic algorithms", Algorithms and Complexity: New Directions and Recent Results, J. F. Traub, Ed., Academic Press, New York, NY, pp. 21-39.

[16] Rivest, R. L., A. Shamir and L. Adleman [1978]. "A method for obtaining digital signatures and public-key cryptosystems", Comm. ACM, 21, February 1978, pp. 120-126.

[18] Sedgewick, R. [1975]. Quicksort, Stanford University Computer Science Department Report STAN-CS-75-492, May 1975.

[19] Solovay, R. and V. Strassen [1977]. "A fast Monte-Carlo test for primality", SIAM J. Comput., 6, March 1977, pp. 84-85.

[20] Weide, B. W. [1978]. Statistical Methods in Algorithm Design and Analysis, Ph.D. Thesis, Carnegie-Mellon University, August 1978.

[21] Yao, A. C. and F. F. Yao [1976]. "The complexity of searching an ordered random table", Proceedings of the Seventeenth Annual Symposium on the Foundations of Computer Science, October 1976, IEEE, pp. 222-227.

Departments of Computer Science and Mathematics, Carnegie-Mellon University, Pittsburgh, Pennsylvania 15213

This research was supported in part by the Office of Naval Research under Contract N0014-76-C-0370.

Discussant's Report on "Probabilistic Analysis of Algorithms," by Jon Bentley

It was a pleasure to listen to Jon Bentley's talk; it gave a useful introduction to and overview of the probabilistic analysis of algorithms. Especially interesting was the discussion of Bentley's work on the traveling salesman problem.

As Bentley pointed out, there are two fundamentally different approaches in this area. One type of algorithm, which is often called a coin-flipping algorithm, provides its own randomness, perhaps through the use of a random number generator. Algorithms of this type can sometimes be designed which work well on even the worst-case input. The primality test of Solovay and Strassen [SS77] and Rabin [Ra76] is a famous example. A second type of approach is to assume some distribution of inputs, and analyze the average behavior of an algorithm under this assumption. This is sometimes called the distributional approach. (Yao [Ya77] presents some interesting relationships between these two approaches.)

When both types of algorithms are available, the randomized approach has some significant advantages. It eliminates the need to make a (possibly unjustified) assumption about the distribution of inputs. It eliminates the risk that the one problem which a hapless user wishes to solve will be one of the rare ones on which the algorithm takes a very long time. Sometimes an algorithm which is usually assumed to rely on randomness in its input can be modified to provide its own randomness; for example, Carter and Wegman have shown that hashing can be modified in this way [CW77]. Thus researchers might be tempted to seek fast coin-flipping algorithms for problems such as boolean satisfiability, graph coloring, etc. It is important to note, however, that this is an extremely ambitious undertaking, since if one NP-complete problem has a solution by a polynomial time coin-flipping algorithm, then all of them do. (See [AM77].)

Thus for many problems we will probably need to content ourselves with distributional solutions. This gives rise to an important question: when doing our analysis, what sort of distribution shall we assume? One approach, which Bentley has used, is to actually gather statistics about real problem instances in an attempt to obtain a realistic distribution. This is a very useful approach, and can be used to tune the algorithm to the input. But what can we do if we do not have a good way of obtaining statistics about the nature of the input in advance? One approach which has been employed and seems promising is to investigate the <u>distributional robustness</u> of algorithms, i.e., to investigate the sensitivity of algorithms to changes in the input distribution. For example, consider the problem of sorting n numbers which are drawn independently according to a probability density function f. If f is the uniform distribution, it is well-known that address calculation sorting yields an expected time of O(n); see [Kn73, Section 5.2.1] for a thorough analysis of this algorithm. If f is "sufficiently smooth," the expected complexity continues to be linear [Kn73, Section 5.2.1, exercise 38, attributed to Karp]. Weide [We78] has shown that the algorithm can be made to work in expected linear time provided that the density function has finite support and is bounded except that it contains finitely many delta functions. Devroye [De79] has shown that it can be made to work for certain density functions which have poles and/or infinite support. For some of these distributions the version of the algorithm discussed in [Kn73] would take more than linear time. Another example of this sort of analysis of distributional robustness is the work of Willard on searching [Wi80].

He shows that a slight deviation from a uniform input distribution can destroy the O(log log n) expected time of interpolation search. He then shows how to modify the algorithm to guarantee O(log log n) expected time over a much broader class of distributions.

It is interesting to note that sometimes probabilistic approaches can be used to prove results whose statement does not involve probability. [ES74] provides a number of interesting examples of this. In [AKLL79] this sort of approach is used to prove the existence of short universal traversal sequences.

References

[AKLL79] R. Aleliunas, R. M. Karp, R. J. Lipton, and L. Lovász (1979), "Random Walks, Universal Traversal Sequences, and the Complexity of Maze Problems," Proceedings 20th Annual Symposium on Foundations of Computer Science, pp. 218-223.

[AM77] L. Adleman and K. Manders (1977), "Reducibility, Randomness, and Intractibility," Proceedings Ninth Annual ACM Symposium on Theory of Computing, pp. 151-163.

[CW77] J. L. Carter and M. N. Wegman (1977), "Universal Classes of Hash Functions," Proceedings Ninth Annual ACM Symposium on Theory of Computing, pp. 106-112.

[De79] L. Devroye (1979), "Average Time Behavior of Distributive Sorting Algorithms," Technical Report No. SOCS 79.4.

[ES74] P. Erdös and J. Spencer (1974), Probabilistic Methods in Combinatorics, Academic Press, New York.

[Kn73] D. Knuth (1973), The Art of Computer Programming, Vol. 3: Sorting and Searching, Addison-Wesley, Reading, Mass.

[Ra76] M. O. Rabin (1976), "Probabilistic Algorithms," in Algorithms and Complexity: New Directions and Recent Results, J. F. Traub, ed., Academic Press, New York.

[SS77] R. Solovay and V. Strassen (1977), "A Fast Monte-Carlo Test for Primality," SIAM J. Comput., 6:1, pp. 84-85.

[We78] B. W. Weide (1978), Statistical Methods in Algorithm Design and Analysis, Ph.D. Thesis, Carnegie-Mellon University, Pittsburgh, Pennsylvania; appeared as CMU Computer Science Report CMU-CS-78-142.

[Wi80] D. E. Willard (1982), "A Log Log N Search Algorithm for Nonuniform Distributions," Proceedings of the Symposium on Applied Probability-Computer Science: The Interface, Birkhäuser-Boston, Boston, MA.

[Ya77] A. Yao (1977), "Probabilistic Computations: Toward a Unified Measure of Complexity," Proceedings of the Eighteenth Annual Symposium on Foundations of Computer Science, pp. 222-227.

Discussant: Dr. George S. Lueker, Department of ICS, University of California, Irvine, CA 92717.

POINT PROCESS METHOD IN QUEUEING THEORY

Peter Franken

Abstract

In this paper we shall study a stochastic point process approach to problems in queueing theory. It will appear that many results previously obtained under special assumptions of independence or special distributional assumptions can be obtained under considerably weaker conditions by using the point process theory. In this way we will demonstrate that such results have considerably greater generality than previously shown.

Section 2 provides some of the ground work in point process theory needed for our task. In Section 3 these methods are used to study various problems related to time-stationary and customer-stationary queues. In particular we discuss M/G/s/r queues, Little-type results, extensions to Takács virtual waiting time formulas and some inequalities related to busy periods.

1. Introduction

Recently there has been considerable interest in queueing models in which there may be dependence among the interarrival times and service times. For several important basic questions - such as the existence of stationary distributions, the relationships among different stationary distributions (e.g., time-stationary and arrival-stationary), and model continuity - the general theory of point processes is very useful. As representative papers, we would mention König and Matthes (1963), Franken (1976), Miyazawa (1979), and König and Schmidt (1980). A comprehensive discussion of the point process approach to queueing is given in the monograph by Franken, König, Arndt and Schmidt (1981), abr. FKAS(1981).

The purpose of this paper is to present some basic ideas of the point process approach and to illustrate how they can be applied by considering a few examples. In Sections 2 and 3 we discuss the point process approach applied to some standard queueing systems under quite weak assumptions concerning the sequence of interarrival times and service times. The presentation is essentially based on chapters 1, 2 and 4 in FKAS(1981). However, there are some differences. In particular, we reduced some mathematical formalities in order to make this paper easier to read. For the same reason, we restrict attention to queues with FCFS discipline although related results can be obtained for other queueing disciplines; see FKAS(1981).

Extending the well-known Kendall notation, we use the symbol G/G/s/r to denote a FCFS queueing system with s servers in parallel ($s \leq \infty$), r waiting places ($r \leq \infty$), and a stationary, metrically transitive sequence $([\alpha_n,\beta_n], -\infty < n < \infty)$ of interarrival times α_n and associated service times β_n, in the case $s = \infty$, we write $G/G/\infty$. We note that for some queueing systems under consideration, the assumption of metric transitivity of $([\alpha_n,\beta_n])$ can be relaxed somewhat, see Franken (1981). Furthermore, we assume $0 < a = E\alpha_n < \infty$ and $0 < b = E\beta_n < \infty$ hold. We use the notation:

$$A(t) = P(\alpha_n \leq t);$$

$$B(t) = P(\beta_n \leq t);$$

$$\rho = b/a.$$

No assumption about independence in the sequence $([\alpha_n,\beta_n])$ is made in general. However, in some special cases the sequence $([\alpha_n,\beta_n])$ will also have some of the following properties:

(i) (α_n) and (β_n) are independent sequences;

(ii) (α_n) is a sequence of i.i.d. random variables;

(iii) (β_n) is a sequence of i.i.d. random variables.

If we have properties (i) and (ii), we will call this a GI/G/s/r queue; if properties (i) and (iii), this is G/GI/s/r; if properties (i), (ii), and (iii), this is GI/GI/s/r. Finally, we use M instead of GI if

$$A(t) = 1 - e^{-\lambda t} \text{ or } B(t) = 1 - e^{-\mu t}.$$

In Section 2 we rigorously define what is meant by the term "queue in steady state". For this purpose we discuss stationary complete arrival processes in Section 2.1. The word "complete" indicates here that we consider not only the arrival epochs but also the associated service times. We state the theorem about the one-to-one correspondence between stationary sequences of interarrival and service times and stationary marked point processes in which the arrival epochs are the points and service times the marks. This result allows us to introduce the notions of synchronous and stationary complete arrival processes generated by the same stationary sequence of interarrival and service times. Finally we discuss the asymptotic behavior of these processes.

In Section 2.2 we treat the time behavior of G/G/s/r queues with FCFS queueing discipline. Under some additional conditions on the underlying sequence of interarrival times and service times, we construct stationary processes and sequences which describe the system behavior in continuous time and at arrival or departure epochs, respectively. This leads to the notions of time-stationary, arrival-stationary, and departure-stationary distributions of system characteristics such as waiting time, queue size, etc.. We point out that these different distributions describe the same system in steady state, but from different points of view depending on the choice of the time origin. Proofs in Section 2 are omitted; the emphasis is on interpretation of the statements. For the proofs we refer to Port and Stone (1973), FKAS(1981), and Streller (1980).

In Section 3 we prove several relationships among the different stationary distributions of the queueing system characteristics and their moments. In particular, we discuss M/G/s/r queues, Little-type formulas, extensions of Takács formulas, and the mean busy period.

2. Queues in Steady State

In this section we rigorously define what is meant by a "queue in steady state" and introduce some stationary processes and sequences describing the time behavior of such queues.

2.1 Stationary Arrival Processes

In the study of "steady-state" queues it is convenient to think of arrivals to an empty system beginning at time $-\infty$. The sequence $([t_n,\beta_n], -\infty < n < \infty)$ of arrival instants, t_n, and service times, β_n, is called the complete arrival process (abbr. CAP). The word "complete" indicates that we consider not only the arrival times, but also the associated service times.

We treat CAP's as a special case of marked point processes. Consider a metric space K (in the following we need spaces of the form $K = R_+^n$, $n \leq \infty$, $R_+ = [0,\infty)$).

2.1 Definition. A sequence $\phi = ([t_n,k_n])$ of $(R \times K)$-valued r.v.'s with the property

$$t_n \leq t_{n+1} \text{ for every } n; \lim_{n\to\pm\infty} t_n = \pm\infty \tag{2.1}$$

is called a marked point process (abbr. MPP) with the mark space K. A sequence (t_n) satisfying (2.1) is called a point process.

Thus, a CAP is a MPP with the mark space $K = [0,\infty)$. (There are other types of queues, e.g., priority queues, for which β_n is a vector.) Because we do not assume that t_n is strictly less than t_{n+1} for every n, batch arrivals are included. However, since the service times β_n are indexed, it follows that the customers in a batch are ordered. In this way, the notion of the service time of the last arrival before t is well defined.

Since we are primarily interested in the steady-state behavior of the queueing system under consideration, we can assume, without loss of generality, that the underlying CAP is in steady state. However, there are two possible ways to define "CAP in steady state" corresponding to the well-known notions of ordinary and stationary renewal processes. We introduce the following generalization of an ordinary renewal process.

2.2 Definition. Let $\xi = ([\alpha_n,k_n])$ be a stationary sequence of $(R_+ \times K)$-valued r.v.'s satisfying $a = E\alpha_1 < \infty$. The MPP $\phi(\xi) = ([t_n,k_n])$ where

$$t_1 = 0;\ t_n = \sum_{j=1}^{n-1} \alpha_j \text{ for } n \geq 2;\ t_n = -\sum_{j=0}^{n} \alpha_{-j} \text{ for } n \leq 0;$$

is called the synchronous MPP generated by the stationary sequence $\xi = ([\alpha_n,k_n])$.

For any (synchronous or asynchronous) MPP $\phi = ([t_n,k_n])$ we define the shift operator S_u, $u \in R$, by

$$S_u\phi = ([t_n-u,k_n]). \tag{2.2}$$

In this way the marks are shifted along with corresponding points. Thus, we have the following generalization of a stationary renewal process.

2.3 Definition. A MPP $\bar{\phi} = ([\bar{t}_n, \bar{k}_n])$ is called stationary if the shifted MPP $S_u\bar{\phi}$ has the same distribution for all u. The expectation $\lambda = E\bar{N}(1)$ is called the intensity of $\bar{\phi}$, where $\bar{N}(1) = \#([0,1) \cap (\bar{t}_n))$ denotes the number of points of MPP $\bar{\phi}$ in [0,1).

For abreviation we use the notations (ξ,P) for the "stationary sequence $\xi = ([\alpha_n,k_n])$ with the probability distribution P" and $(\bar{\phi},\bar{P})$ for the "stationary MPP $\bar{\phi} = ([\bar{t}_n,\bar{\beta}_n])$ with the probability distribution $\bar{P}$".

At first glance, definitions 2.2 and 2.3 may seem to deal with quite different objects. However, the following propositions show that there is a close relation between stationary sequences and stationary MPP, similar to the relation between ordinary and stationary renewal processes with the same underlying lifetime distribution.

2.4 Proposition. There exists a one-to-one mapping between the families $\{(\xi,P),\ 0 < a(P) = E_P\alpha_1 < \infty\}$ and $\{(\bar{\phi},\bar{P}),\ 0 < \lambda(\bar{P}) < \infty\}$. In particular for a given (ξ,P) the corresponding $(\bar{\phi},\bar{P})$ is determined by the formula

$$\bar{P}(\bar{\phi} \in C) = \frac{1}{a(P)} \int_0^\infty P(\{\alpha_1 > t\} \cap \{S_t\phi(\xi) \in C\})dt, \tag{2.3}$$

or, equivalently,

$$E_{\bar{P}}f(\bar{\phi}) = \frac{1}{a(P)} E_P \int_0^{\alpha_1} f(S_t\phi(\xi))dt. \tag{2.4}$$

(for every $f \geq 0$). Furthermore,

$$\lambda(\bar{P}) = 1/a(P). \tag{2.5}$$

In view of (2.3) a stationary MPP $\bar{\phi}$ has, with probability 1, no points at 0. Thus, we can number the points $\bar{t}_n$ in the following way

$$\cdots \leq \bar{t}_0 < 0 < \bar{t}_1 \leq \cdots \tag{2.6}$$

From (2.3) and (2.6) we obtain the formula

$$\bar{P}(t_1 < x) = \frac{1}{a(P)} \int_0^x P(\alpha_1 > t)dt. \tag{2.6a}$$

which is well-known for renewal processes.

If (ξ,P) satisfies an additional condition we can prove a limit theorem which, in particular, states that $\bar{P}$ arises as a limiting distribution for $\phi(\xi)$. Recall that the sequence ξ is called metrically transitive if $P(A) = 0$ or $P(A) = 1$ for all shift invariant sets $A \subseteq \prod_{-\infty}^{+\infty} (R_+ \times K)$.

<u>2.5 Proposition</u>. If, in addition to the conditions of Proposition 2.4, (ξ,P) is metrically transitive, then for every function $f \geq 0$.

$$\lim_{T\to\infty} \frac{1}{T} \int_0^T f(S_u\phi(\xi))du = E_{\bar{P}}\, f(\bar{\phi}) \quad P\text{-a.s.}, \tag{2.7}$$

$$\lim_{n\to\infty} \frac{1}{n} \sum_{j=1}^{n} f(S_{\bar{t}_j}(\bar{\phi})) = E_P\, f(\phi(\xi)) \quad \bar{P}\text{-a.s.}. \tag{2.8}$$

So we can speak about the synchronous MPP $\phi(\xi)$ and the stationary MPP $\bar{\phi}(\xi)$ generated by (ξ,P). Note that, in view of definition (2.2), $\phi(\xi)$ is actually a function of ξ. The notation $\bar{\phi}(\xi)$ means only that $\bar{P}$ is generated by P according to proposition 2.4; however, $\bar{\phi}$ is not a function of ξ.

Now we can define what is meant by a "CAP in steady state". We start with a stationary, metrically transitive sequence $\xi = ([\alpha_n,\beta_n])$ of interarrival and service times with the property $a(P) = E_P\alpha_1 < \infty$. In view of propositions 2.4 and 2.5, the synchronous and the stationary MPP's $\phi(\xi)$ and $\bar{\phi}(\xi)$ can be considered as two different but equivalent descriptions of the same underlying complete arrival process in steady state. The model $\phi(\xi)$ corresponds to the origin being an arbitrary arrival epoch; thus, we also call $\phi(\xi)$ the <u>arrival-stationary</u> CAP. The model $\bar{\phi}(\xi)$ corresponds to the origin being an arbitrary point on the time axis; thus, we call $\bar{\phi}(\xi)$ the time-stationary CAP.

Consider $\xi = ([\alpha_n, \beta_n])$ with the mean interarrival time $a = E_P\alpha_1 < \infty$ and the mean service time $b = E_P\beta_1 < \infty$. Let $\lambda = 1/a$ be the intensity of $\bar{\phi}(\xi)$; λ is also called the arrival rate. Denote by $N(t)$ and $\bar{N}(t)$ the counting processes associated with $\phi(\xi)$ and $\bar{\phi}(\xi)$, respectively: $N(t) = \#([0,t) \cap (t_n))$, $\bar{N}(t) = \#([0,t) \cap (\bar{t}_n))$. From (2.7) and (2.8) we obtain the following relationships

$$\lim_{t\to\infty} N(t)/t = \lambda, \quad P\text{-a.s.,} \tag{2.9}$$

$$\lim_{n\to\infty} \frac{1}{n} \sum_{j=1}^{n} \bar{\alpha}_j = a, \quad \bar{P}\text{-a.s.,} \tag{2.10}$$

$$\lim_{n\to\infty} \frac{1}{n} \sum_{j=1}^{n} \bar{\beta}_j = b, \quad \bar{P}\text{-a.s..} \tag{2.11}$$

Recall, that (2.9) is well-known in the case of renewal processes. On the other hand using the Birkhof-Khinchin ergodic theorem we have

$$\lim_{t\to\infty} \bar{N}(t)/t = \lambda, \quad \bar{P}\text{-a.s.} \tag{2.12}$$

$$\lim_{n\to\infty} \frac{1}{n} \sum_{j=1}^{n} \alpha_j = a, \quad P\text{-a.s.,} \tag{2.13}$$

$$\lim_{n\to\infty} \frac{1}{n} \sum_{j=1}^{n} \beta_j = b, \quad P\text{-a.s..} \tag{2.14}$$

Thus, the mean interarrival time a (mean service time b) arises as the expectation of a generic interarrival time α (service time β) in the arrival-stationary model $\phi(\xi)$ and as a limiting sample path customer-average in both models. The arrival rate λ arises as the expectation of the number of arrivals in a time unit in the time-stationary model $\bar{\phi}(\xi)$ and as limiting sample path time-averages in both models.

Consider a stationary sequence (ξ,P) and the corresponding stationary MPP $(\bar{\phi}(\xi),\bar{P})$. For a set $Y \subseteq K$ with $P(k_1 \in Y) > 0$ consider the stationary MPP $\bar{\phi}_Y$ with the mark space Y which arises from $\bar{\phi}$ by deletion of all points $[\bar{t}_n,\bar{k}_n]$ with $\bar{k}_n \notin Y$. Then the intensity λ_Y of $\bar{\phi}_Y$ is related to λ by

$$\lambda_Y = \lambda P(k_1 \in Y). \qquad (2.15)$$

The following so-called Campbell's Theorem is very useful in queueing theory (see Section 3.1).

2.7 Proposition. Consider (ξ,P) and the corresponding $(\bar{\phi},\bar{P})$. For every function $f \geq 0$ on $R \times K$,

$$E_{\bar{P}} \sum_{n=-\infty}^{+\infty} f(\bar{t}_n,\bar{k}_n) = \lambda \int_{-\infty}^{+\infty} \int_K f(t,y)dt\, P(k_1 \in dy). \qquad (2.16)$$

For $u < t$ we define the restriction of an arbitrary MPP $\phi = \{[t_n,k_n]\}$ to $[u,t)$ by

$$_{[u,t)}\phi = \{[t_n,k_n]:[t_n,k_n] \in \phi \text{ and } t_n \in [u,t)\}. \qquad (2.17)$$

2.2 Stationary Queueing Processes

Now we discuss the time behavior of FCFS queues G/G/s/r. First, we have to define the appropriate state space Z. The state of the system at an arbitrary epoch will be given by an $(s+r)$-dimensional vector $\underset{\sim}{z}$. The first s components of this vector are the residual service times at each server arranged in descending order (the residual service time at an idle server is equal to 0). The last r components are the required service times of customers presently in the queue, arranged in the order of their arrival. If there are j customers waiting $(j < r)$ the last $r - j$ components are zero. Such a detailed description allows us to describe the system behavior by simple stochastic equations in discrete or continuous time.

We consider first the arrival-stationary CAP $\phi(\xi)$, $\xi = ([\alpha_n,\beta_n])$. There exists a function $h_{s,r}$ such that the system states $\underset{\sim}{z}_n = (z_{n,1},\ldots)$ immediately before the arrival of the nth customer satisfy the following recurrence equation

$$\underset{\sim}{z}_{n+1} = h_{s,r}(\underset{\sim}{z}_n, \alpha_n, \beta_n), \quad -\infty < n < \infty . \tag{2.18}$$

For example, for loss systems G/G/s/0 we have

$$h_{s,0}(\underset{\sim}{z}_n, \alpha_n, \beta_n) = R(\underset{\sim}{z}_n - \alpha_n \underset{\sim}{i} + \beta_n \underset{\sim}{e} 1_{(z_{n,s}=0)})_+,$$

where $\underset{\sim}{i} = (1,\ldots,1)$, $\underset{\sim}{e} = (0,\ldots,0,1)$, $\underset{\sim}{x}_+ = (\max(x_1,0),\ldots,\max(x_s,0))$ and R denotes the operator of rearranging the vector's components in descending order. However, the exact form of the function $h_{s,r}$ is not important in the following.

There also exists a function $\hat{h}_{s,r}$ such that the system states $\underset{\sim}{z}(u)$ and $\underset{\sim}{z}(t)$ at the two epochs u and t, $u < t$, satisfy the equation

$$\underset{\sim}{z}(t) = \hat{h}_{s,r}(\underset{\sim}{z}(u), {}_{[u,t)}\phi) \tag{2.19}$$

where ${}_{[u,t)}\phi$ is defined in (2.17).

For an arbitrary $\underset{\sim}{d} \in Z$ define

$$\underset{\sim}{z}_1(\underset{\sim}{d}) = \underset{\sim}{d}, \quad \underset{\sim}{z}_{n+1}(\underset{\sim}{d}) = h_{s,r}(\underset{\sim}{z}_n(\underset{\sim}{d}), \alpha_n, \beta_n). \tag{2.20}$$

Similarly for $t > 0$ define

$$\underset{\sim}{z}(\underset{\sim}{d}, t) = \hat{h}_{s,r}(\underset{\sim}{d}\, {}_{[0,t)}\phi). \tag{2.21}$$

The sequence $(\underset{\sim}{z}_n(\underset{\sim}{d}))$ describes the transient behavior of the system at arrival epochs with the initial state $\underset{\sim}{d}$ and arrival-stationary CAP $\phi(\xi)$. The process $(\underset{\sim}{z}(\underset{\sim}{d},t))$ defined by (2.21) has the analogous meaning for the continuous-time behavior.

Now we are interested in stationary solutions of (2.18), because such a solution can be considered as a model of the arrival-stationary behavior of the system.

<u>2.8 Proposition</u>. Consider a G/G/s/r system with the arrival-stationary CAP $\phi(\xi)$, where $\xi = ([\alpha_n, \beta_n])$ is a stationary, metrically transitive sequence with finite a and b. Under an additional assumption A(s,r), there exists a stationary sequence $(\underset{\sim}{z}_n)$ which satisfies (2.18).

Moreover, there exists a function $g_{s,r}$ such that for every n

$$\underset{\sim}{z}_n = g_{s,r}(([\alpha_j,\beta_j]),\ j < n). \tag{2.22}$$

Under a further assumption, B(s,r), we have that for every $\underset{\sim}{d} \in Z$

$$(\underset{\sim}{z}_{n+k}(\underset{\sim}{d}),\ k \geq 1) \underset{n\to\infty}{\Longrightarrow} (\underset{\sim}{z}_k), \tag{2.23}$$

where $\underset{\sim}{z}_n(\underset{\sim}{d})$ is defined by (2.20) and $\Longrightarrow$ denotes the weak convergence of random sequences (in fact, we have here convergence in variation).

The relationship (2.23) means that starting from an arbitrary initial condition $\underset{\sim}{d}$ the system tends to the steady state. Obviously, (2.23) ensures the uniqueness of the stationary sequence $(\underset{\sim}{z}_n)$ satisfying (2.18).

What are the conditions denoted by the symbols A(s,r) and B(s,r)? There is not yet a complete answer to this question for some systems considered. The most recent results can be found in Borovkov (1980), FKAS(1981) and the references there. Here we will give a few examples.

We do not need additional assumptions A(∞) and B(∞) in the case G/G/∞. For loss systems G/G/s/0, it can be shown, without additional assumptions, that at least one stationary solution of (2.18) also exists. However, this solution may not be unique and it may not satisfy (2.22). Because the condition A(s,0) and B(s,0) are quite complicated in general, we illustrate them only. For systems G/GI/s/0 the condition $P(\beta_1 < \alpha_1) > 0$ is sufficient for the validity of all assertations in proposition 2.8. Thus, the condition $B(x) = P(\beta_1 < x) > 0$ for all $x > 0$ is sufficient too. For GI/GI/s/0 systems we have a weaker sufficient condition, $P(\beta_1 < s\alpha_1) > 0$.

For G/G/1/∞ queues both condition A(1,∞) and B(1,∞) hold with $\rho < 1$. For G/G/s/∞ the condition $\rho < s$, ensures the existence of a stationary solution of (2.18) which satisfies (2.22) and (2.23) for

$\underset{\sim}{d} = 0$ (but, in general, not for all $\underset{\sim}{d} \in Z$). However, for GI/GI/s/∞ the condition $\rho < s$ ensures the validity of both assertions in proposition 2.8.

In the following we assume for the system considered that at least the condition A(s,r) is fulfilled. Then we can interpret $(\underset{\sim}{z}_n)$ as the model of the system behavior in steady state (observed at arrival epochs only). Thus, we call $(\underset{\sim}{z}_n)$ the arrival-stationary state process (sequence) of the system under consideration. From $(\underset{\sim}{z}_n)$ we can easily derive other variables of interest such as the actual work load v_n, the number of customers in the system ℓ_n, and the number of customers in the queue q_n;

$$v_n = \sum_{j=1}^{s+r} z_{n,j}; \quad \ell_n = \sum_{j=1}^{s+r} 1_{\{z_{n,j}>0\}}; \quad q_n = \sum_{j=s+1}^{s+r} 1_{\{z_{n,j}>0\}}. \tag{2.24}$$

For $s = 1$ the actual waiting time w_n is equal to $w_n = z_{n,1}+\dots+z_{n,r+1}$. For $s > 1$ we have to use the Kiefer-Wolfowitz representation for the determination of w_n from $\underset{\sim}{z}_n$. The sequences (v_n), (w_n), (ℓ_n), (q_n) are stationary because $(\underset{\sim}{z}_n)$ is stationary. In view of (2.22) the distributions of $(\underset{\sim}{z}_n)$, (v_n), (w_n), (ℓ_n), (q_n) are determined by the distributions P of $([\alpha_n,\beta_n])$. Thus, we can use the notation:

$V(x) = P(v_1 < x)$ - the (stationary) distribution function of the actual work,

$V = E_P v_1$ - the (stationary) mean actual work load,

$W(x) = P(w_1 < x)$ - the (stationary) distribution function of the actual waiting time,

$W = E_P w_1$ - the (stationary) mean actual waiting time,

$p_k = P(\ell_1 = k)$ - the (stationary) probability that an arriving customer will find k customers in the system,

$B = P(\ell_1 = s + r)$ - the (stationary) blocking or loss probability (if $s + r < \infty$).

Denote by

$$n(t) = \max\{j: t_j < t\} \tag{2.25}$$

the (random) number of the last customer in the CAP $\phi(\xi)$ to arrive before t. The process

$$\underset{\sim}{z}(t) = \hat{h}_{s,r}(\underset{\sim}{z}_{n(t)}, t - t_{n(t)}, \beta_{n(t)}) \tag{2.26}$$

satisfies (2.19) and thus describes the behavior of the system in continuous time. We point out that the process $(\underset{\sim}{z}(t))$ is non-stationary although $(\underset{\sim}{z}_n)$ is stationary. However, we can construct a stationary process $(\bar{\underset{\sim}{z}}(t))$ which describes the behavior of the system in continuous time if we start from the time-stationary CAP $(\bar{\phi}(\xi),\bar{P})$ generated by the same (ξ,P). In this case we have the following continuous-time analogue of Proposition 2.8, where conditions A(s,r) and B(s,r) are the same.

<u>2.9 Proposition</u>. Consider a G/G/s/r system with time-stationary CAP $(\bar{\phi}(\xi),\bar{P})$, where ξ satisfies the general assumption in Proposition 2.8. Under the condition A(s,r), there exists a stationary process $(\bar{\underset{\sim}{z}}(t))$ satisfying equation (2.19) with $\bar{\phi}$ instead of ϕ. Moreover, there exists a function $\hat{g}_{s,r}$ such that for every t

$$\bar{\underset{\sim}{z}}(t) = \hat{g}_{s,r}({}_{(-\infty,t)}\bar{\phi}). \tag{2.27}$$

Under the additional assumption B(s,r), we have for every $\underset{\sim}{d} \in Z$

$$(\underset{\sim}{z}(\underset{\sim}{d},t+u),\ u \geq 0) \underset{t\to\infty}{\Longrightarrow} (\bar{\underset{\sim}{z}}(u),\ u \geq 0). \tag{2.28}$$

We call $(\bar{\underset{\sim}{z}}(t))$ the <u>time-stationary state process</u> of the system considered. From $(\bar{\underset{\sim}{z}}(t))$ we can derive the time-stationary processes $(\bar{v}(t))$, $(\bar{w}(t))$, $(\bar{\ell}(t))$ and $(\bar{q}(t))$ of <u>work load</u>, <u>virtual waiting time</u>, number of <u>customers</u> in the <u>system</u> and in the <u>queue</u> in a way similar to that used in deriving (v_n), etc. from $(\underset{\sim}{z}_n)$. In view of (2.27) the distributions of these processes are completely determined by the distribution $\bar{P}$ of $\bar{\phi}(\xi)$. The most relevant quantities are:

$\bar{V}(x) = \bar{P}(\bar{v}(0) < x)$ - the (stationary) distribution function of the virtual work load,

$\bar{V} = E_{\bar{P}}\bar{v}(0)$ - the (stationary) mean virtual work load,

$\bar{W}(x) = \bar{P}(\bar{w}(0) < x)$ - the (stationary) distribution function of the virtual waiting time,

$\bar{W} = E_{\bar{P}}\bar{w}(0)$ - the (stationary) mean virtual waiting time,

$\bar{p}_k = \bar{P}(\bar{\ell}(0) = k)$, $k \geq 0$ - the (time-stationary) distribution of the number of customers in the queue,

$\bar{L} = E_{\bar{P}}\bar{\ell}(0)$ and $\bar{Q} = E_{\bar{P}}\bar{q}(0)$ - the (time-stationary) mean number of customers in the system and in the queue, respectively,

$\bar{D} = \bar{P}(\bar{\ell}(0) \geq s)$ - the time-stationary delay (waiting) probability if $r = \infty$.

Define

$$\bar{\underset{\sim}{z}}_n = \begin{cases} \bar{\underset{\sim}{z}}(\bar{t}_n - 0), & \text{if } \bar{\alpha}_{n-1} > 0 \\ h_{s,r}(\bar{\underset{\sim}{z}}_{n-1}, \bar{\alpha}_{n-1}, \bar{\beta}_{n-1}), & \text{otherwise.} \end{cases}$$

Furthermore define $\bar{w}_n$, $\bar{\ell}_n$ $\cdots$ from $\bar{\underset{\sim}{z}}_n$ in the same way as w_n, ℓ_n $\cdots$ were defined from $\underset{\sim}{z}_n$. We emphasize that the MPP's

$$([\bar{t}_n, (\bar{\beta}_n, \bar{\underset{\sim}{z}}_n)]); \quad ([\bar{t}_n, (\bar{\beta}_n, \bar{w}_n)]), \cdots$$

are stationary, but the sequences $(\bar{\beta}_n)$, $(\bar{z}_n)$, $(\bar{w}_n)$, $\cdots$ are not stationary in general.

Proposition 2.5 leads immediately to the following important result.

<u>2.10 Proposition</u>. There are the following relationships between the processes $(\underset{\sim}{z}(t))$ and $(\bar{\underset{\sim}{z}}(t))$: for every function $f \geq 0$

$$\lim_{T\to\infty} \frac{1}{T} \int_0^T f(\underset{\sim}{z}(t))dt = E_{\overline{P}}\, f(\overline{\underset{\sim}{z}}(0)), \quad P\text{-a.s.}, \tag{2.29}$$

$$\lim_{n\to\infty} \frac{1}{n} \sum_{j=1}^{n} f(\overline{\underset{\sim}{z}}_j) = E_P\, f(\underset{\sim}{z}_1), \quad \overline{P}\text{-a.s.}. \tag{2.30}$$

In particular for every $A \subseteq Z$ we have

$$\lim_{T\to\infty} \frac{1}{T} \int_0^T 1_{\{z(t)\varepsilon A\}} dt = \lim_{T\to\infty} \frac{1}{T} \int_0^T P(\underset{\sim}{z}(t)\varepsilon A)dt = \overline{P}(\overline{\underset{\sim}{z}}(0)\varepsilon A), \quad P\text{-a.s.} \tag{2.31}$$

$$\lim_{n\to\infty} \frac{1}{n} \sum_{j=1}^{n} 1_{\{\overline{\underset{\sim}{z}}_j \varepsilon A\}} = \lim_{n\to\infty} \frac{1}{n} \sum_{j=1}^{n} \overline{P}\{\overline{\underset{\sim}{z}}_j \varepsilon A\} = P(\underset{\sim}{z}_1 \varepsilon A), \quad \overline{P}\text{-a.s.}.$$

It is impossible to simultaneously define a stationary sequence describing the steady-state behavior at arrival epochs and a continuous-time stationary process describing the steady-state behavior of the same system as measureable functions $\phi(\xi)$ or $\overline{\phi}(\xi)$, respectively. (However, there exist several papers in which that has been assumed erroneously, e.g., Little (1961)). Nevertheless, Proposition 2.10 shows that both can be introduced (even in the sense of Proposition 2.10) starting either from the arrival-stationary CAP $\phi(\xi)$ or from the time-stationary CAP $\overline{\phi}(\xi)$. Every probability or expectation in the time-stationary (arrival-stationary) model arises as a limiting time-average (arrival-average) in the arrival-stationary (time-stationary) model.

2.11 Remark. In order to define the so-called departure-stationary distributions, we start with the time-stationary model. We call $\overline{t}^*$ a departure epoch if $\overline{\ell}(\overline{t}^* - 0) - \overline{\ell}(\overline{t}^* + 0) + \overline{N}(\overline{t}^* + 0) - \overline{N}(\overline{t}^*) > 0$. If $s + r < \infty$, we consider the departure epochs of both the served and the unserved customers. If there are batch departures, the customers in a batch will be ordered according to the order of their arrivals. Now we consider the MPP $\overline{\phi}^* = ([\overline{t}_n^*, (\overline{z}_n^*, \overline{\ell}_n^*)])$, where $\overline{t}_n^* \cdots \leq \overline{t}_0^* < 0 < \overline{t}_1^* \leq \cdots,$ are the departure epochs; $\overline{\underset{\sim}{z}}_n^* = \overline{z}(\overline{t}_n^* + 0)$ if $\overline{t}_n^* < \overline{t}_{n+1}^*$ and $\overline{\underset{\sim}{z}}_n^* = \overline{\underset{\sim}{z}}_{n+1}^*$ otherwise;

$$\bar{\ell}^*_n = \sum_{j=1}^{s+2} 1_{\{z_{n,j}>0\}}$$

if $\bar{t}^*_n < \bar{t}^*_{n+1}$ and $\bar{\ell}^*_n = \bar{\ell}^*_{n+1} + 1_{\{\bar{\ell}^*_{n+1} < s+r\}}$. The MPP $\bar{\phi}^*$ is stationary by definition; its distribution we denote by $\bar{P}^*$. According to the Proposition 2.4, $(\bar{\phi}^*,\bar{P}^*)$ uniquely determines a stationary sequence (ξ^*,P^*), $\xi^* = ([\alpha^*_n,(\underset{\sim}{z}^*_n,\ell^*_n)])$. The sequence ξ^* describes the steady state behavior of the system immediately after departure epochs if the origin is taken as an arbitrary departure epoch. Similar to v_n, w_n, ℓ_n,... and $V(x)$, V; $W(x)$, W; p_k,... we introduce v^*_n, w^*_n, ℓ^*_n,... and $V^*(x)$, V^*, $W^*(x)$, W^*; p^*_k,... . Consequently we call $(\underset{\sim}{z}^*_n)$ the departure-stationary state process, $V^*(x)$, $W^*(x)$, p^*_k... _departure-stationary distributions_ of the work load, the waiting time, the number of customers in the system, ... We will also use the term _customer-stationary_ when we discuss arrival-stationary and departure-stationary distributions together.

2.12 Remark. The notion of MPP is closely related to the notion of _stochastic processes_ with _embedded point process_ (abbr. PEP), see FKAS (1981), Streller (1980) and references there. Typical examples of PEPs are the processes $(\underset{\sim}{z}(t))$ and $(\bar{\underset{\sim}{z}}(t))$ with (t_n) and $(\bar{t}_n)$ as embedded points respectively. The intensity conservation principle (e.g., FKAS(1981) and König and Schmidt (1980)) is based on the interpretation of queueing processes as PEPs. In particular PEPs are useful in reliability analysis of stationary, redundant systems (Franken and Streller (1980)).

3. Relationships Between Time- and Customer-Stationary Distributions

In this section we consider queueing systems in steady state, using the notation introduced in Section 2.2.

3.1 Little Formulas

We first show that the point process approach provides very simple proofs of Little-type formulas, e.g., FKAS(1981), Franken (1981).

3.1 Theorem. For a $G/G/s/\infty$ queue,

$$\bar{Q} = \lambda W \tag{3.1}$$

$$\bar{L} = \lambda(W + \beta_1) \tag{3.2}$$

$$\bar{V} = \lambda E_P(w_1\beta_1 + \frac{1}{2}\beta_1^2). \tag{3.3}$$

Proof. Consider the stationary MPP $([\bar{t}_n,(\bar{\beta}_n,\bar{w}_n)])$ with mark space $K = R_+ \times R_+$ where $\bar{w}_n = \bar{w}(\bar{t}_n - 0)$. Define

$$f_1(t,\beta,w) = \begin{cases} 1 & \text{if } t < 0,\ w > -t \\ 0 & \text{otherwise.} \end{cases}$$

For the time-stationary queue length $\bar{q}(0)$, we have

$$\bar{q}(0) = \sum_n f_1(\bar{t}_n,\bar{\beta}_n,\bar{w}_n).$$

Using Proposition 2.7 with $f = f_1$, we get (3.1). To get (3.2) and (3.3), use the functions

$$f_2(t,\beta,w) = \begin{cases} 1 & \text{if } t < 0,\ w+\beta > -t \\ 0 & \text{otherwise,} \end{cases}$$

and

$$f_3(t,\beta,w) = \begin{cases} \beta & \text{if } t < 0,\ w > -t \\ \beta + w + t & \text{if } t < 0,\ w < -t < \beta + w \\ 0 & \text{otherwise.} \end{cases}$$

We point out that formulas (3.1) to (3.3) are relationships between expectations, where these expectations are with respect to different probabilities P and $\bar{P}$ (as indicated in Section 2.2). The queue $M/G/s/\infty$ is an exception; see Theorem 3.4. In the queueing literature Little-type formulas are usually relationships between sample-path averages or their expectations (for non-stationary as well as stationary queues, cf., e.g., Stidham (1974)). We point out that Theorem 3.1 can be recast in this form. Using Proposition 2.10 we can rewrite (3.1) as

$$\lim_{T\to\infty} \frac{1}{T} \int_0^T q(t)\,dt = \lim_{T\to\infty} \frac{N(T)}{T} \cdot \lim_{n\to\infty} \frac{1}{n} \sum_{j=1}^{n} w_j \quad \text{P-a.s..} \tag{3.4}$$

Moreover, if the assertion (2.23) is valid for the system under consideration, we have

$$\lim_{T\to\infty} \frac{1}{T} \int_0^T q(t,\underset{\sim}{d})\,dt = \lim_{T\to\infty} \frac{N(T)}{T} \lim_{n\to\infty} \frac{1}{n} \sum_{j=1}^{n} w_j(\underset{\sim}{d}) \quad \text{P-a.s.} \tag{3.5}$$

where $q(t,\underset{\sim}{d})$ denotes the queue length at time t if the system with CAP $\phi(\xi)$ starts in the state $\underset{\sim}{d}$ at time 0 (the waiting times $w_j(\underset{\sim}{d})$ are defined similarly). Relationships similar to (3.4) and (3.5) are also valid for queues with time-stationary CAP $\overline{\phi}(\xi)$ and an arbitrary initial state. For detailed discussion of Little formulas we refer to FKAS (1981), Section 4.2, and Franken (1981).

3.2 Theorem. For a G/G/s/r queue in steady state,

$$E_{\overline{P}}[\min[s,\overline{\ell}(0)]] = \lambda(1-B)b_s, \tag{3.6}$$

where the left side denotes the mean number of busy servers at an arbitrary time epoch, $B = P(\ell_1 = s+r)$ is the blocking probability and b_s is the mean service time of an arbitrary served customer, cf., (3.7).

Proof. We consider the stationary MPP $([\overline{t}_n,\overline{\beta}_n,\overline{\ell}_n])$ and delete all points with the mark $\overline{\ell}_n = s+r$. The thinned MPP $([\overline{t}^s_n,\overline{\beta}^s_n])$ where $\overline{t}^s_n$ are the arrival epochs of served customers and $\overline{\beta}^s_n$ are their service times, is also stationary. According to (2.15) the intensity of this MPP is equal to $\lambda(1-B)$. Denote by $\xi^s = ([\alpha^s_n,\beta^s_n])$ the stationary sequence which corresponds to $([\overline{t}^s_n,\overline{\beta}^s_n])$ according to Proposition 2.4 and denote by P^s

the distribution of ξ^s. Then $P^s(\cdot) = P(\circ | \ell_1 < r+s)$, where P is the distribution of the basic sequence ξ (cf., e.g., Franken (1981)). We obtain

$$b_s = E_P(\beta_1 | \ell_1 < r+s) \tag{3.7}$$

Thus, $b \neq b_s$ in general (examples are given in Disney and Franken (1981)). However, in G/GI/s/r queues β_1 and ℓ_1 are independent in view of (2.22) and we have $b = b_s$.

We get (3.6) by applying (2.16) to the MPP $([\bar{t}_n^s, \bar{\beta}_n^s])$ and the function

$$f(t,\beta) = \begin{cases} 1 & \text{if } t < 0,\ \beta > -t \\ 0 & \text{otherwise.} \end{cases}$$

3.3 Corollary. (a) For a G/G/s/∞ queue,

$$E_{\bar{P}}(\min(s,\bar{\ell}(0))) = \lambda b = \rho. \tag{3.8}$$

(b) For a G/G/1/∞ queue,

$$\bar{D} = \rho. \tag{3.9}$$

(c) For a G/GI/s/r queue,

$$E_{\bar{P}}(\min(s,\bar{\ell}(0))) = \rho(1 - B). \tag{3.10}$$

Theorem 3.2 and Corollary 3.3 lead to some inequalities for the blocking and delay probabilities B and $\bar{D}$, see Disney and Franken (1981).

3.2 Queues with Poisson Arrivals

There are many studies for M/GI/s/r queues that prove that the time- and arrival-stationary distributions coincide. We are able to prove the same result for the more general class of M/G/s/r queues using the approach explained in Section 2.2.

3.4 Theorem. In M/G/s/r queues the random state vectors $\underset{\sim}{z}_1$ and $\bar{\underset{\sim}{z}}(0)$ are identically distributed.

Proof. Using (2.3) it is easy to show that for M/G/s/r queues the restrictions ${}_{(-\infty,0)}\phi(\xi)$ and ${}_{(-\infty,0)}\bar{\phi}(\xi)$ are identically distributed. In view of (2.22), $\underset{\sim}{z}_1$ is a function of ${}_{(-\infty,0)}\phi(\xi)$ and, in view of (2.27), $\bar{\underset{\sim}{z}}(0)$ is a function of ${}_{(-\infty,0)}\bar{\phi}(\xi)$. This proves the theorem.

From Theorem 3.4 it follows that in M/G/s/r queues, the time- and arrival-stationary distributions of all interesting variables (waiting time, queue length) coincide. Further related results are given in FKAS(1981).

3.3 Number of Customers in the System

First recall that p_j, $\bar{p}_j$ and p_j^* are the arrival-stationary, time-stationary and departure-stationary probabilities of j customers in the system.

3.5 Theorem. For a G/G/s/r queue,

$$p_j = p_j^* \text{ for all } j. \tag{3.11}$$

Proof. We consider the time-stationary model. Then we have two stationary MPPs $([\bar{t}_n,\bar{\ell}_n])$ and $([\bar{t}_n^*,\bar{\ell}_n^*])$ with $\bar{\ell}_n^* = \bar{\ell}(\bar{t}_n^* + 0)$. If we delete all points with marks different from j in these processes we get stationary MPPs where intensities are equal to λp_j and λp_j^*, respectively according to (2.15). Denote by $\bar{N}_j(t)$ and $\bar{N}_j^*(t)$ counting processes corresponding to these MPPs. In view of (2.12) we have

$$\lim_{t\to\infty} \bar{N}_j(t)/t = \lambda p_j, \quad \lim_{t\to\infty} \bar{N}_j^*(t)/t = \lambda p_j^*,$$

On the other hand, $|\bar{N}_j(t) - \bar{N}_j^*(t)| \leq 1$ for each t. Thus, we have (3.11).

The relationships between $\bar{p}_j$ and p_j are quite complicated in general, see FKAS, Section 4.3. However, the following result can be proved easily.

3.6 Theorem. For G/GI/1/0 and G/M/s/r queues,

$$\bar{p}_j \min(j,s) = \rho p_{j-1}, \quad 1 \leq j \leq s + r. \tag{3.12}$$

Proof. For G/GI/1/0 queues (3.12) coincides with (3.10). In G/M/s/r queues no arrival and departure epochs coincide. We consider the time-stationary model. Denote by $G_j(x)$ the distribution function of $X_j = \int_0^1 1_{\{\bar{\ell}(t)=j\}}dt$. Then

$$E_{\bar{P}}X_j = \bar{p}_j$$

$$E_{\bar{P}}\, \bar{N}^*_{j-1}(1) = E_{\bar{P}} \int_0^1 u\, \mu \min(j,s)dG_j(u) = \mu \min(j,s)E_{\bar{P}}X_j. \tag{3.13}$$

The formula (3.13) is based on the fact that in the random time X_j the departures form a Poisson process with intensity $\mu \min(j,s)$. On the other hand, as was shown in the proof of Theorem 3.5, the left side in (3.13) is equal to $\lambda p^*_{j-1} = \lambda p_{j-1}$. This finishes the proof.

From (3.6) we can conclude that (3.12) is in general not true for G/G/1/0 queues. In FKAS(1981), Section 4.3 it is shown that in general (3.13) is not valid for GI/GI/s/r queues.

Consider a r.v. $X \geq 0$ with the distribution function F and finite mean EX. Denote by X_R the r.v. with distribution function $F_R(x) = (EX)^{-1} \int_0^x (1 - F(u))du$. For r.v.'s X,Y with distribution functions F and G, $X \overset{st}{\leq} Y$ iff $F(x) \geq G(x)$ for all x (stochastic order). If $X \overset{st}{\leq} Y$, then $f(X) \overset{st}{\leq} f(Y)$ for every non-decreasing function f.

Recall that a distribution function F of a r.v. X is called NBUE (NWUE) if $X_R \overset{st}{\leq} (\overset{st}{\geq})X$.

In addition to Theorem 3.5 and 3.6, we prove the following useful inequalities between $\bar{\ell}(0)$ and ℓ_1, see also FKAS(1981), Disney and Franken (1981), König and Schmidt (1980).

<u>3.7 Theorem.</u> For GI/G/s/r queues,

$$\bar{\ell}(0) \overset{st}{\leq} \min(\ell_1 + 1, s + r). \tag{3.14}$$

If, in addition, the distribution function $A(t)$ of the interarrival times is NBUE (NWUE), then we have

$$\bar{\ell}(0) \overset{st}{\geq} (\overset{st}{\leq}) \ell_1. \tag{3.15}$$

The proof is based on the following lemma which is itself of interest.

3.8 Lemma. For GI/G/s/r queues the r.v.'s z_1 and $\bar{z}_0 = \bar{z}(\bar{t}_0 - 0)$ are identically distributed.

Proof. From construction it follows that the MPP's $\psi = ([t_n, z_n])$ and $\bar{\psi} = ([\bar{t}_n, \bar{z}_n])$ are related in the same way as the underlying CAP ϕ and $\bar{\phi}$. Thus, we can apply formulas (2.3)-(2.5) to ψ and $\bar{\psi}$ to get for every $C \in Z$

$$P(\bar{z}_0 \in C) = \lambda \int_0^\infty P(\{\alpha_1 > t\} \cap \{z_0(S_t\psi) \in C\}dt.$$

For $t < \alpha_1$ the mark z_0 of the first negative point in $S_t\psi$ coincides with $z_1(\psi) = z_1$. Moreover for GI/G/s/r queues z_1 and α_1 are independent in view of (2.22). Thus,

$$\bar{P}(\bar{z}_0 \in C) = P(z_1 \in C)\ \lambda \int_0^\infty P(\alpha_1 > t)dt = P(z_1 \in C).$$

Proof of Theorem 3.7. For $t < 0$ denote by $\bar{N}^*(t)$ ($N^*(t)$) the number of departures in the interval $[t,0)$ in the time-stationary (arrival-stationary) model. Since $\bar{\ell}_0 \overset{d}{=} \ell_1$ in view of Lemma 3.8 ($X \overset{d}{=} Y$ denotes that the r.v.'s X and Y are identically distributed) and $\bar{N}^*(t)$ is nonnegative we obtain

$$\begin{aligned} \bar{\ell}(0) &= \min(\bar{\ell}_0 + 1, s + r) - \bar{N}^*(\bar{t}_0) \\ &\overset{st}{\leq} \min(\bar{\ell}_0 + 1, s + r) \\ &\overset{d}{=} \min(\ell_1 + 1, s + r), \end{aligned} \tag{3.16}$$

Now we prove (3.15) if $A(x)$ is NBUE (the NWUE case is similar).

Using (2.6a) we obtain $-\bar{t}_0 \overset{st}{\leq} -t_0$. Using Lemma 3.8 and the stationarity of (z_n) and (ℓ_n) we have $\ell_1 \overset{d}{=} \ell_0 \overset{d}{=} \bar{\ell}_0$ and $\underset{\sim}{z}_1 \overset{d}{=} \underset{\sim}{z}_0 \overset{d}{=} \bar{\underset{\sim}{z}}_0$. Since there are no arrivals in $(\underline{t}_0,0)$ and $(t_0,0)$ and $\underset{\sim}{z}_0 \overset{d}{=} \bar{\underset{\sim}{z}}_0$,

$$N^*(t_0) - N^*(t_0 - \bar{t}_0) \overset{d}{=} \bar{N}^*(\bar{t}_0).$$

Thus,

$$\bar{N}^*(\bar{t}_0) \overset{st}{\leq} N^*(t_0).$$

Finally,

$$\begin{aligned} \bar{\ell}(0) &\overset{d}{=} \min(\bar{\ell}_0 + 1, s + r) - \bar{N}^*(\bar{t}_0) \\ &\overset{st}{\leq} \min(\ell_0 + 1, s + r) - N^*(t_0) \\ &= \ell_1 \quad \text{(by definition).} \end{aligned}$$

The proofs show that the result of Theorem 3.7 is also true for queues with balking and reneging, for example.

3.4 Busy Cycles

For a rigorous description of a generic busy cycle, we need a new distribution P', namely the conditional distribution

$$P'(\cdot) = P(\cdot \mid \ell_1 = 0),$$

see FKAS(1981) and Franken (1981). We omit here a discussion of conditions which ensure $P(\ell_1 = 0) > 0$ and thus the existence of infinitely many busy cycles with finite mean length.

We denote by η, ν, and χ the length of the busy period, the number of customer served during a busy period and the length of the idle period, respectively.

__3.9 Theorem.__ For a G/G/s/r queue with $p_0 = P(\ell_1 = 0) > 0$,

$$E_{P'}(\eta + \chi) = \lambda^{-1} E_{P'}\nu, \tag{3.18}$$

$$p_0 = (E_{P'}\nu)^{-1} \tag{3.19}$$

$$\bar{p}_0 = E_{P'}\chi / E_{P'}(\chi + \eta). \tag{3.20}$$

We refer to FKAS(1981) for proofs and to Disney and Franken (1981) for discussion of the mean busy period in single-server queues.

3.10 Theorem. For a $G/G/1/\infty$ queue,

$$\bar{W}(x) = 1 - \rho + \lambda\, E_P \min\{\beta_1, (x-w_1)_+\}), \tag{3.21}$$

$$\bar{P}(\bar{z}_1(0) < x) = 1 - \rho + \rho B_R(x), \tag{3.22}$$

where $\bar{z}_1(0)$ is the stationary residual service time of the customer in service at an arbitrary time epoch ($\bar{z}_1(0) = 0$ if the system is empty),

$$B_R(x) = b^{-1} \int_0^x (1 - B(t))dt, \text{ and } y_+ = \max(y,0) \text{ for } y \in R.$$

In particular, for the $G/GI/1/\infty$ queue,

$$\bar{W}(x) = 1 - \rho + \lambda(B_R * W)(x). \tag{3.23}$$

Proof. We use the formula (2.4) for CAP's $\phi(\xi)$ and $\bar{\phi}(\xi)$ with $f(\phi) = 1_{\{w(0) \geq x\}}$ for each fixed x. We have

$$\bar{P}(\bar{w}(0) \geq x) = \lambda\, E_P(\int_0^{\alpha_1} 1_{\{w(t) \geq x\}} dt)$$

$$= \lambda\, E_P((w_1 + \beta_1 - x)_+ - (w_2 - x)_+).$$

The last formula is equivalent to (3.21) because w_1 and w_2 are identically distributed.

By (2.22), w_1 is a function of (α_n, β_n), $n < 1$. Thus, for $G/GI/1/\infty$ queues w_1 and β_1 are independent. Formula (3.23) follows in this case from (3.21) by simple calculations. The proof of (3.22) is similar to that of (3.21) and will be omitted.

Formulas (3.22) and (3.23) were proved in Takács (1963) for $GI/GI/1/\infty$ queues with non-arithmetic distribution function A(t). See FKAS(1981) for a comprehensive discussion of this topic (also for G/G/s/r queues) and further references.

4. Acknowledgements

I am grateful to Prof. R. L. Disney, Prof. R. D. Foley and Dr. W. Whitt for helpful discussions during the preparation of the paper and to Ms. Paula L. Kirk for her excellent typing of various drafts of the mansucript.

5. References

[1] Borovkov, A. A. (1980), Asymptotic Methods in Queueing Theory, Nauka, Moscow, (in Russian).

[2] Disney, R. L. and Franken, P. (1981), "Further Comments on Some Queueing Inequalities," (to appear Elektron. Informationsverab. Kybernet).

[3] Franken, P. (1976), "Einige Anwendungen der Theorie zufälliger Punktprocesse in der Bedienungstheorie," Math. Nachr., 70, pp. 303-319.

[4] Franken, P. and Streller, A. (1980), "Reliability Analysis of Complex Repairable Systems by Means of Marked Point Processes," J. Appl. Probab., 17, pp. 154-167.

[5] Franken, P., König, D., Arndt, U. and Schmidt, U. (1981), Queues and Point Processes, Akademie-Verlag, Berlin, and Wiley-Intersciences, New York.

[6] Franken, P. (1981), "The Point Process Approach to Queueing Theory and Related Topics," Stochastic Process. Appl. (to appear).

[7] König, D., and Matthes, K. (1963), "Verallgemeinerungen der Erlangschen Formeln I," Math. Nachr., 26, pp. 45-56.

[8] König, D., and Schmidt, V. (1980), "Imbedded and Non-imbedded Stationary Characteristics of Queueing Systems with Varying Service Rate and Point Processes," J. Appl. Probab., 17, pp. 753-767.

[9] Little, J. (1961), "A Proof for the Queuing Formula $L = \lambda W$," Oper. Res., 9, pp. 383-387.

[10] Miyazawa, M. (1979), "A Formal Approach to Queueing Processes in the Steady State and Their Applications," J. Appl. Probab., 16, pp. 322-347.

[11] Stidham, S. (1974), "A Last Word on $L = \lambda W$," Oper. Res., 22, pp. 417-421.

[12] Streller, A. (1980), "Stochastic Processes with an Embedded Point Process," Math. Operationsforsch. Statist. Ser. Statist. (to appear).

[13] Takács, L. (1963), "The Limiting Distribution of the Virtual Waiting Time and the Queue Size for a Single Server Queue with Recurrent Input and General Service Times," Sankhyā, Ser. A, 25, pp. 91-100.

Humboldt-University, Department of Mathematics, 1086 Berlin, German Democratic Republic.

ERROR MINIMIZATION IN DECOMPOSABLE STOCHASTIC MODELS

P. J. Courtois

1. Introduction

In the last few years, growing attention has been paid to the analysis of stochastic models by decomposition and aggregation techniques. Various applications have been developed, especially in relation with studies of information traffic in complex models of switching systems, computer systems, or data communication networks.

But decomposition yields approximative results only, and a useful determination of the error remains a difficulty in practice. This paper describes recent attempts at estimating the error made when the Simon-Ando aggregation technique is applied to the calculation of the equilibrium probability vector of nearly decomposable stochastic matrices. A consequence of this analysis, potentially useful in practice, is to show how one can construct aggregates which keep the error at a minimum. In this context an important result, recently obtained by H. Vantilborgh [9], is introduced.

2. Notations

Scalars and row vectors are represented by lower case letters and matrices by upper case. The symbols $\underline{1}$ and $\underline{0}$ represent row vectors, all components of which are one and zero, respectively. To alleviate the notation, vector and matrix orders will be omitted whenever context allows it. We use the vector norms $\|x\|_1 = \sum_i |x_i|$ and $\|x\|_\infty = \max_i(|x_i|)$, and the subordinate matrix norms under left vector multiplication, i.e., $\|A\|_1 = \max_i \sum_j |a_{ij}|$, and $\|A\|_\infty = \max_j \sum_i |a_{ij}|$. A^T is used to denote the transpose of A, and $u \perp v$ means that the vector u is orthogonal to v. We use the classical order relation, and write $e = O(h^n)$ if $\lim_{h\to 0} h^{-n}e = k$ where k is some constant.

3. Nearly Completely Decomposable Systems

We are concerned with finite non-negative stochastic matrices which, after an appropriate arrangement of the states, can be written under the form

$$Q = \begin{bmatrix} Q_{11} & Q_{12} & \cdots & Q_{1N} \\ Q_{21} & Q_{22} & \cdots & \\ \vdots & & & \\ Q_{N1} & \cdot\cdot & & Q_{NN} \end{bmatrix}$$

where the elements of the matrices Q_{IJ}, $I \neq J$, are small compared to those of the diagonal matrices Q_{II}, $I = 1,\ldots,N$. These matrices are said to be nearly completely decomposable. They were first studied by Simon and Ando [6], in the context of economic models. Later, they were found useful also in the study of queueing networks and probabilistic models for computer systems [2]. They appear quite naturally in complex structures organized as sets of subsystems which interact weakly with one another.

If Q is irreducible, its eigenvalue one is simple; the associated left eigenvector, let us call it ν, is unique and satisfies:

$$\nu Q = \nu \tag{3.1}$$

If Q is acyclic and ν is normalized so that $\|\nu\|_1 = 1$, ν can be interpreted as the vector of steady-state probabilities of the Markov

Chain defined by Q.

The aggregation procedure proposed by Simon and Ando to calculate an approximation x of ν consists of the following five steps:

1) A diagonal stochastic matrix

$$Q^* = \text{diag}(Q_1^*, Q_2^*, \ldots, Q_N^*),$$

where each Q_I^* is stochastic, irreducible and acyclic is constructed from Q by distributing the rowsum $\sum_{J \neq I} \sum_j q_{i_I j_J}$ (where $q_{i_I j_J}$ denotes the (i,j) element of Q_{IJ}) over the elements of the i_Ith row of Q_{II}, $I = 1, \ldots, N$. The Q_I^* are often called aggregates. One can write

$$Q = Q^* + \varepsilon C \tag{3.2}$$

where $\varepsilon = \max_{i,I} \left(\sum_{J \neq I} \sum_j q_{i_I j_J} \right)$ is small and can be interpreted as the maximum degree of coupling between subsystems Q_{II}; C has rowsums equal to zero, and is such that $\max_{i,I} \sum_{J \neq I} \sum_j c_{i_I j_J} = 1$.

2) The steady-state vectors ν_I^* of the aggregates Q_I^*, $I = 1, \ldots, N$ are computed: $\nu_I^* Q_I^* = \nu_I^*$, and

3) An $N \times N$ stochastic aggregative matrix P of transition probabilities between aggregates is obtained as

$$P = [P_{IJ}] = [\nu_I^* Q_{IJ} \, \underline{1}^T]. \tag{3.3}$$

4) The steady-state vector X of P, X = XP, is computed, and finally

5) the vector $x = [X_1 \nu_1^*, X_2 \nu_2^*, \ldots, X_N \nu_N^*]$ is used as an approximation to ν.

A physical justification for this procedure lies in the theorems proven by Simon and Ando [6]. These theorems show that, for ν sufficiently small, the dynamic behavior of a nearly completely decomposable system Q defined by (3.2) consists of two periods; a short-term period where each subsystem Q_{II} approaches an internal equilibrium approximately independently from the other subsystems; and a long-term period during which the entire system reaches its equilibrium under the influence of the weak couplings between subsystems and during which the internal equilibria of these subsystems remains approximately in a constant ratio.

The obvious computational advantage of this procedure is to reduce the computation of a large eigenvector problem to the computation of $(N+1)$ smaller ones. But the broadest and most promising field of application lies beyond the realm of matrix calculus. More generally, the principle of near complete decomposability gives guidelines as to how complex systems can be analyzed in terms of separate subsystems, possibly by different types of techniques, the results being ultimately amalgamated into a macro model of the entire system (heterogeneous aggregation [2]). Besides, this principle and the Simon-Ando theorem mentioned above relate a partitioning of the state space of a model to its time behavior. This mapping suggests a way by which the local equilibria of the aggregates can be used as an approximation of the short-term transient behavior of the entire model; these approximations may be of great help in view of the difficulty in obtaining the transient solution of the simplest mathematical models.

4. The Error of Aggregation

In practice, however, this approach presents two types of difficulties. First, the Simon-Ando decomposition yields approximative results only, and a tight determination of the error has remained a problem. The difference $(\nu - x)$ has been shown to be in $O(\varepsilon)$ [2], but the constant of proportionality has not been bounded a priori. Secondly, there is a degree of freedom left in the construction of the aggregates Q_I^*. More precisely, there is an infinity of ways to redistribute the off diagonal block rowsums over the rows of the matrices Q_I's. And one can verify numerically that the constant of proportionality is sensitive to the way this is done.

In the remainder of this paper, we deal with these two problems. We are not able to propose an a priori tight error bound yet; but, following [9], we emphasize the possibility of constructing aggregates which keep the error at a minimum, allowing the analyst, in many practical situations, to get around the absence of a tight a priori error bound.

The error of aggregation $(\nu - x)$ can be made more explicit in the following way. Let $\beta_I = \sum_{i\varepsilon I} \nu_{i_I}$ be the exact steady-state probability of being in aggregate I. Then,

$$\nu_{i_I} - x_{i_I} = \nu_{i_I} - X_I \nu^*_{i_I} \quad = \beta_I \nu^*_{i_I} - X_I \nu^*_{i_I} + \nu_{i_I} - \beta_I \nu^*_{i_I}$$

$$= (\beta_I - X_I)\, \nu^*_{i_I} + \beta_I (\beta_I^{-1} \nu_{i_I} - \nu^*_{i_I})$$

The error of aggregation which affects the probability of being in a given state i_I consists therefore of two weighted components: an <u>aggregate error</u> $(\beta_I - X_i)$ affecting the estimation X_I of the probability β_I being in subsystem I, and a <u>marginal error</u> affecting the estimation $\nu^*_{i_I}$ of the conditional probability $\beta_I^{-1} \nu_{i_I}$.

These two errors have been shown to be of $O(\varepsilon)$ [2]. Moreover, a necessary and sufficient condition for $(\nu_{i_I} - x_{i_I})$ to be null is simply that the marginal error be null. Indeed, if $(\beta_I^{-1}\ {}_{i_I} - \nu^*_{i_I}) = 0$, the aggregative matrix constructed at step 3) of the Simon-Ando procedure becomes

$$P_{exact} = [\beta_I^{-1} \nu_I \; Q_{ij} \; \underline{1}^T];$$

The steady-state vector of P_{exact} is precisely β so that the difference $(\beta - X)$ is also null.

5. The Marginal Error

Minimizing the marginal error is thus vital in the Simon-Ando approximation. In this section we establish an expression for this error and discuss the possibilities of bounding it.

We consider one subsystem only, say Q_1:

$$Q = \begin{bmatrix} Q_1 & E \\ F & G \end{bmatrix}, \tag{5.1}$$

G representing the remaining (N - 1) other subsystems and their mutual interactions. Let $\nu = [\nu_1, \nu_2]$ be a consistent partition for the steady-state vector of Q; we then have

$$\nu_1 Q_1 + \nu_2 F = \nu_1 \tag{5.2}$$

$$\nu_1 E + \nu_2 G = \nu_2 \tag{5.3}$$

The Simon-Ando approximation uses an approximation ν_1^* of ν_1; ν_1^* is the steady-state vector of an aggregate Q_1^* obtained from Q_1 as:

$$Q_1^* = Q_1 + s,$$

where S has same rowsums as E:

$$S\underline{1}^T = E\underline{1}^T \tag{5.4}$$

Since $\nu_1^*(Q_1 + S) = \nu_1^*$, one obtains easily from (5.2), (5.3) that

$$[\nu_1 - \nu_1^*][Q_1 + E(I-G)^{-1}F - I] = \nu_1^*[S - E(I-G)^{-1}F] \tag{5.5}$$

which one would like to solve for $[\nu_1 - \nu_1^*]$.

Note that $(I-G)^{-1}$ exists since G has no eigenvalue equal to one, otherwise Q would not be irreducible.

The rowsums of $E(I-G)^{-1}F$ are also equal to those of E:

$$E(I-G)^{-1}F\,\underline{1}^T = E(I-G)^{-1}(I-G)\underline{1}^T = E\underline{1}^T. \tag{5.6}$$

Thus, $Q_1 + E(I-G)^{-1}F$ is a stochastic matrix and has an eigenvalue equal to one, so that $[Q_1 + E(I-G)^{-1}F - I]$ is singular.

Fortunately, we can solve this system by means of the theorem which is proven in the appendix. The conditions to use this theorem are indeed verified, i.e.:

(i) The vector $[\nu_1 - \nu_1^*]$ is orthogonal to $\underline{1}$ when the same normalization is taken for ν_1 and ν_1^*;

(ii) the matrix $[Q_1 + E(I-G)^{-1}F - I]$ is of nullity one since, by assumption, the vector ν_1 is unique up to a multiplicative constant.

If we apply the theorem for the particular case when the vector h (defined in the appendix) is equal to $\underline{1}$, we obtain for (5.5) the unique solution:

$$[\beta_1^{-1}\nu_1 - \nu_1^*] = \nu_1^*[S - E(I-G)^{-1}F](Q_1 + E(I-G)^{-1}F - I)^- . \qquad (5.7)$$

The matrix $(Q_1 + E(I-G)^{-1}F - I)^-$ is a generalized inverse (see note 2 of the appendix), the construction of which is specified by the theorem.

Expression (5.7) was first established by Zarling [10] who used it to derive an upper bound for the marginal error. Let us discuss briefly the validity of this bound. Let

$$M^- = (Q_1 + E(I-G)^{-1}F - I)^- ; \qquad (5.8)$$

from (5.7), we obtain:

$$\|\beta_1^{-1}\nu_1 - \nu_1^*\|_1 = \|\nu_1^* [S - E(I-G)^{-1}F]\ M^-\|_1 . \qquad (5.9)$$

The vector $\nu_1^*[S - E(I-G)^{-1}F]$ is orthogonal to $\underline{1}$ by virtue of (5.4) and (5.6). Thus we have

$$\|\beta_1^{-1}\nu_1 - \nu_1^*\|_1 \le \frac{1}{2} \nu_1^*[S - E(I-G)^{-1}F]\|_1 \max_{k,\ell} \sum_j |m_{kj}^- - m_{j\ell}^-| \qquad (5.10)$$

since the maximum of (5.9) occurs for some vector $\nu_1^*[S - E(I-G)^{-1}F]$ with exactly one positive element and one negative element, say the elements in positions k and ℓ respectively.

Moreover, we obtain from (5.6):

$$\|\nu_1^*[S - E(I-G)^{-1}F]\|_1 \leq \|\nu_1^* S\|_1 + \|\nu_1^* E(I-G)^{-1}F\|_1 = 2\|\nu_1^* E\|_1 ,$$

and thus,

$$\|\beta_1^{-1}\nu_1 - \nu_1^*\|_1 \leq \|\nu_1^* E\|_1 \max_{k,\ell} \sum_j |\bar{m}_{kj} - \bar{m}_{\ell j}| . \tag{5.11}$$

This is the essential formulation of Zarling's bound. It shows clearly how the error depends on the coupling matrix E weighted by the equilibrium vector ν_1^*; $\|\nu_1^* E\|_1$ was called in [2] the effective degree of coupling between the aggregate $Q_1 + S$ and the remainder of the system.

This bound (5.11) is impossible to evaluate in practice since it requires the calculation of $(I-G)^{-1}$. A method is given in [10] to obtain an approximation of (5.11). But this approximation, like the bound itself in fact, is not tight enough to be useful in practice. Numerical examples given in [3] show that this approximation can be as much as two orders of magnitude larger than the error actually made by the Simon-Ando approximation.

A first reason for this important discrepancy is that (5.11) is independent from the matrix S, while as we shall see, the error is very much dependent on the particular choice of S. In other words, (5.11) covers all possible choices of S, including the worst possible ones.

It is shown in [10] that the approximation of (5.11) which is proposed can be effectively reached for a worst choice of S, F and G. But it is only reached in the limit as $\varepsilon \to 0$, a limiting case where the actual error is zero. Moreover it requires that F have a special structure with given columns being equal to $\underline{0}$. This is a situation which is never met in many models like queueing network models for example. A second reason is that (5.11) bounds the sum of the absolute errors over all states of an aggregate, while, in practice, one is interested in the error affecting individual states only, and thus in the bound $\| \beta_1^{-1}\nu_1 - \nu_1^* \|_\infty$ which, on the average, can be about n times smaller, n being the size of the aggregate.

One could think that the same approach as above taken to bound $\| \beta_1^{-1}\nu_1 - \nu_1^* \|_1$ could also be followed to bound $\| \beta_1^{-1}\nu_1 - \nu_1^* \|_\infty$. But this is not the case. Since $[\beta_1^{-1}\nu_1 - \nu_1^*] \perp \underline{1}$, one will necessarily obtain

$$\| \beta_1^{-1}\nu_1 - \nu_1^* \|_\infty \leq \frac{1}{2} \| \beta_1^{-1}\nu_1 - \nu_1^* \|_1 ,$$

the equality being verified whenever the vector has exactly two non-zero elements. An improvement of a factor 1/2 only could thus be expected; this gain is negligible in view of the fact that Zarling's bound is proportional to the order of the aggregate.

6. A Class of Ideal Aggregates

The freedom we have in choosing S is an handicap in the search for error bounds since any general bound will inevitably cover the worst cases. But, in practice, this freedom may be turned into an advantage as it makes possible the construction of aggregates which minimize the error.

Indeed, since the rowsums of $E(I-G)^{-1}F$ are equal to those of E (see 5.6), a possible choice for S is:

$$S^{id} = E(I-G)^{-1}F. \tag{6.1}$$

This *ideal* S^{id}, as (5.7) clearly shows, nullifies the marginal error $[\beta_1^{-1}\nu_1 - \nu_1^*]$, and thus also the error of aggregation.

An ideal aggregate should therefore be constructed as

$$Q_1^{*id} = Q_1 + S^{id} = Q_1 + E(I-G)^{-1}F . \tag{6.2}$$

Zarling [10] must be credited for having been the first to catch sight of this possibility.

Unfortunately, the construction of this ideal aggregate requires also the computation of $(I-G)^{-1}$; at least in most of the cases where G is arbitrary. This computation which amounts to the resolution of the whole original system except for one subsystem only, must be considered, by hypothesis, as being prohibitive.

However, when the original matrix Q has a special structure, equation (2) gives a first indication as to how an optimal choice for S could be approximated. This is based on the observation that an element $(E(I-G)^{-1}F)_{ij}$ is non-zero iff both the ith row of E and the jth column of F have at least one non-zero element. Thus, in the construction of an ideal aggregate for Q_1, only the transition probabilities between a state i from which Q_1 can be left and a state j through which Q_1 can be entered from the remainder of the system should be incremented by a quantity $(S^{id})_{ij}$. When the original system Q is sparse for example, and if F has only one non-zero column j, all columns of S^{id} are also zero except the jth which, because of (5.6) reduces to E $\underline{1}$. Exact aggregation is possible in this particular case without computing $(I-G)^{-1}$.

- Markov matrices corresponding to queueing networks which have the product form property have also a special sparse structure which makes ideal aggregation possible without the inversion of $(I-G)$; inversions of submatrices of the routing matrix only are necessary [8], [4], a method sometimes referred to as the short-circuit approach.

7. Conformable Aggregates

But special sparse structures as those discussed hereabove are scanty in practice. One is therefore led to consider algorithms which attempt to approximate the optimal choice S^{id} for S without actually computing $(I-G)^{-1}$.

Let us consider again the equation

$$[\beta_1^{-1}\nu_1 - \nu_1^*] = \nu_1^*[S - E(I-G)^{-1}F]\, M^- .$$

For all permissible choices of S, i.e., for all matrices S which satisfy (5.4), we have that:

- the corresponding vector ν_1^* remains of order $O(\varepsilon^0)$;
- M^- remains unchanged since it depends on Q_1 and Q only;
- the difference $(\beta_1^{-1}\nu_1 - \nu_1^*)$ remains of $O(\varepsilon)$ (cfr. [2]).

Consequently, a choice of S such that

$$S - E(I-G)^{-1}F = O(\varepsilon^k),\ k \geq 2 \tag{7.1}$$

will cause $[\beta_1^{-1}\nu_1 - \nu_1^*]$ to be also in $O(\varepsilon^k)$, $k \geq 2$. In other words, the error can be made arbitrarily small by choosing S sufficiently close to $E(I-G)^{-1}F$, ν_1^* and M^- remaining of constant magnitude.

An aggregate ${}^{(k)}Q_1^* = Q_1 + S$ where S satisfies (7.1) will be called a conformable aggregate of order k. Since E and F are given, the main problem to construct a conformable aggregate of order $k \geq 2$ is to construct an approximation A of $(I - G)$ such that

$$EA^{-1}F - E(1-G)^{-1}F = O(\varepsilon^k),\ k \geq 2. \tag{7.2}$$

Unfortunately, classical results on inverses of perturbed matrices (see e.g., [7]) don't help very much in the determination of useful conditions which A should satisfy to achieve (7.2); repeated use of norm inequalities seems to overestimate the restrictions A should comply with.

8. A Conformable Aggregate of Order 2

However, aggregation procedures which yield an $O(\varepsilon^2)$-accuracy exist and have already been proposed in [1] and [2]. More recently, H. Vantilborgh [9] found a new interesting way of constructing conformable aggregates of order 2. Although mathematically equivalent to these former approaches, it throws a new light on the process of aggregation, and offers more advantages from a practical stand point. We present here this result in a way which relates it directly to the original Simon-Ando theorems. Let

$$Q = \sum_{I=1}^{N} \lambda_{1_I} Z(1_I) + \sum_{I=1}^{N} \sum_{i \neq 1}^{n(I)} \lambda_{i_I} Z(i_I) \tag{8.1}$$

be the spectral decomposition of Q where the λ_{1_I}, $I = 1,\ldots,N$ are the N dominant eigenvalues of Q. Suppose that an arbitrary permissible aggregation of Q yields Q*. The spectral decomposition of Q* is:

$$Q^* = \sum_{I=1}^{N} Z^*(1_I) + \sum_{I=1}^{N} \sum_{i \neq 1}^{n(I)} \lambda^*_{i_I} Z^*(i_I).$$

From the first theorem of Simon and Ando (see theorem 1.1 in [2], e.g.), we obtain

$$Q \simeq \sum_{I=1}^{N} \lambda_{1_I} Z(1_I) + Q^* - \sum_{I=1}^{N} Z^*(1_I).$$

By the second Simon-Ando theorem (see th. 1.2 in [2]), we obtain for a block Q_{KL} of Q $(K,L = 1,\ldots,N)$:

$$Q_{KL} \simeq \sum_{I=1}^{N} \lambda_{1_I} \underline{1}^T_{n(K)} \alpha_{KL} \nu^*_L + Q^*_{KL} - \sum_{I=1}^{N} Z^*_{KL}(1_I)$$

which by Simon-Ando third theorem (see equation 1.28 in [2]) yields

$$Q_{KL} \simeq \underline{1}^T_{n(k)} P_{KL} \nu^*_L + Q^*_{KL} - \sum_{I=1}^{N} Z^*_{KL}(1_I) ,$$

where P_{KL} is the (K,L)-element of the aggregative matrix P. For $K \neq L$, we have that

$$Q^*_{KL} = Z^*_{KL}(I_I) = 0, \quad I = 1,\ldots,N$$

and for K = L:

$$Z^*_{KK}(1_I) = 0, \quad I \neq K$$

$$Z^*_{KK}(1_K) = \underline{1}^T_{n(k)} \nu^*_K$$

This yields:

$$Q_{KL} \simeq \underline{1}^T_{n(K)} P_{KL} \nu^*_L , \quad K \neq L$$

$$Q_{KK} \simeq \underline{1}^T_{n(K)} (P_{KK}-1) \nu^*_K + Q^*_{KK} .$$

If we now define the matrices

$$\hat{P} = [P_{KL}], \quad K,L = 2,\ldots,N$$

$$\hat{Q}^* = [Q^*_{KK}], \quad K = 2,\ldots,N$$

$$U = \begin{bmatrix} \underline{1}^T_{n(2)} & & \\ & \ddots & \\ & & \underline{1}^T_{n(N)} \end{bmatrix} \text{ and } V^* = \begin{bmatrix} \nu^*_2 & & \\ & \ddots & \\ & & \nu^*_N \end{bmatrix},$$

we obtain an approximation of G which is given by

$$G \simeq U(\hat{P} - I)V^* + \hat{Q}^* . \tag{8.2}$$

Or,

$$(I - G) \simeq A = U(I - \hat{P})V^* + (I - \hat{Q}^*). \tag{8.3}$$

It is not difficult to verify that corresponding elements of $(I - G)$ and A can differ by at most ε.

The inverse of A can be obtained in the following way. The matrices U and V enjoy the following properties:

$$V^*U = I, \quad (I - \hat{Q}^*)U = 0 \tag{8.4}$$

If one introduces the fundamental matrix R^* of the stochastic matrix $\hat{Q}^*$ (see e.g., theorem 4.3.1 in [5]):

$$R^* = (I - (\hat{Q}^* - UV^*))^{-1} \tag{8.5}$$

which enjoys the properties (see th. 4.3.3 in [5]):

$$R\hat{Q}^* = \hat{Q}^*R, \; RU = U, \; V^*R = V^*,$$

one can verify that

$$A[U(I - \hat{P})^{-1}V^* + R - UV^*] = [U(I - \hat{P})^{-1}V^* + R - UV^*]A = I,$$

and thus that

$$A^{-1} = U(I-\hat{P})^{-1}V^* + R - UV^* \ . \tag{8.6}$$

The first term of this inverse A^{-1} has an important property which was first established by H. Vantilborgh [9].

For $\varepsilon \to 0$, G tends to a stochastic matrix so that $(I-G)$ is singular at the point $\varepsilon = 0$. It is shown in [9] that this point $\varepsilon = 0$ is a zero of order 1 of $(I-G)$, i.e., that the Laurent expansion of $(I-G)^{-1}$ at the point $\varepsilon = 0$ is given by

$$(I-G)^{-1} = \sum_{i=-1}^{\infty} \varepsilon^i Y_i \tag{8.7}$$

Moreover, the coefficient Y_{-1} is precisely given by

$$Y_{-1} = U(I-\hat{P})^{-1} V^* \ .$$

Consequently,

$$EU(I-\hat{P})^{-1} V^*F - E(I-G)^{-1}F = O(\varepsilon^2) \ . \tag{8.8}$$

The rowsums of $EU(I-\hat{P})^{-1} V^*F$ are equal to those of E; i.e.,

$$EU(I-\hat{P})^{-1} V^*F \ \underline{1}^T = EU(I-\hat{P})^{-1} \begin{bmatrix} P_{21} \\ \vdots \\ P_{N1} \end{bmatrix} = EU \ \underline{1}^T = E \ \underline{1}^T \ , \tag{8.9}$$

since P is a $N \times N$ stochastic matrix.

Therefore, $EU(I-\hat{P})^{-1} V^*F$ is a permissible choice for S, and

$$^{(2)}Q_1^* = Q_1 + EU(I-\hat{P})^{-1} V^*F \tag{8.10}$$

is a conformable aggregate of order 2 for subsystem Q_1.

9. Applications

In terms of the matrix Q blocks, expression (8.10) can be rewritten as:

$$ {}^{(2)}Q_1^* = Q_1 + [Q_{12}\, \underline{1}^T, \ldots, Q_{1N}\, \underline{1}^T][1 - \hat{P}]^{-1} \begin{bmatrix} \nu_2^* \; Q_{21} \\ \vdots \\ \nu_N^* \; Q_{N1} \end{bmatrix} . \tag{9.1} $$

This type of conformable aggregate is easily constructed as it depends on aggregate characteristics only. An aggregation procedure which yields an $O(\varepsilon^2)$ approximation of the steady-state vector of Q would consist of two steps. First an arbitrary permissible decomposition is used to obtain a set of aggregate equilibrium vectors ν_1^* and a matrix P of transition probabilities between aggregates. In a second step, a conformable matrix ${}^{(2)}Q^*$ is constructed; and steps 2) and 5) of the aggregation procedure of section 3) are repeated to obtain an $O(\varepsilon^2)$-approximation of the steady-state vector of the original matrix.

This procedure requires, for each aggregate, the inversion of a matrix $(1 - \hat{P})$. However, because we already know the equilibrium vector of P, these inversions can be avoided in the following way. As expression (9.1) shows, the equilibrium vector of ${}^{(2)}Q_1^*$ - which is an $O(\varepsilon^2)$-approximation of $\beta_1^{-1}\nu_1$ - is parallel to the first block of the equilibrium vector of the $(n_1 + N - 1) \times (n_1 + N - 1)$ stochastic matrix

$$ Q' = \underbrace{\left[\begin{array}{c|ccc} Q_1 & Q_{12}\, \underline{1}^T & \cdots & Q_{1N}\, \underline{1}^T \\ \hline \nu_2^* \; Q_{21} & & & \\ \vdots & & \hat{P} & \\ \nu_N^* \; Q_{N1} & & & \end{array} \right]}_{n_1 \quad\quad N-1} \begin{array}{l} \} \, n_1 \\ \} \, N-1 \end{array} \tag{9.2} $$

This matrix is itself nearly completely decomposable since it can be written as

$$ \begin{bmatrix} Q_1^* & 0 \\ 0 & \hat{P}^* \end{bmatrix} + O(\varepsilon) \; . $$

Thus, if we consider the above matrix as being nearly completely decomposable into two aggregates Q_1^* and $\hat{P}^*$ only, we can apply again the above reasoning to this two aggregate-matrix. The consequence of this is that an $O(\varepsilon^2)$-approximation of the first block of the equilibrium vector of Q' is itself parallel to the first block of the equilibrium vector of the $(n_1+1) \times (n_1+1)$ stochastic matrix

$$Q'' = \begin{bmatrix} Q_1 & E\,\underline{1}^T \\ \hat{x}F & 1-\hat{x}\,F\,\underline{1}^T \end{bmatrix}, \tag{9.3}$$

where $\hat{x} = (1-X_1)^{-1}\,[X_2\nu_2^*, X_3\nu_3^*, \ldots, X_N\nu_N^*]$ is a normalized equilibrium vector for the aggregate G*. Therefore, an $O(\varepsilon^2)$-approximation of $\beta_1^{-1}\nu_1$ can be also obtained as the equilibrium vector of the conformable aggregate

$${}^{(2)}Q_1^* = Q_1 + E\,\underline{1}^T(\hat{x}\,F\,\underline{1}^T)^{-1}\,\hat{x}\,F\,, \tag{9.4}$$

where the vector $\hat{x}$ is directly available from the $O(\varepsilon)$-approximation x obtained at the first step of the approximation procedure.

10. Conformable Aggregates of Order k, $k \geq 1$

This construction of ideal and conformable aggregates has a probabilistic interpretation which is intuitively appealing. Expression (6.2)

$$Q_1^{*id} = Q_1 + S^{id} + Q_1 + E(I-G)^{-1}F$$

shows that an ideal aggregate is constructed by adding to the probability q_{ij} of a one-step transition from state i to state j of Q_1, a probability $[E(I-G)^{-1}F]_{ij}$. This last probability is known in Markov chain theory (see e.g., [5]) as the taboo probability of leaving state i for state j without returning to any state of Q_1. The second term of the right-hand side of (9.1) is an $O(\varepsilon^2)$-approximation of this taboo probability obtained by condensing the subsystem G to a set of (N - 1) aggregated states. The second term in (9.4) is an $O(\varepsilon^2)$-approximation of the taboo probability obtained by reducing G to a single aggregated state.

Besides, expression (9.4) can be understood in the following way. An element (i,j) of a conformable aggregate of order 2 is given by

$$^{(2)}q^*_{ij} = q_{ij} + [E\, \underline{1}^T]_i \quad (\hat{x}\, F\, \underline{1}^T)^{-1}\, [\hat{x}\, F]_j .$$

Thus, this element is obtained by adding to the one-step transition probability q_{ij}, a fraction of the non diagonal block row sum $[E\, \underline{1}^T]_i$ which is directly proportional to the probability $(\hat{x}\, F\, \underline{1}^T)^{-1}\, [\hat{x}\, F]_j$ of entering the subset of states Q_1 precisely through state j. In other words, the probability q_{ij} must be incremented by a fraction of the probability of leaving the aggregate through state i which is proportional to the probability of entering the aggregate through state j. This is the most intuitive interpretation to which Vantilborgh's beautiful rule for constructing conformable aggregate of order 2 boils down.

This rule suggests also a general heuristic to construct aggregates without knowledge of the probability vector x. If one assumes, as a $O(\varepsilon)$-approximation, that all states of the remaining system G are equiprobable, i.e., that $\hat{x} = (N - x_1)^{-1}\underline{1}$, then (9.4) becomes

$$^{(1)}Q^*_1 = Q_1 + E\, \underline{1}^T(\underline{1}\, F\, \underline{1}^T)^{-1}\, \underline{1}\, F.$$

This construction was already proposed, on different heuristic grounds, by Zarling. One could call this a conformable aggregate of order 1. It implies a return to the aggregate with equal probability from all other states of the system. With this type of construction, an error:

$$10^{-5} < \|\nu_1 \beta_1^{-1} - X_1 \nu^*_1\|_\infty < 10^{-4}$$

has been obtained [9] for the example dealt with in the appendix of [2] where $\varepsilon = 10^{-3}$.

Finally, it is shown in [9], that conformable aggregates of any order can be constructed by an iterative application of (9.4): at each iteration a new vector $\hat{x}$ is calculated to obtain conformable aggregates of the next superior order. This iterative procedure converges and yields, in k iterations, an error $\|\nu - x\|_\infty$ of order $O(\varepsilon^k)$, for any $k \geq 1$.

11. Concluding Remarks

The error of aggregation is sensitive to the structure of the aggregates. To be of practical use, error bounds should therefore take this structure into account. The existence of such bounds is still an open question. It is however possible to construct aggregates which reduce considerably the error; and by an iterative construction of such aggregates, to reduce the error to an arbitrarily small order of magnitude.

Such "conformable aggregates" which were first proposed by Vantilborgh [9], are of great help whenever a matrix representation of the system under analysis is available. Otherwise, they will provide guidelines and heuristics as to how ideal aggregates should be approached.

- Conformable aggregates raise interesting questions. A conformable aggregate has not necessarily the same structure as the subsystem from which it is derived. For example, a conformable aggregate in a network of queues is no longer a single server queue. Further work is thus required to see how the above results could be exploited in these models.

12. Appendix

The following theorem generalizes to matrices of nullity one, a theorem originally proven in [10] for matrices with rowsums equal to zero.

Theorem. Let A be a $n \times n$ matrix of nullity one, and u a vector of length n such that $u\,A = 0$, $u \neq \underline{0}$; let

$$x\,A = k \tag{3}$$

be a compatible system of linear equations, and let h be any vector of length n with at least one nonzero element, say h_n:

$$h = [h' \quad h_n].$$

If, after a rearrangement, if necessary, of the equations,

$$A = \left[\begin{array}{c|c} A_1 & e^T \\ \hline b & d \end{array}\right],$$

where A_1 is of order $(n-1)$ and non-singular, and if $h\,u^T \neq 0$, then the unique solution $y \perp h$ of (3) is given by $y = k\,A^-$, where

$$A^- = \begin{bmatrix} Z & -Z\,h_n^{-1}\,h'^T \\ \underline{0} & 0 \end{bmatrix}$$

with $Z = (A_1 - h_n^{-1}\,h'^T\,b)^{-1}$.

Proof. Since A is of nullity one, a vector $z \neq \underline{0}$ exists such that $A\,z^T = \underline{0}$; z is unique up to a multiplicative constant. By a permutation, if necessary, of the columns of A, i.e., of the equations of (3), we can make $z_n \neq 0$. Then a post-multiplication of both sides of $x\,A = k$ by the non-singular matrix

$$\begin{bmatrix} I_{n-1} & \vdots & \\ & \vdots & z^T \\ \hline \underline{0} & \vdots & \end{bmatrix}$$

yields

$$x \begin{bmatrix} A_1 & \underline{0}^T \\ \hline b & 0 \end{bmatrix} = [k_1 \ldots k_{n-1}, \; k\, z^T].$$

Since the system (3) is compatible, $k\, z^T \equiv 0$; thus,

$$[x_1 \ldots x_{n-1}]\, A_1 + x_n b = [k_1 \ldots k_{n-1}].$$

If we divide both sides of $x\, h^T = 0$ by $h_n \neq 0$, we also have

$$x_n = -h_n^{-1}\, [x_1 \ldots x_{n-1}]\, h'^T \; ; \tag{4}$$

thus,

$$[x_1 \ldots x_{n-1}][A_1 - h_n^{-1}\, h'^T b] = [k_1 \ldots k_{n-1}] \tag{5}$$

Let us now show that $(A_1 - h_n^{-1} h'^T b)$ is invertible. Since A_1 is not singular, we can express b in function of A_1: $b = g\, A_1$; as A is of nullity one, we then have also $d = g\, e^T$. Thus,

$$(A_1 - h_n^{-1}\, h'^T b) = (I - h_n^{-1}\, h'^T g)\, A_1;$$

taking determinants, we obtain

$$\begin{aligned} |A_1 - h_n^{-1}\, h'^T b| &= |I - h_n^{-1}\, h'^T g|\; |A_1| \\ &= (1 - h_n^{-1}\, h' g^T)\; |A_1| \\ &= h\, f^T\, |A_1|. \end{aligned}$$

with $f = [g, -1]$. Moreover, we have the $f = cu$ with c being any nonzero scalar because

$$\begin{aligned} f\, A &= [g, -1]\, A \\ &= [g\, A_1 - b,\; g\, e^T - d] = \underline{0}. \end{aligned}$$

Thus, $h f^T = chu^T \neq 0$ by hypothesis; also $|A_1| \neq 0$ by hypothesis so that $(A_1 - h_n^{-1} h'^T b)$ is invertible.

The proof is now completed if we post-multiply both sides of (5) by $Z = [A_1 - h_n^{-1} h'^T b]^{-1}$ and substitute in (4).

Note 1. The theorem remains valid for whatever element of h is $\neq 0$; this non-zero element can always be brought into the last position by a rearrangement, if necessary, of the unknowns, i.e., of the rows of A.

Note 2. One can verify that the matrices A and A^- have the property that $AA^-A = A$ and $A^-AA^- = A^-$. For this reason A^- is called a "reflexive g-inverse", or a "g_2-inverse". One has

$$AA^-A = \begin{bmatrix} A_1 Z & -A_1 Z\, h_n^{-1}\, h'^T \\ bZ & b\, Z\, h_n^{-1}\, h'^T \end{bmatrix} \begin{bmatrix} A_1 & e^T \\ b & d \end{bmatrix},$$

$$= \begin{bmatrix} A_1 Z(A_1 - h_n^{-1}\, h'^T b) & A_1 Z(e^T - h_n^{-1}\, h'^T d) \\ b\, Z(A_1 - h_n^{-1}\, h'^T b) & b\, Z(e^T - h_n^{-1}\, h'^T d) \end{bmatrix};$$

from the proof of the theorem, we have

$$b = gA_1,\ d = g\, e^T \text{ and } Z = A_1^{-1}(I - h_n^{-1}\, h'^T g)^{-1},$$

which, by substitution in the matrix above yield $AA^-A = A$. The verification of $A^-AA^- = A^-$ is straightforward.

13. References

[1] Courtois, P. J., Louchard, G., (1976), Approximation of Eigencharacteristics in Nearly Completely Decomposable Stochastic Systems. Stochastic Process. Appl., 4, 283-296.

[2] Courtois, P. J. (1977), Decomposability, Queueing and Computer System Applications, Academic Press, New York.

[3] Courtois, P. J. (1977), The Error of Aggregation. Comments on Zarling's Analysis. MBLE Report R345.

[4] Courtois, P. J. (1978), Exact Aggregation in Queueing Networks. Proc. 1st AFCET-SMF meeting on Appl. Math., Ecole Polytechn. Palaiseau (France), Tome I, 35-51.

[5] Kemeny, J. G., Snell, J. L. (1960), Finite Markov Chains. Van Nostrand Inc.

[6] Simon, H. A., Ando, A. (1961), Aggregation of variables in dynamic systems, Econometrica, 29, 111-138.

[7] Stewart, G. W. (1973), Introduction to Matrix Computations. Academic Press.

[8] Vantilborgh, H. (1978), Exact Aggregation in Exponential Queueing Networks. J. Assoc. Comput. Mach., 25, 620-629.

[9] Vantilborgh, H. (1980), Aggregation with an Error of $O(\varepsilon^2)$, submitted for publication to J. Assoc. Comput. Mach.

[10] Zarling, R. L. (1976), Numerical Solution of Nearly Decomposable Queueing Network. Ph.D. dissertation, Department of Computer Science, North Carolina University, Chapel Hill.

Philips Research Laboratory, 2, av. Van Becelaere, 1170-BRUSSELS, BELGIUM.

COMPUTATIONAL METHODS FOR PRODUCT FORM QUEUEING NETWORKS

Charles H. Sauer

1. Extended Abstract

A major objective of computing systems (including computer communication systems) development in the last two decades has been to promote sharing of system resources. Sharing of resources necessarily leads to contention, i.e., queueing, for resources. Contention and queueing for resources are typically quite difficult to quantify when estimating system performance. A major research topic in computing systems performance in the last two decades has been solution and application of queueing models. These models are usually networks of queues because of the interactions of system resources. For general discussion of queueing network models of computing systems, see Sauer and Chandy [SAUE81a] and recent special issues of *Computing Surveys* (September 1978) and *Computer* (April 1980).

In the fifties and early sixties Jackson showed (see JACK63) that a certain class of networks of queues has a *product form* solution in the sense that

$$P(\vec{n}_1,\ldots,\vec{n}_M) = \frac{X_1(\vec{n}_1)\ldots X_M(\vec{n}_M)}{G(\vec{N})}$$

where $P(\vec{n}_1,\ldots,\vec{n}_M)$ is the joint queue length distribution in a network with M queues, $X_m(\vec{n}_m)$, $m = 1,\ldots,M$, is a factor obtained from the marginal queue length distribution of queue m in isolation and $G(\vec{N})$ is a normalizing constant. Computer scientists and others have extended

Jackson's result to encompass networks with heterogeneous cusotmers, with both open and closed routing chains, with some queueing disciplines other than first-come-first-served, with general service time distributions for some of those other disciplines and with some state dependent routing mechanisms. The most important results for computer scientists are found in BASK75, CHAN77, KOBA75 and TOWS80.

The existence of a product form solution makes tractable the solution of networks with large numbers of queues and/or large populations, but the computational methods are non-trivial unless the normalizing constant itself is easily obtained. There have been many computational algorithms proposed for product form networks. No single algorithm is entirely satisfactory for all applications. This talk considers four exact algorithms, additional methods useful with those algorithms and heuristic methods which have been proposed as inexpensive alternatives to exact algorithms.

Most of the attention on computational methods has focused on closed networks, i.e., networks with customer populations of fixed size. There are at least two reasons for this emphasis: (1) Closed networks seem the most useful in computer system modeling. (2) Numerical solution of open networks (those with sources and sinks) is trivial if arrival rates are constant. Arrival processes must be Poisson in open networks known to have a product form solution, but the product form solution allows arrival rates dependent on network populations. The algorithms for closed networks can be extended to open networks with population dependent arrival rates and to mixed (closed and open routing chain) networks, as discussed in this talk.

The oldest satisfactory algorithm, and still the most important algorithm, is the "Convolution" algorithm, first proposed in Buzen's Ph.D. Thesis (see BUZE73). The Convolution algorithm obtains the normalizing constant, $G(\vec{N})$, where $\vec{N}$ is the vector of populations of the different types of customers, from the factors $X_m(\vec{n}_m)$, m = 1,...,M, in a manner similar to the convolution of discrete probability distributions. Performance measures are then obtained from expressions involving these factors and the normalizing constant.

A fundamental problem with the Convolution algorithm, and two of the other three algorithms discussed in this talk, is that the normalizing constant, $G(\vec{N})$, may have a value outside of floating point range even though the network parameters and performance measures have values well within floating point range. This problem can usually, but not always, be eliminated by appropriate use of scaling values in determining the factors $X_m(\vec{n}_m)$, m = 1,...,M [REIS78b, CHAN79, LAM80, SAUE81a]. In addition to this problem, Convolution and one of the other algorithms have a problem with numerical instability for $G(\vec{N})$ near the lower limit (in magnitude) of floating point range.

Next to Convolution, the most important algorithm is the Mean Value Analysis of Reiser and Lavenberg [REIS78a, REIS80]. Mean Value Analysis avoids the use of normalizing constants and the numerical problems associated with them. This is done by use of recursive relationships between mean queueing times in the given network and performance measures in the corresponding network with one customer removed. Little's Rule may then be applied to obtain throughputs and mean queue lengths in the given network. For single server fixed rate queues and infinite server queues this is quite simply done; the only performance measures which need to be considered are mean queueing times, throughputs and mean queue lengths. However, for other kinds of queues, marginal queue length distributions must be obtained. As originally proposed in REIS78a, Mean Value Analysis experiences severe numerical problems in determining marginal queue length distributions

[CHAN79, SAUE81a]. Reiser later provides an alternate approach for determining queue length distributions with Mean Value Analysis, but this approach will be much more computationally expensive than Convolution [REIS80]. (For networks with only single server fixed rate and infinite server queues, Mean Value Analysis and Convolution require essentially the same computational effort.) Mean Value Analysis has recently been extended to be as general as Convolution [SAUE81b].

The Local Balance Algorithm for Normalizing Constants (LBANC) uses equations which are derived in a manner similar to those of Mean Value Analysis, but which explicitly involve the same normalizing constants as Convolution [CHAN79, SAUE81a]. Though LBANC has the same problem as Convolution that the normalizing constant may be outside of floating point range, LBANC does not suffer the numerical instability of Convolution for normalizing constants near the lower limit (in magnitude) of floating point range. LBANC and Convolution can be used together in a hybrid manner, with LBANC used for single server fixed rate and infinite server queues and Convolution used for other queues. LBANC also has special advantages in implementations on programmable calculators.

The algorithm to Coalesce Computation of Normalizing Constants (CCNC) is especially appropriate for programmable calculator implementations where networks have only single server fixed rate and infinite server queues [CHAN79, SAUE81a].

All four of these algorithms may be very expensive in both time and memory for networks with large heterogeneous customer populations. Reiser and Lavenberg proposed heuristic modifications of Mean Value Analysis which provide inexpensive approximate solutions for such networks [REIS78a]. Others have proposed refinements of those heuristics; the most promising refinements seem to be those of Chandy and Neuse [CHAN80].

2. References

BASK75 F. Baskett, K. M. Chandy, R. R. Muntz, and F. Palacios-Gomez, "Open, Closed, and Mixed Networks of Queues with Different Classes of Customers", J. Assoc. Comput. Mach., 22, 2 (April 1975).

BUZE73 J. P. Buzen, "Computational Algorithms for Closed Queueing Networks with Exponential Servers," Comm. ACM, 16, 9 (Sept. 1973) pp. 527-531.

CHAN77 K. M. Chandy, J. H. Howard and D. F. Towsley, "Product Form and Local Balance In Queueing Networks," J. Assoc. Comput. Mach., 24, 2 pp. 250-263 (April 1977).

CHAN79 K. M. Chandy and C. H. Sauer, "Computational Algorithms for Product Form Queueing Networks," RC-7950, IBM Research, Yorktown Heights, N.Y. (November 1979). Comm. ACM., 23, 10 (October 1980).

CHAN80 K. M. Chandy and D. Neuse, "Fast Accurate Heuristic Algorithms for Queueing Network Models of Computing Systems," TR-157, Dept. of Computer Sciences, University of Texas at Austin (September 1980).

JACK63 J. R. Jackson, "Jobshop-like Queueing Systems," Management Sci., 10, pp. 131-142 (1963).

KOBA75 H. Kobayashi and M. Reiser, "On Generalization of Job Routing Behavior in a Queueing Network Model," IBM Research Report RC-5252 (1975).

LAM80 S. S. Lam, "Dynamic Scaling and Growth Behavior of Queueing Network Normalization Constants," Technical Report TR-148, Department of Computer Sciences, University of Texas at Austin (December 1980).

REIS78a M. Reiser and S. S. Lavenberg, "Mean Value Analysis of Closed Multichain Queueing Networks," IBM Research Report RC-7023, Yorktown Heights, NY (March 1978). J. Assoc. Comput. Mach., 27, 2 (April 1980) pp. 313-322.

REIS78b M. Reiser and C. H. Sauer, "Queueing Network Models: Methods of Solution and their Program Implementation," in K. M. Chandy and R. T. Yeh, Editors, Current Trends in Programming Methodology, Volume III: Software Modeling and Its Impact on Performance. Prentice-Hall (1978) pp. 115-167.

REIS80 M. Reiser, "Mean-Value Analysis and Convolution Method for Queue-Dependent Servers in Closed Queueing Networks," IBM Research Report RZ-1009 (Zurich, March 1980).

SAUE81a C. H. Sauer and K. M. Chandy, Computer System Performance Modeling, Prentice-Hall, Englewood Cliffs, NJ (1981).

SAUE81b C. H. Sauer, "Computational Algorithms for State-Dependent Queueing Networks," IBM Research Report RC-8698 (February 1981).

TOWS80 D. F. Towsley, "Queueing Network Models with State-Dependent Routing," J. Assoc. Comput. Mach., 27, 2 (April 1980) pp. 323-337.

IBM Thomas J. Watson Research Center, Yorktown Heights, New York 10598

NETWORKS OF QUEUES, I
Richard Muntz, Chairman

S. S. Lavenberg
M. Reiser
W. J. Stewart

CLOSED MULTICHAIN PRODUCT FORM QUEUEING NETWORKS WITH LARGE POPULATION SIZES

S. S. Lavenberg

Abstract

We consider closed multichain product form queueing networks which have at least one infinite server service center. When modeling interactive computer systems an infinite server service center typically represents the collection of interactive terminals. We establish conditions under which performance measures for such a network converge, as the population sizes of the chains increase, to performance measures for a network in which the infinite server service center is replaced by a Poisson source. Then, for an important class of such networks, we obtain large population approximations for throughputs, mean queue sizes and mean response times which are based on the asymptotic result and on the equations of mean-value analysis. The approximations are typically several orders of magnitude less expensive to compute than are the exact values. Numerical results are presented which indicate that the approximations are useful for closed networks with 100 or more customers.

1. Introduction

Closed queueing networks with multiple routing chains which have product form solution, first introduced in Baskett, et. al [2], are commonly used in the performance modeling of computer systems (e.g., see Graham [4], Sauer and Chandy [9]). However, the computational complexity of algorithms for computing performance measures for such networks (e.g., the convolution algorithm in Reiser and Kobayashi [7] or the mean-value analysis algorithm in Reiser and Lavenberg [8]) is proportional to the product of the population sizes for each chain. Hence, computational complexity is a limiting factor in the use of such networks for large population sizes. The asymptotic behavior of such networks as the population sizes increase is of interest since asymptotic results may suggest useful approximations for large finite population sizes.

Pittel [6] investigated the asymptotic behavior as the population sizes increase of a class of closed multichain product form networks with capacity constraints. He showed that the set of saturated service centers and the asymptotic queue size distributions for the nonsaturated service centers are determined by the solution of a nonlinear programming problem. Even if the capacity constraints are removed, a nonlinear programming problem must be solved. Here we restrict our attention to multichain product form networks without capacity constraints which have at least one IS (infinite server) service center. When modeling interactive multiprogrammed computer systems such a service center typically represents the collection of interactive terminals. We first consider the asymptotic behavior of such networks as the population sizes increase for all chains which visit an IS service center and the mean service demands at the IS service center increase in proportion to the population sizes. (In [6] all mean service demands were held fixed.) We show that the joint queue size distribution of the rest of the network and performance measures based on the queue size distribution converge to the joint queue size distribution and performance measures of a network in which the IS service center is replaced by a Poisson source. If all chains visit the IS service center then the asymptotic network is open, while if some chains do not visit the IS service center then the asymptotic network is mixed, i.e., it has both open and closed chains. Then, for an important class of such networks, we obtain large population approximations for throughputs, mean queue sizes and mean response times. The approximations are based on the asymptotic result and on the equations of mean-value analysis. The approximations require that for each closed chain a certain nonlinear equation in a single unknown be solved. This equation is shown to have a unique solution.

Other approximations have been proposed for closed networks based on the equations of mean-value analysis. These approximations require the solution of simultaneous nonlinear equations. Bard [1] obtained an approximation which requires the solution of Pittel's [6] asymptotic equations. Reiser and Lavenberg [8] and Schweitzer [10] suggested approximations designed to work better than Bard's for small population sizes. However, the uniqueness of the solution of the equations they obtain has not been established, nor has the convergence of the iterative solution methods they used. Chandy and Neuse [3] have improved Schweitzer's approximation by employing it in the first step of a multistep procedure. Error bounds have not been obtained for any of these approximations or for the approximation we present. The establishment of error bounds is a difficult open problem.

In Section 2 we present the known results for closed multichain product form networks that we will use in the remainder of the paper. We establish the asymptotic results in Section 3 and we develop the approximations in Section 4. In Section 4 we also present examples in which we numerically compare the exact and approximate values of performance measures. The approximate values are typically several orders of magnitude less expensive to compute than are the exact values. The approximations are found to be very accurate (errors do not exceed a few percent) for throughputs and, at moderate asymptotic traffic intensities, for mean queue sizes and mean response times. The approximations are least accurate (errors can exceed 20%) for mean queue sizes at service centers with high asymptotic traffic intensities. However, even if the approximations are not highly accurate, they are still useful when the exact values are expensive to compute.

2. Preliminaries

We consider closed multichain queueing networks whose stationary state probabilities have the product form given below in (2.1). Each routing chain contains a fixed number of probabilistically identical customers each of whom proceeds through a subset of the service centers according to an irreducible Markov chain. We adopt the following notation. (All vectors are row vectors.)

L: number of service centers

R: number of routing chains

K_r: population size of chain r

$\underset{\sim}{K} = (K_1,\ldots,K_R)$: population vector

$\tau_{r,\ell}$: mean service demand of a chain r customer at service center ℓ

$\mu_\ell(k)$: service rate of service center ℓ as a function of the number of customers at the center

$S(r)$: set of service centers visited by chain r

$R(\ell)$: set of chains which visit service center ℓ

P_r: stochastic routing matrix of chain r

$\underset{\sim}{\Theta}_r = (\Theta_{r,\ell}: \ell \in S(r))$: stationary probability vector for P_r, i.e., the unique probability vector which satisfies $\underset{\sim}{\Theta}_r = \underset{\sim}{\Theta}_r P_r$

$k_{r,\ell}$: number of chain r customers in service center ℓ, $r \in R(\ell)$

k_ℓ: number of customers in service center ℓ

$\underset{\sim}{k}_\ell = (k_{r,\ell}: r \in R(\ell))$: state of service center ℓ

$\underset{\sim}{k} = (\underset{\sim}{k}_1,\ldots,\underset{\sim}{k}_L)$: state of network

$A(\underset{\sim}{K}) = \{\underset{\sim}{k}: \sum_{\ell \varepsilon S(r)} k_{r,\ell} = K_r,\ r = 1,\ldots,R\}$: set of states of network

$p(\underset{\sim}{k},\underset{\sim}{K})$: stationary probability of network being in state $\underset{\sim}{k} \varepsilon A(\underset{\sim}{K})$

$\lambda_{r,\ell}(\underset{\sim}{K})$: throughput of chain r customers at service center ℓ

The product form referred to above is given by

$$p(\underset{\sim}{k},\underset{\sim}{K}) = \left[\prod_{\ell=1}^{L} \alpha_\ell(\underset{\sim}{k}_\ell)\right] / g(\underset{\sim}{K}) \quad \underset{\sim}{k} \varepsilon A(\underset{\sim}{K}) \tag{2.1}$$

where

$$\alpha_\ell(\underset{\sim}{k}_\ell) = \left[\prod_{j=1}^{k_\ell} \frac{1}{\mu_\ell(j)}\right] k_\ell! \prod_{r \varepsilon R(\ell)} \frac{(\Theta_{r,\ell}\tau_{r,\ell})^{k_{r,\ell}}}{k_{r,\ell}!} \tag{2.2}$$

and $g(\underset{\sim}{K})$ is a normalizing constant. (We adopt the convention that an empty product, such as the first product in (2.2) with $k_\ell = 0$, is equal to one.) Furthermore, if $K_r \geq 1$

$$\lambda_{r,\ell}(\underset{\sim}{K}) = \Theta_{r,\ell} \frac{g(\underset{\sim}{K}-\underset{\sim}{e}_r)}{g(\underset{\sim}{K})}, \tag{2.3}$$

where $\underset{\sim}{e}_r$ is a unit vector whose r-th component is equal to one.

Networks which have the product form solution in (2.1) include those in which service centers are of the following types:

(i) FCFS: customers are served in order of arrival; multiple servers are allowed; service demands for customers in different chains have the same exponential distribution.

(ii) PS: all customers are served simultaneously by a single server, each customer receiving an equal share of the service rate (processor sharing); service demands for customers in different chains can have different phase-type distributions.

(iii) LCFSPR: customers are served by a single server according to the last-come first-served preemptive resume queueing discipline; service demands for customers in different chains can have different phase-type distributions.

(iv) IS: there is effectively an infinite number of servers so that the amount of time a customer spends in the service center depends only on its service demand and not on the number of customers in the center; service demands for customers in different chains can have different phase-type distributions.

The product form in (2.1) also holds for other types of service centers such as those defined in Kelly [5]. Furthermore, this product form holds if for each chain a set of classes is defined at each service center it visits. Customers in the same chain change class as they proceed through the service centers and the routing is according to an irreducible Markov chain over the classes instead of over the service centers. The service demand at a PS, LCFSPR or IS service center can depend on the class of the customer as well as on the chain. In the case of classes the quantities $\Theta_{r,\ell}$ and $\tau_{r,\ell}$ in (2.2) are defined as follows: Let P^*_r denote the stochastic routing matrix over the classes of chain r and let $\underset{\sim}{\Theta}^*_r$ denote the unique probability vector which satisfies $\underset{\sim}{\Theta}^*_r = \underset{\sim}{\Theta}^*_r P^*_r$. Let $\Theta^*_{r,c}$ denote the component of $\underset{\sim}{\Theta}^*_r$ corresponding to class c and let $\tau^*_{r,c}$ denote the mean service demand of a chain r customer in class c. Finally let $C_r(\ell)$ denote the set of classes for chain r at service center ℓ. Then,

$$\Theta_{r,\ell} = \sum_{c \varepsilon C_r(\ell)} \Theta^*_{r,c} \tag{2.4}$$

and

$$\tau_{r,\ell} = \sum_{c \varepsilon C_r(\ell)} \Theta^*_{r,c} \tau^*_{r,c} / \Theta_{r,\ell} \tag{2.5}$$

3. Asymptotic Results

We will consider networks which have at least one IS service center. Suppose that service center L is an IS service center so that for all $k \geq 1$

$$\mu_L(k) = k. \tag{3.1}$$

Let $K_r = [a_r K]$, $r \in R(L)$, where K is a positive integer, $a_r > 0$, $r \in R(L)$, $\sum_{r\in R(L)} a_r = 1$ and [x] denotes the integer part of x. Furthermore, let

$$\lambda_r = K_r/\tau_{r,L} \qquad r \in R(L). \tag{3.2}$$

We are interested in the asymptotic behavior of the stationary state probabilities for the subnetwork consisting of service centers $1,\ldots,L-1$ and of the throughputs $\lambda_{r,\ell}(\underset{\sim}{K})$ as $K \to \infty$ with a_r and λ_r fixed for all $r \in R(L)$. Since λ_r is fixed it follows that $\tau_{r,L} \to \infty$ for all $r \in R(L)$.

We let $\overline{R(L)}$ denote the complement of R(L) and we adopt the following additional notation:

$\mu_\ell = \lim_{k\to\infty} \mu_\ell(k)$: asymptotic service rate of service center ℓ (we assume this limit exists and is positive)

$\underset{\sim}{k}^* = (\underset{\sim}{k}_1,\ldots,\underset{\sim}{k}_{L-1})$: state of the subnetwork

$n_r = \sum_{\ell\in S(r),\ell\neq L} k_{r,\ell}$: number of chain r customers in the subnetwork

$A^*(\underset{\sim}{K}) = \{\underset{\sim}{k}^*: n_r \leq K_r,\ r \in R(L);\ n_r = K_r,\ r \in \overline{R(L)}\}$: set of states of subnetwork

$q(\underset{\sim}{k}^*,\underset{\sim}{K})$: stationary probability that the subnetwork is in state $\underset{\sim}{k}^* \in A^*(\underset{\sim}{K})$ (we let $q(\underset{\sim}{k}^*,\underset{\sim}{K}) = 0$ if $\underset{\sim}{k}^* \notin A^*(\underset{\sim}{K})$).

Note that if $\underset{\sim}{k}^* \in A^*(\underset{\sim}{K})$ then

$$q(\underset{\sim}{k}^*,\underset{\sim}{K}) = p((\underset{\sim}{k}^*,\underset{\sim}{k}_L),\underset{\sim}{K}) \tag{3.3}$$

where

$$\underset{\sim}{k}_L = (K_r - n_r:\ r \in R(L)).$$

For simplicity we first present asymptotic results in the case where all chains visit service center L so that $R(L) = \{1,\ldots,R\}$ and $\overline{R(L)}$ is empty. These results which are given below in (3.6), (3.9) and (3.10) follow directly from Theorem 1 which is stated later in this section and which holds in the general case where all chains need not visit service center L.

Let $\underset{\sim}{0}$ denote a vector all of whose components are equal to 0. For each $\ell \neq L$ let

$$\rho_{r,\ell} = \lambda_r \Theta_{r,\ell} \tau_{r,\ell} / \Theta_{r,L} \qquad r \in R(\ell) \tag{3.4}$$

and

$$\rho_\ell = \sum_{r \in R(\ell)} \rho_{r,\ell} . \tag{3.5}$$

The quantity ρ_ℓ / μ_ℓ is the asymptotic traffic intensity for service center ℓ. If for each $\ell \neq L$, $\rho_\ell < \mu_\ell$ then for all $\underset{\sim}{k}^* \geq \underset{\sim}{0}$

$$\lim_{K\to\infty} q(\underset{\sim}{k}^*, \underset{\sim}{K}) = \prod_{\ell=1}^{L-1} [\delta_\ell(\underset{\sim}{k}_\ell)/g_\ell] \tag{3.6}$$

where

$$\delta_\ell(\underset{\sim}{k}_\ell) = \left[\prod_{j=1}^{k_\ell} \frac{1}{\mu_\ell(j)}\right] k_\ell! \prod_{r \in R(\ell)} \frac{(\rho_{r,\ell})^{k_{r,\ell}}}{k_{r,\ell}!} \tag{3.7}$$

and

$$g_\ell = 1 + \sum_{k=1}^{\infty} \left[\prod_{j=1}^{k} \frac{1}{\mu_\ell(j)}\right] \rho_\ell^k < \infty . \tag{3.8}$$

Also, if for each $\ell \neq L$, $\rho_\ell < \mu_\ell$ then

$$\lim_{K\to\infty} \lambda_{r,\ell}(\underset{\sim}{K}) = \lambda_r \Theta_{r,\ell} / \Theta_{r,L} \qquad r \in R(\ell) . \tag{3.9}$$

If for some $\ell \neq L$, $\rho_\ell > \mu_\ell$ then for all $\underset{\sim}{k}^* \geq \underset{\sim}{0}$

$$\lim_{K\to\infty} q(\underset{\sim}{k}^*, \underset{\sim}{K}) = 0 . \tag{3.10}$$

Thus, if $\rho_\ell < \mu_\ell$ for each $\ell \neq L$ then as K increases the stationary state probability distribution for the subnetwork converges to that for a stable open network consisting of service centers $1,\ldots,L-1$ and having R routing chains. Chain r arrivals at the open network are Poisson with rate λ_r and the mean number of visits a chain r customer makes to service center ℓ while in the open network is equal to $\Theta_{r,\ell}/\Theta_{r,L}$. The throughputs for the subnetwork also converge to the throughputs for the open network and the throughputs for service center L converge to the open network arrival rates. If $\rho_\ell > \mu_\ell$ for some $\ell \neq L$ then the subnetwork becomes unstable as K increases.

We next consider the general case where all chains need not visit service center L. This case is more complex notationally but the results are similar. For each $\ell \neq L$ we partition $R(\ell)$ into two subsets, $R^0(\ell)$ and $R^1(\ell)$, where $R^0(\ell) = R(\ell) \cap R(L)$ and $R^1(\ell) = R(\ell) \cap \overline{R(L)}$. Thus, chains in $R^0(\ell)$ visit service center L and chains in $R^1(\ell)$ do not. We write each of the previously defined vectors $\underset{\sim}{K}$, $\underset{\sim}{k}_\ell$, $\ell \neq L$ and $\underset{\sim}{k}^*$ as the concatenation of two vectors, the first of which has components for those chains which visit service center L and the second of which has components for the remaining chains. Thus, we write $\underset{\sim}{K} = (\underset{\sim}{K}^0,\underset{\sim}{K}^1)$, $\underset{\sim}{k}_\ell = (\underset{\sim}{k}_\ell^0,\underset{\sim}{k}_\ell^1)$, $\ell \neq L$, and $\underset{\sim}{k}^* = (\underset{\sim}{k}^0,\underset{\sim}{k}^1)$ where for example $\underset{\sim}{K}^0 = \{K_r\colon r \in R(L)\}$ and $\underset{\sim}{K}^1 = \{K_r\colon r \in \overline{R(L)}\}$. We also let $k_\ell^i = \sum_{r \in R^i(\ell)} k_{r,\ell}$, $i = 0,1$ so that $k_\ell = k_\ell^0 + k_\ell^1$. Note that as $K \to \infty$, $\underset{\sim}{K}^0 \to \underset{\sim}{\infty}$ while $\underset{\sim}{K}^1$ is fixed. If $\underset{\sim}{K}^0 = \underset{\sim}{\infty}$ then $A^*(\underset{\sim}{K})$, the previously defined set of states of the subnetwork, is given by $A^*(\underset{\sim}{\infty},\underset{\sim}{K}^1) = \{\underset{\sim}{k}^*\colon n_r = K_r,\ r \in \overline{R(L)}\}$. Finally, for each $\ell \neq L$ we let

$$\rho_{r,\ell} = \begin{cases} \lambda_r \Theta_{r,\ell}\tau_{r,\ell}/\Theta_{r,L} & r \in R^0(\ell) \\ \\ \Theta_{r,\ell}\tau_{r,\ell} & r \in R^1(\ell) \end{cases} \tag{3.11}$$

and

$$\rho_\ell^i = \sum_{r \in R^i(\ell)} \rho_{r,\ell} \qquad i = 0,1. \tag{3.12}$$

We prove the following theorem in Appendix A.

<u>Theorem 1</u>. If for each $\ell \neq L$, $\rho_\ell^0 < \mu_\ell$ then for all $\underset{\sim}{k}^* \in A^*(\underset{\sim}{\infty},\underset{\sim}{K}^1)$

$$\lim_{K\to\infty} q(\underset{\sim}{k}^*,\underset{\sim}{K}) = \left[\prod_{\ell=1}^{L-1} \delta_\ell(\underset{\sim}{k}_\ell)\right] / G(\underset{\sim}{K}^1) \tag{3.13}$$

where $\delta_\ell(\underset{\sim}{k}_\ell)$ is given by (3.7) and $G(\underset{\sim}{K}^1)$ is a normalizing constant given by

$$G(\underset{\sim}{K}^1) = \sum_{\underset{\sim}{k}^1 : n_r = K_r,\, r\in\overline{R(L)}} \prod_{\ell=1}^{L-1} \beta_\ell(\underset{\sim}{k}_\ell^1) \tag{3.14}$$

where

$$\beta_\ell(\underset{\sim}{k}_\ell^1) = \left[\prod_{j=1}^{k_\ell^1} \frac{1}{\mu_\ell(j)}\right] k_\ell^1! \left[\prod_{r\in R^1(\ell)} \frac{(\rho_{r,\ell})^{k_{r,\ell}}}{k_{r,\ell}!}\right] g_\ell(k_\ell^1) \tag{3.15}$$

and

$$g_\ell(k_\ell^1) = 1 + \sum_{k=1}^{\infty} \left[\prod_{j=k_\ell^1+1}^{k_\ell^1+k} \frac{1}{\mu_\ell(j)}\right] \frac{(k_\ell^1+k)!}{k_\ell^1!\,k!} (\rho_\ell^0)^k < \infty . \tag{3.16}$$

Also, if for each $\ell \neq L$, $\rho_\ell^0 < \mu_\ell$ then

$$\lim_{K\to\infty} \lambda_{r,\ell}(\underset{\sim}{K}) = \begin{cases} \lambda_r \Theta_{r,\ell}/\Theta_{r,L} & r \in R^0(\ell) \\ \Theta_{r,\ell} G(\underset{\sim}{K}^1 - \underset{\sim}{e}_r)/G(\underset{\sim}{K}^1) & r \in R^1(\ell) \text{ and } K_r \geq 1. \end{cases} \tag{3.17}$$

If for some $\ell \neq L$, $\rho_\ell^0 > \mu_\ell$ then for all $\underset{\sim}{k}^* \in A^*(\underset{\sim}{\infty},\underset{\sim}{K}^1)$

$$\lim_{K\to\infty} q(\underset{\sim}{k}^*,\underset{\sim}{K}) = 0 . \tag{3.18}$$

Thus, if $\rho_\ell^0 < \mu_\ell$ for each $\ell \neq L$ then as K increases the stationary state probability distribution for the subnetwork converges to that for a stable mixed (open and closed) network consisting of service centers $1,\ldots,L-1$ and having R routing chains. For $r \in R(L)$ chain r is open, arrivals at the network are Poisson with rate λ_r and the mean number of visits a chain r customer makes to service center ℓ while in the network is equal to $\Theta_{r,\ell}/\Theta_{r,L}$. For $r \in \overline{R(L)}$ chain r is closed and there are K_r customers in the network. The throughputs for the open and closed chains are given in (3.17). If $\rho_\ell^0 > \mu_\ell$ for some $\ell \neq L$ then the subnetwork becomes unstable as K increases.

We conclude this section by considering the asymptotic behavior of performance measures for the subnetwork. Let $\underset{\sim}{m}(\underset{\sim}{K})$ denote the stationary state vector for the closed subnetwork, i.e., $\underset{\sim}{m}(\underset{\sim}{K})$ is a random vector whose probability distribution is given by $\{q(\underset{\sim}{k}^*,\underset{\sim}{K}): \underset{\sim}{k}^* \in A^*(\underset{\sim}{K})\}$. Let $\underset{\sim}{m}$ denote the stationary state vector for the stable mixed network. We will consider performance measures which can be expressed as the expected value of a function of the stationary state vector. Such performance measures as queue size probabilities, moments of queue size and utilizations can be expressed in this way. We prove the following corollary to Theorem 1 in Appendix A.

<u>Corollary 1</u>. Let $f(\underset{\sim}{k}^*)$ be a function defined for all $\underset{\sim}{k}^* \in A^*(\underset{\sim}{\infty},\underset{\sim}{K}^1)$. If $\rho_\ell^0 < \mu_\ell$ for each $\ell \neq L$ and if $E[|f(\underset{\sim}{m})|] < \infty$ then

$$\lim_{K\to\infty} E[f(\underset{\sim}{m}(\underset{\sim}{K}))] = E[f(\underset{\sim}{m})]. \tag{3.19}$$

Thus, if $\rho_\ell^0 < \mu_\ell$ for each $\ell \neq L$ then as K increases performance measures for the subnetwork converge to performance measures for the stable mixed network.

4. Approximation of Performance Measures

In almost all networks of practical interest either (i) $\mu_\ell(j) = \mu_\ell$ for all $j \geq 1$ (single fixed rate server), (ii) $\mu_\ell(j) = j$ for all $j \geq 1$ (IS service center) or (iii) for some finite $J > 0$, $\mu_\ell(j)$ is arbitrary for all $j \leq J$ and $\mu_\ell(j) = \mu_\ell$ for all $j > J$ (limited queue size dependence). The infinite sums g_ℓ in (3.8) and $g_\ell(k_\ell^1)$ in (3.16) are easy to evaluate in these cases. For example in case (i) $g_\ell(k_\ell^1) = 1/(1-\rho_\ell^0/\mu_\ell)^{k_\ell^1+1}$ and in case (ii) $g_\ell(k_\ell^1) = e^{\rho_\ell^0}/k_\ell^1!$. In these cases the computational complexity of algorithms (e.g., convolution [7] and mean-value analysis [8]) which compute mean queue sizes, mean waiting times, throughputs and utilizations for closed networks is proportional to the product of the chain populations. If K_r is large for all $r \in R(L)$ it is tempting to approximate performance measures for a closed network by performance measures for the asymptotic network provided the asymptotic network is stable. If all chains visit service center L so that the asymptotic network is open the computational effort for the asymptotic network is negligible. If some chains do not visit service center L so that the asymptotic network is mixed the computational effort will still be reduced substantially. (The convolution algorithm for mixed networks is discussed in [7] and the computational effort is proportional to the product of the closed chain populations. Mean-value analysis has not been presented in the literature for mixed networks, although it can be extended to handle mixed networks.) Unfortunately, as we shall see in the numerical examples to be presented later in this section, performance measures for the closed network can converge quite slowly to their asymptotic values. Thus, it is desirable to obtain approximations to performance measures for the closed network which are better than the asymptotic approximations but which are still easy to compute.

In the remainder of this section we restrict our attention to closed networks in which all chains visit service center L so that the asymptotic network is open and in which every other service center either has a single unit rate server, i.e., $\mu_\ell(k) = 1$ for all k, or is an

IS service center, i.e., $\mu_\ell(k) = k$ for all k. We assume for ease of notation that service centers $1,\ldots,L_1$, where $L_1 \leq L-1$, have a single unit rate server, that service centers $L_1+1,\ldots,L$ are IS and that all chains visit all service centers. We further assume that $L_1 \geq 1$ so that not all service centers are IS. (If all service centers are IS then performance measures can be obtained with negligible computational effort.) We will obtain a simple approximation for $\lambda_r(\underset{\sim}{K}) = \lambda_{r,L}(\underset{\sim}{K})$, the chain r throughout at service center L, $N_{r,\ell}(\underset{\sim}{K})$, the mean number of chain r customers in service center ℓ, $\ell \neq L$, and $T_r(\underset{\sim}{K})$, the mean response time for chain r (the response time is defined to be the time a customer spends in the subnetwork from when it leaves service center L until it next returns there). Note from Little's formula that

$$T_r(\underset{\sim}{K}) = N_r(\underset{\sim}{K})/\lambda_r(\underset{\sim}{K}) \, , \tag{4.1}$$

where $N_r(\underset{\sim}{K})$ is the mean number of chain r customers in the subnetwork, and also that

$$T_r(\underset{\sim}{K}) = \frac{K_r}{\lambda_r(\underset{\sim}{K})} - \frac{K_r}{\lambda_r} \tag{4.2}$$

where we have substituted K_r/λ_r for $\tau_{r,L}$. Combining (4.1) and (4.2) it follows that

$$\lambda_r(\underset{\sim}{K}) = \left(1 - \frac{N_r(\underset{\sim}{K})}{K_r}\right)\lambda_r \, . \tag{4.3}$$

This immediately suggests the following approximation for $\lambda_r(\underset{\sim}{K})$ if the asymptotic network is stable:

$$\lambda_r(\underset{\sim}{K}) \approx \left(1 - \frac{N_r(\underset{\sim}{\infty})}{K_r}\right)\lambda_r \tag{4.4}$$

where $N_r(\underset{\sim}{\infty})$ is the asymptotic value of $N_r(\underset{\sim}{K})$. However, we now develop a better approximation based on mean-value analysis [8].

It follows from the mean-value analysis equations (e.g., see (3.1)-(3.4) in [8]) that if $K_r \geq 1$ then

$$N_{r,\ell}(\underset{\sim}{K}) = \lambda_r(\underset{\sim}{K})(\theta_{r,\ell}/\theta_{r,L})\tau_{r,\ell}[1 + N_{,\ell}(\underset{\sim}{K}-e_r)] \qquad \ell = 1,\ldots,L_1 \tag{4.5}$$

where $N_{,\ell}(\underset{\sim}{K})$ is the mean number of customers in service center ℓ, and

$$N_{r,\ell}(\underset{\sim}{K}) = \lambda_r(\underset{\sim}{K})(\theta_{r,\ell}/\theta_{r,L})\tau_{r,\ell} \qquad \ell = L_1 + 1,\ldots,L - 1. \tag{4.6}$$

Substituting (4.3) into (4.5) and (4.6) it follows that if $K_r \geq 1$ then

$$N_{r,\ell}(\underset{\sim}{K}) = \rho_{r,\ell}\left(1 - \frac{N_r(\underset{\sim}{K})}{K_r}\right)\begin{cases}[1 + N_{,\ell}(\underset{\sim}{K}-\underset{\sim}{e}_r)] & \ell = 1,\ldots,L_1 \\ 1 & \ell = L_1 + 1,\ldots,L - 1,\end{cases} \tag{4.7}$$

where $\rho_{r,\ell}$ is given by (3.4). We assume in what follows that $\rho_\ell < 1$, $\ell = 1,\ldots,L_1$, so that the asymptotic network is stable and the asymptotic performance measures we are considering are finite. It follows from Corollary 1 that $N_{,\ell}(\underset{\sim}{K})$ converges to a finite value so that for large K

$$N_{,\ell}(\underset{\sim}{K}-\underset{\sim}{e}_r) \approx N_{,\ell}(\underset{\sim}{K}) \,. \tag{4.8}$$

It follows from (4.7) that

$$[\lim_{K\to\infty} N_{r,\ell}(\underset{\sim}{K})]/[\lim_{K\to\infty} N_{s,\ell}(\underset{\sim}{K})] = \rho_{r,\ell}/\rho_{s,\ell} \tag{4.9}$$

so that for large K

$$N_{r,\ell}(\underset{\sim}{K})/N_{,\ell}(\underset{\sim}{K}) \approx \rho_{r,\ell}/\rho_\ell \,. \tag{4.10}$$

Substituting the approximations given in (4.8) and (4.10) into (4.7) we obtain for large K the approximation

$$N_{r,\ell}(\underset{\sim}{K}) \approx \frac{\rho_{r,\ell}\left(1 - \frac{N_r(\underset{\sim}{K})}{K_r}\right)}{1-\rho_\ell\left(1 - \frac{N_r(\underset{\sim}{K})}{K_r}\right)} \qquad \ell = 1,\ldots,L_1 \,. \tag{4.11}$$

Finally, summing (4.11) for $\ell = 1,\ldots,L_1$ and summing (4.7) for $\ell = L_1+1,\ldots,L-1$ we have that for large K

$$N_r(\underset{\sim}{K}) \approx \sum_{\ell=1}^{L_1} \frac{\rho_{r,\ell}\left(1-\frac{N_r(\underset{\sim}{K})}{K_r}\right)}{1-\rho_\ell\left(1-\frac{N_r(\underset{\sim}{K})}{K_r}\right)} + \sum_{\ell=L_1+1}^{L-1} \rho_{r,\ell}\left(1-\frac{N_r(\underset{\sim}{K})}{K_r}\right) . \tag{4.12}$$

Thus, $N_r(\underset{\sim}{K})/K_r$ is an approximate solution of the equation $f_r(x) = 0$, where

$$f_r(x) = K_r x - \sum_{\ell=1}^{L_1} \frac{\rho_{r,\ell}(1-x)}{1-\rho_\ell(1-x)} - \sum_{\ell=L_1+1}^{L-1} \rho_{r,\ell}(1-x) . \tag{4.13}$$

Since $f_r(0) < 0$, $f_r(1) > 0$ and $f_r'(x) > 0$ for all x, the equation $f_r(x) = 0$ has a unique solution $x_r(K_r)$ and $0 < x_r(K_r) < 1$. Thus, for large K

$$N_r(\underset{\sim}{K})/K_r \approx x_r(K_r) . \tag{4.14}$$

Substituting the approximation given in (4.14) into (4.11) for $\ell = 1,\ldots,L_1$, into (4.7) for $\ell = L_1+1,\ldots,L-1$, and into (4.3) and (4.1) yields for large K the approximations

$$N_{r,\ell}(\underset{\sim}{K}) \approx \begin{cases} \dfrac{\rho_{r,\ell}[1-x_r(K_r)]}{1-\rho_\ell[1-x_r(K_r)]} & \ell = 1,\ldots,L_1 \\ \\ \rho_{r,\ell}[1-x_r(K_r)] & \ell = L_1+1,\ldots,L-1 \end{cases} \tag{4.15}$$

$$\lambda_r(\underset{\sim}{K}) \approx [1-x_r(K_r)]\lambda_r \tag{4.16}$$

$$T_r(\underset{\sim}{K}) \approx \frac{K_r x_r(K_r)}{[1-x_r(K_r)]\lambda_r} . \tag{4.17}$$

It is straightforward to show that $x_r(K_r)$ is a strictly decreasing function of K_r and that

$$\lim_{K_r\to\infty} x_r(K_r) = 0 \tag{4.18}$$

$$\lim_{K_r\to\infty} K_r x_r(K_r) = \sum_{\ell=1}^{L_1} \frac{\rho_{r,\ell}}{1-\rho_\ell} + \sum_{\ell=L_1+1}^{L-1} \rho_{r,\ell} = N_r(\underset{\sim}{\infty}). \tag{4.19}$$

It follows from (4.18) and (4.19) that the approximations given in (4.15)-(4.17) yield the exact results in the limit as $K \to \infty$.

Next we present examples in which we numerically compare values of $\lambda_r(\underset{\sim}{K})$, $N_{r,\ell}(\underset{\sim}{K})$ and $T_r(\underset{\sim}{K})$ with the approximate values obtained from (4.15)-(4.17). The form of the closed networks we consider is shown in Figure 1. Service center 1 is PS, service centers $2,\ldots,L-1$ are FCFS and service center L is IS. The service rates at service centers $1,\ldots,L-1$ are constants equal to one. Thus, $L_1 = L-1$. A chain r customer which completes service at service center 1 next enters service center ℓ, $\ell = 2,\ldots,L$, with probability $p_{r,\ell} > 0$ where $\sum_{\ell=2}^{L} p_{r,\ell} = 1$. All other branching is deterministic and is shown in Figure 1. Note that all chains visit service center L so that the asymptotic network is open. It is easy to show that

$$\rho_{r,\ell} = \begin{cases} \lambda_r \tau_{r,\ell}/p_{r,L} & \ell = 1 \\ \lambda_r \tau_{r,\ell} p_{r,\ell}/p_{r,L} & \ell = 2,\ldots,L-1. \end{cases} \tag{4.20}$$

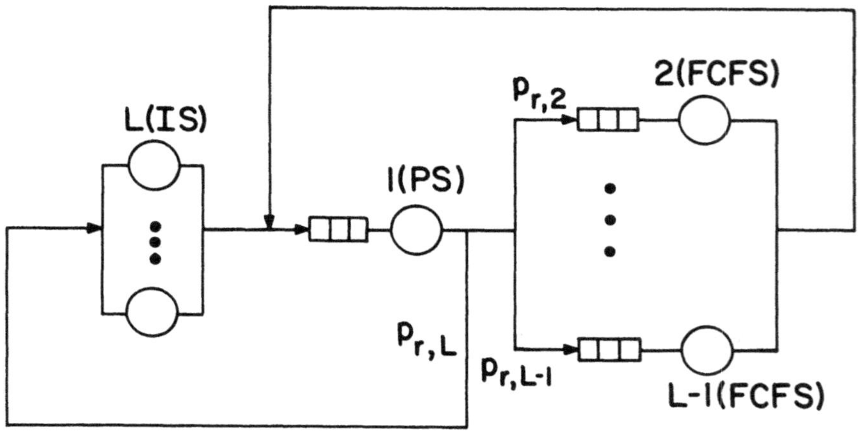

Figure 1. Form of Networks for Examples

Table 1 contains numerical results for a network with $R = 1$, $L = 4$ and parameter values

$$p_2 = .72,\ p_3 = .08,\ p_4 = .2$$

$$\tau_1 = 1,\ \tau_2 = .694,\ \tau_3 = 6.25 \qquad (4.21)$$

(Since there is only one chain we have dropped the chain subscript r. The single remaining subscript, if any, refers to the service center.) For these parameter values

$$\rho_1 = 2\rho_\ell = 5\lambda \quad \ell = 2,3. \qquad (4.22)$$

Values of exact and approximate performance measures are given in Table 1 for 3 values of λ and, for each value of λ, for several values of K. Since $N_2(K) = N_3(K)$, only $N_2(K)$ is given. The rows with $K = \infty$ contain the values of the asymptotic performance measures. If the exact and approximate values differ by more than 1 percent the percentage difference is given in parentheses next to the approximate value. Table 2 contains numerical results for a network with $R = 1$, $L = 10$ and parameter values

$$p_2 = p_3 = p_4 = p_5 = .18,\ p_6 = p_7 = p_8 = p_9 = .02,\ p_{10} = .2$$

$$\tau_1 = 1,\ \tau_2 = \tau_3 = \tau_4 = \tau_5 = 2.78,\ \tau_6 = \tau_7 = \tau_8 = \tau_9 = 25. \qquad (4.23)$$

For these parameter values

$$\rho_1 = 2\rho_\ell = 5\lambda \quad \ell = 2,\dots,9. \qquad (4.24)$$

Since $N_2(K) = N_\ell(K)$, $\ell = 3,\dots,9$, only $N_2(K)$ is given. Finally Table 3 contains numerical results for a network with $R = 2$, $L = 4$ and parameter values

$$p_{1,2} = .72,\ p_{1,3} = .08,\ p_{1,4} = .2$$

Table 1. Performance Measure Values for a Network with R = 1, L = 4 and Parameter Values Given in (4.21)

λ	ρ_1	K	$\lambda(K)$	$N_1(K)$	$N_2(K)$	T(K)
				Exact Values		
.1	.5	50	.0969	.909	.317	15.9
		100	.0984	.951	.325	16.3
		200	.0992	.974	.329	16.5
		400	.0996	.987	.331	16.6
		∞	.1	1	.333	16.7
.15	.75	50	.140	2.10	.533	22.5
		100	.145	2.42	.562	24.5
		200	.147	2.66	.579	25.9
		400	.149	2.81	.589	26.9
		∞	.150	3	.6	28
.18	.9	50	.163	3.42	.674	29.3
		100	.169	4.49	.726	35.1
		200	.174	5.61	.762	41.1
		400	.176	6.65	.786	46.6
		600	.177	7.17	.795	49.4
		∞	.18	9	.818	59.1
				Approximate Values		
.1	.5	50	.0968	.939(3)	.319	16.3(3)
		100	.0984	.968(2)	.326	16.5(1)
		200	.0992	.984(1)	.330	16.6
		400	.0996	.992	.331	16.6
.15	.75	50	.140	2.32(10)	.537	24.3(8)
		100	.144	2.60(7)	.565	25.8(5)
		200	.147	2.78(5)	.581	26.8(3)
		400	.148	2.88(2)	.590	27.4(2)
.18	.9	50	.161(-1)	4.07(19)	.670	33.7(15)
		100	.168	5.24(17)	.724	39.8(13)
		200	.173	6.38(14)	.761	45.7(11)
		400	.176	7.33(10)	.786	50.6(9)
		600	.177	7.77(8)	.795	52.8(7)

Table 2. Performance Measure Values for a Network with $R=1$, $L=10$ and Parameter Values Given in (4.23)

λ	ρ_1	K	$\lambda(K)$	$N_1(K)$	$N_2(K)$	T(K)
				Exact Values		
.1	.5	50	.0935	.851	.303	35.0
		100	.0965	.917	.317	35.8
		200	.0982	.956	.325	36.2
		400	.0991	.977	.329	36.4
		∞	.1	1	.333	36.7
.15	.75	50	.133	1.81	.490	43.1
		100	.140	2.20	.536	46.3
		200	.145	2.51	.565	48.6
		400	.147	2.72	.581	50.1
		∞	.15	3	.6	52.0
.18	.9	50	.153	2.73	.607	49.7
		100	.163	3.81	.683	56.8
		200	.170	5.01	.736	64.1
		400	.174	6.18	.771	70.8
		600	.176	6.79	.785	74.2
		∞	.18	9	.818	86.4
				Approximate Values		
.1	.5	50	.0934	.876(3)	.305	35.5(1)
		100	.0965	.933(2)	.318	36.0
		200	.0982	.965	.325	36.3
		400	.0991	.982	.329	36.5
.15	.75	50	.132	1.95(8)	.494	44.6(3)
		100	.140	2.34(6)	.539	47.4(2)
		200	.145	2.61(4)	.566	49.4(2)
		400	.147	2.79(3)	.582	50.6
.18	.9	50	.151(-1)	3.11(14)	.608	52.7(6)
		100	.162	4.32(13)	.683	60.3(6)
		200	.170	5.59(12)	.737	67.7(6)
		400	.174	6.75(9)	.771	74.2(5)
		600	.176	7.31(8)	.785	77.2(4)

$p_{2,2} = .72,\ p_{2,3} = .08,\ p_{2,4} = .2$

$\tau_{1,1} = 1,\ \tau_{1,2} = 1.39,\ \tau_{1,3} = 12.5$

$$\tau_{2,1} = 2,\ \tau_{2,2} = 1.39,\ \tau_{2,3} = 12.5 \qquad (4.25)$$

For these parameter values

$\rho_{1,1} = \rho_{1,\ell} = 5\lambda_1 \qquad \ell = 2,3$

$$\rho_{2,1} = 2\rho_{2,\ell} = 10\lambda_2 \qquad \ell = 2,3 \qquad (4.26)$$

Since $N_{r,2}(\underset{\sim}{K}) = N_{r,3}(\underset{\sim}{K})$, $r = 1,2$, only $N_{r,2}(\underset{\sim}{K})$ is given.

The exact finite population size values in the tables were computed using an APL program which implements the mean-value analysis algorithm given in [8]. The program was run under VS APL on an IBM 370/168. The execution time to compute a row of exact vaules in Tables 1 and 2 ranged from approximately 1 second when $K = 50$ to approximately 9 seconds when $K = 600$. The execution times to compute a pair of rows of exact values in Table 3 was approximately 12 seconds when $\underset{\sim}{K} = (50,25)$, 45 seconds when $\underset{\sim}{K} = (100,50)$ and 200 seconds when $\underset{\sim}{K} = (200,100)$. (The execution time was not obtained when $\underset{\sim}{K} = (400,200)$ but would be approximately 4 times that when $\underset{\sim}{K} = (200,100)$). The execution time to compute a row of approximate values in Tables 1 and 2 was approximately .02 seconds and to compute a pair of rows of approximate values in Table 3 was approximately .04 seconds. ($x_r(K_r)$ was computed using Newton's method). Thus, the approximate values are very inexpensive to compute compared to the exact values.

We now discuss the results in the tables. Note that the throughputs converge more quickly to their asymptotic values than do the mean queue sizes and mean response times. The slowest convergence is for the mean queue size at service center 1, the service center with the highest

Table 3. Performance Measure Values for a Network with R = 2, L = 4 and Parameter Values Given in (4.25)

r	λ_r	$\rho_{r,1}$	K_r	$\lambda_r(\underset{\sim}{K})$	$N_{r,1}(\underset{\sim}{K})$	$N_{r,2}(\underset{\sim}{K})$	$T_r(\underset{\sim}{K})$
				Exact Values			
1	.05	.25	50	.0488	.464	.379	25.1
2	.025	.25	25	.0242	.456	.188	34.4
1			100	.0494	.481	.389	25.5
2			50	.0246	.476	.194	35.2
1			200	.0497	.490	.395	25.7
2			100	.0248	.488	.197	35.6
1			∞	.05	.5	.4	26
2			∞	.025	.5	.2	36
1	.075	.375	50	.0711	1.11	.735	36.2
2	.0375	.375	25	.0348	1.06	.360	51.2
1			100	.0729	1.25	.787	38.8
2			50	.0360	1.22	.389	55.6
1			200	.0739	1.36	.819	40.5
2			100	.0367	1.34	.407	58.7
1			∞	.075	1.5	.857	42.9
2			∞	.0375	1.5	.429	62.9
1	.09	.45	50	.0829	1.86	1.03	47.3
2	.045	.45	25	.0400	1.75	.499	68.7
1			100	.0858	2.41	1.15	55.0
2			50	.0419	2.31	.564	82.1
1			200	.0875	2.96	1.24	62.2
2			100	.0432	2.88	.613	95.3
1			400	.0886	3.46	1.30	68.5
2			200	.0439	3.41	.646	107
1			∞	.09	4.5	1.38	80.8
2			∞	.045	4.5	.692	131

(Table 3 continued on next page)

(Table 3 continued)

r	λ_r	$\rho_{r,1}$	K_r	$\lambda_r(\underset{\sim}{K})$	$N_{r,1}(\underset{\sim}{K})$	$N_{r,2}(\underset{\sim}{K})$	$T_r(\underset{\sim}{K})$
				Approximate Values			
1	.05	.25	50	.0488	.476(3)	.384(1)	25.2
2	.025	.25	25	.0242	.467(2)	.189	35.0(2)
1			100	.0494	.487(1)	.392	25.6
2			50	.0246	.483(1)	.194	35.5
1			200	.0497	.494	.396	25.9
2			100	.0248	.491	.197	35.7
1	.075	.375	50	.0709	1.22(10)	.757(3)	38.5(6)
2	.0375	.375	25	.0347	1.14(8)	.362	53.6(5)
1			100	.0728	1.34(7)	.802(2)	40.4(4)
2			50	.0360	1.28(5)	.390	57.3(3)
1			200	.0738	1.41(4)	.828(1)	41.5(2)
2			100	.0367	1.38(3)	.408	59.8(2)
1	.09	.45	50	.0820(-1)	2.29(23)	1.07(4)	53.8(14)
2	.045	.45	25	.0397	1.94(11)	.492(-1)	73.5(7)
1			100	.0853	2.89(20)	1.18(3)	61.7(12)
2			50	.0417	2.52(9)	.558(-1)	87.2(6)
1			200	.0873	3.44(16)	1.26(2)	68.4(10)
2			100	.0431	3.10(8)	.608	100.(5)
1			400	.0885	3.86(12)	1.32(2)	73.4(7)
2			200	.0439	3.60(6)	.643	111.(4)

asymptotic traffic intensity. The convergence of all performance measures is slower for larger asymptotic traffic intensities. (For $\rho_1 = .9$ in Table 1, $N_1(600)$ is only within 20% of its asymptotic value). The approximate values are much closer to the exact values than are the asymptotic values. The percentage difference between the exact and approximate values increases with decreasing population sizes and/or increasing asymptotic traffic intensities. The difference is greatest for the mean queue sizes at service center 1 and is least for the throughputs and the other mean queue sizes. In summary, the approximations appear to be very accurate for the throughputs and for the mean queue sizes at service centers whose asymptotic traffic intensity is .5 or less. The approximations are least accurate for the mean queue sizes at service centers with high asymptotic traffic intensities (e.g., .9 or more). However, even if the approximations are not highly accurate, they are still useful when the exact values are expensive to compute.

The approximations we have developed can be extended to the case where the asymptotic network is mixed rather than open. We present this extension in Appendix B. It is worth investigating improving our approximations (in a similar manner to that used by Chandy and Neuse [3] to improve Schweitzer's [10] approximation) while still avoiding having to solve simultaneous nonlinear equations. Further investigation is also required to obtain useful approximations for networks in which some service centers have limited queue size dependent service rates.

5. Acknowledgement

I wish to thank P. Heidelberger and C. H. Sauer for their helpful coments.

6. Appendix A

Proof of Theorem 1: Using (2.1), (2.2), (3.1) and (3.3), and replacing $\tau_{r,L}$ by K_r/λ_r for all $r \in R(L)$, it follows that

$$q(\underset{\sim}{k}^*,\underset{\sim}{K}) = \phi_{A^*(\underset{\sim}{K})}(\underset{\sim}{k}^*)\gamma(\underset{\sim}{n},\underset{\sim}{K}^0)\left[\prod_{\ell=1}^{L-1} \delta_\ell(\underset{\sim}{k}_\ell)\right]/g^*(\underset{\sim}{K}) \qquad \underset{\sim}{k}^* \in A^*(\underset{\sim}{\infty},\underset{\sim}{K}^1) \tag{6.1}$$

where $\phi_{A^*(\underset{\sim}{K})}(\underset{\sim}{k}^*)$ is the indicator function of the set $A^*(\underset{\sim}{K})$, $\underset{\sim}{n} = (n_r\colon r \in R(L))$,

$$\gamma(\underset{\sim}{n},\underset{\sim}{K}^0) = \prod_{r\in R(L)} \frac{K_r!}{(K_r-n_r)!K_r^{n_r}}, \tag{6.2}$$

$\delta_\ell(\underset{\sim}{k}_\ell)$ is given by (3.7), $\rho_{r,\ell}$ is given by (3.11) and $g^*(\underset{\sim}{K})$ is a normalizing constant related to $g(\underset{\sim}{K})$ by

$$g(\underset{\sim}{K}) = g^*(\underset{\sim}{K}) \prod_{r\in R(L)} \frac{(\Theta_{r,L}K_r/\lambda_r)^{K_r}}{K_r!}. \tag{6.3}$$

By definition

$$g^*(\underset{\sim}{K}) = \sum_{\underset{\sim}{k}^*\in A^*(\underset{\sim}{\infty},\underset{\sim}{K}^1)} \phi_{A^*(\underset{\sim}{K})}(\underset{\sim}{k}^*)\gamma(\underset{\sim}{n},\underset{\sim}{K}^0)\left[\prod_{\ell=1}^{L-1} \delta_\ell(\underset{\sim}{k}_\ell)\right] \tag{6.4}$$

It is straightforward to show that for each $\underset{\sim}{k}^* \in A^*(\underset{\sim}{\infty},\underset{\sim}{K}^1)$, $\phi_{A^*(\underset{\sim}{K})}(\underset{\sim}{k}^*)\gamma(\underset{\sim}{n},\underset{\sim}{K}^0)$ is a nondecreasing function of each component of $\underset{\sim}{K}^0$, and hence is a nondecreasing function of K, and

$$\lim_{K\to\infty} \phi_{A^*(\underset{\sim}{K})}(\underset{\sim}{k}^*)\gamma(\underset{\sim}{n},\underset{\sim}{K}^0) = 1. \tag{6.5}$$

Therefore, it follows from the monotone convergence theorem that

$$\lim_{K\to\infty} g^*(\underset{\sim}{K}) = S(\underset{\sim}{K}^1) \tag{6.6}$$

where

$$S(\underset{\sim}{K}^1) = \sum_{\underset{\sim}{k}^*\varepsilon A^*(\underset{\sim}{\infty},\underset{\sim}{K}^1)} \left[\prod_{\ell=1}^{L-1} \delta_\ell(\underset{\sim}{k}_\ell)\right] \tag{6.7}$$

We next evaluate $S(\underset{\sim}{K}^1)$. Let $A^1(\underset{\sim}{K}^1) = \{\underset{\sim}{k}^1: n_r = K_r,\ r \varepsilon \overline{R(L)}\}$. Using the notation of Section 3

$$S(\underset{\sim}{K}^1) = \sum_{\underset{\sim}{k}^1\varepsilon A^1(\underset{\sim}{K}^1)} \sum_{\underset{\sim}{k}^0 \geq \underset{\sim}{0}} \left[\prod_{\ell=1}^{L-1} \delta_\ell(\underset{\sim}{k}_\ell^0,\underset{\sim}{k}_\ell^1)\right]$$

$$= \sum_{\underset{\sim}{k}^1\varepsilon A^1(\underset{\sim}{K}^1)} \prod_{\ell=1}^{L-1} \left[\sum_{\underset{\sim}{k}_\ell^0 \geq \underset{\sim}{0}} \delta_\ell(\underset{\sim}{k}_\ell^0,\underset{\sim}{k}_\ell^1)\right] \tag{6.8}$$

where

$$\sum_{\underset{\sim}{k}_\ell^0 \geq \underset{\sim}{0}} \delta_\ell(\underset{\sim}{k}_\ell^0,\underset{\sim}{k}_\ell^1) = \left[\prod_{j=1}^{k_\ell^1} \frac{1}{\mu_\ell(j)}\right] k_\ell^1! \left[\prod_{r\varepsilon R^1(\ell)} \frac{(\rho_{r,\ell})^{k_{r,\ell}}}{k_{r,\ell}!}\right] g_\ell(k_\ell^1) \tag{6.9}$$

and

$$g_\ell(k_\ell^1) = \sum_{\underset{\sim}{k}_\ell^0 \geq \underset{\sim}{0}} \left[\prod_{j=k_\ell^1+1}^{k_\ell^1+k_\ell^0} \frac{1}{\mu_\ell(j)}\right] \frac{(k_\ell^0+k_\ell^1)!}{k_\ell^1!} \prod_{r\varepsilon R^0(\ell)} \frac{(\rho_{r,\ell})^{k_{r,\ell}}}{k_{r,\ell}!}$$

$$= \sum_{k_\ell^0=0}^{\infty} \left[\prod_{j=k_\ell^1+1}^{k_\ell^1+k_\ell^0} \frac{1}{\mu_\ell(j)}\right] \frac{(k_\ell^0+k_\ell^1)!}{k_\ell^1!\,k_\ell^0!} (\rho_\ell^0)^{k_\ell^0}. \tag{6.10}$$

It follows from the ratio test for convergence of a sum that $g_\ell(k_\ell^1) < \infty$ if $\rho_\ell^0/\mu_\ell < 1$ and $g_\ell(k_\ell^1) = \infty$ if $\rho_\ell^0/\mu_\ell > 1$. The test is inconclusive if $\rho_\ell^0/\mu_\ell = 1$. Thus, if $\rho_\ell^0 < \mu_\ell$ for each $\ell \neq L$ then $S(\underset{\sim}{K}^1) = G(\underset{\sim}{K}^1) < \infty$, where $G(\underset{\sim}{K}^1)$ is the normalizing constant in (3.14), and (3.13) follows from (6.1), (6.5) and (6.6). If $\rho_\ell^0 > \mu_\ell$ for some $\ell \neq L$ then $S(\underset{\sim}{K}^1) = \infty$ and (3.18) follows from (6.1), (6.5) and (6.6).

We now prove (3.17). Using (2.1), (2.2), (3.1) and (3.3), and replacing $\tau_{r,L}$ by K_r/λ_r for all $r \varepsilon R(L)$ it follows that if $s \varepsilon R(L)$ and $K_s \geq 1$ then

$$q(\underset{\sim}{k}^*,\underset{\sim}{K}-\underset{\sim}{e}_s) = \phi_{A^*(\underset{\sim}{K}-\underset{\sim}{e}_s)}(\underset{\sim}{k}^*)\left(\frac{K_s-1}{K_s}\right)^{n_s}\gamma(\underset{\sim}{n},\underset{\sim}{K}^0-\underset{\sim}{e}_s)\left[\prod_{\ell=1}^{L-1}\delta_\ell(\underset{\sim}{k}_\ell)\right]/g_s^*(\underset{\sim}{K})$$

$$\underset{\sim}{k}^* \in A^*(\underset{\sim}{\infty},\underset{\sim}{K}^1) \qquad (6.11)$$

where $g_s^*(\underset{\sim}{K})$ is a normalizing constant related to $g(\underset{\sim}{K}-\underset{\sim}{e}_s)$ by

$$g(\underset{\sim}{K}-\underset{\sim}{e}_s) = g_s^*(\underset{\sim}{K})(\lambda_s/\Theta_{s,L})\prod_{r\in R(L)}\frac{(\Theta_{r,L}K_r/\lambda_r)^{K_r}}{K_r!}\ . \qquad (6.12)$$

It can be shown in the same way as above that

$$\lim_{K\to\infty} g_s^*(\underset{\sim}{K}) = S(\underset{\sim}{K}^1) \qquad (6.13)$$

where the limit is finite if $\rho_\ell^0 < \mu_\ell$ for each $\ell \neq L$. It follows from (2.3), (6.3) and (6.12) that if $s \in R^0(\ell)$ then

$$\lambda_{s,\ell}(\underset{\sim}{K}) = \lambda_s(\Theta_{s,\ell}/\Theta_{s,L})(g_s^*(\underset{\sim}{K})/g^*(\underset{\sim}{K}))\,. \qquad (6.14)$$

The first equality in (3.17) follows by taking the limit of (6.14) as $K \to \infty$. The second equality in (3.17) is even simpler to establish and its proof is omitted.

Proof of Corollary 1: Using (6.1),

$$E[f(\underset{\sim}{m}(\underset{\sim}{K}))] = \left[\sum_{\underset{\sim}{k}^*\in A^*(\underset{\sim}{\infty},\underset{\sim}{K}^1)} \phi_{A^*(\underset{\sim}{K})}(\underset{\sim}{k}^*)f(\underset{\sim}{k}^*)\gamma(\underset{\sim}{n},\underset{\sim}{K}^0)\prod_{\ell=1}^{L-1}\delta_\ell(\underset{\sim}{k}_\ell)\right]/g^*(\underset{\sim}{K})\,. \qquad (6.15)$$

If $\rho_\ell^0 < \mu_\ell$ for each $\ell \neq L$ then $\lim_{K\to\infty} g^*(\underset{\sim}{K}) = G(\underset{\sim}{K}^1) < \infty$ and by assumption

$$E[|f(\underset{\sim}{m})|] = \left[\sum_{\underset{\sim}{k}^*\in A^*(\underset{\sim}{\infty},\underset{\sim}{K}^1)} |f(\underset{\sim}{k}^*)|\prod_{\ell=1}^{L-1}\delta_\ell(\underset{\sim}{k}_\ell)\right]/G(\underset{\sim}{K}^1) < \infty\,. \qquad (6.16)$$

Since $|\phi_{A^*(\underset{\sim}{K})}(\underset{\sim}{k}^*)f(\underset{\sim}{k}^*)\gamma(\underset{\sim}{n},\underset{\sim}{K}^0)| \leq |f(\underset{\sim}{k}^*)|$ for all $\underset{\sim}{k}^* \in A^*(\underset{\sim}{\infty},\underset{\sim}{K}^1)$ it follows from the dominated convergence theorem that the numerator in (6.15) converges to $E[f(\underset{\sim}{m})]G(\underset{\sim}{K}^1)$ which is finite. The denominator in (6.15) converges to $G(\underset{\sim}{K}^1)$ which completes the proof.

7. Appendix B

Here we extend the approximations developed in Section 4 to the case where all chains do not visit service center L. Recall in this case that as $K \to \infty$, $\underset{\sim}{K}^0 \to \underset{\sim}{\infty}$ while $\underset{\sim}{K}^1$ is fixed. In this case if $r \in R(L)$ then (4.3) and (4.7) hold, the approximation given in (4.8) holds for large K and for large K the approximation given in (4.10) becomes

$$N_{r,\ell}(\underset{\sim}{K})/N^0_{,\ell}(\underset{\sim}{K}) \approx \rho_{r,\ell}/\rho^0_\ell \tag{7.1}$$

where $N^0_{,\ell}(\underset{\sim}{K}) = \sum_{r \in R(L)} N_{r,\ell}(\underset{\sim}{K})$. It can then be shown in a similar manner to that used in Section 4 that if $r \in R(L)$ then for large K

$$N_{r,\ell}(\underset{\sim}{K}) \approx \frac{\rho_{r,\ell}\left(1 - \frac{N_r(\underset{\sim}{K})}{K_r}\right)[1 + N^1_{,\ell}(\underset{\sim}{K})]}{1-\rho^0_\ell\left(1 - \frac{N_r(\underset{\sim}{K})}{K_r}\right)} \qquad \ell = 1,\ldots,L_1 \tag{7.2}$$

and

$$N_r(\underset{\sim}{K}) \approx \sum_{\ell=1}^{L_1} \frac{\rho_{r,\ell}\left(1 - \frac{N_r(\underset{\sim}{K})}{K_r}\right)[1 + N^1_{,\ell}(\underset{\sim}{K})]}{1-\rho^0_\ell\left(1 - \frac{N_r(\underset{\sim}{K})}{K_r}\right)} + \sum_{\ell=L_1+1}^{L-1} \rho_{r,\ell}\left(1 - \frac{N_r(\underset{\sim}{K})}{K_r}\right) \tag{7.3}$$

where $N^1_{,\ell}(\underset{\sim}{K}) = \sum_{r \in \overline{R(L)}} N_{r,\ell}(\underset{\sim}{K})$. Thus, $N_r(\underset{\sim}{K})/K_r$ is an approximate solution of the equation $g_r(x) = 0$ where

$$g_r(x) = K_r x - \sum_{\ell=1}^{L_1} \frac{\rho_{r,\ell}(1-x)[1 + N^1_{,\ell}(\underset{\sim}{K})]}{1-\rho^0_\ell(1-x)} + \sum_{\ell=L_1+1}^{L-1} \rho_{r,\ell}(1-x). \tag{7.4}$$

Let $\underset{\sim}{N}^1(\underset{\sim}{K}) = (N^1_{,1}(\underset{\sim}{K}),\ldots, N^1_{,L_1}(\underset{\sim}{K}))$. The equation $g_r(x) = 0$ has a unique solution $x_r(K_r,\underset{\sim}{N}^1(\underset{\sim}{K}))$ and $0 < x_r(K_r,\underset{\sim}{N}^1(\underset{\sim}{K})) < 1$. Thus, for large K if $r \in R(L)$ then

$$N_r(\underset{\sim}{K})/K_r \approx x_r(K_r,\underset{\sim}{N}^1(\underset{\sim}{K})). \tag{7.5}$$

However, $\underset{\sim}{N}^1(\underset{\sim}{K})$ is unknown.

We next show how to recursively compute approximations for $N_{r,\ell}(\underset{\sim}{K})$, denoted $\hat{N}_{r,\ell}(\underset{\sim}{K})$, $r \in \overline{R(L)}$, $\ell = 1,\ldots,L_1$, where the recursion is over all population vectors $\underset{\sim}{J} = (\underset{\sim}{K}^0,\underset{\sim}{J}^1)$ such that $\underset{\sim}{0} \le \underset{\sim}{J}^1 \le \underset{\sim}{K}^1$. It follows directly from the mean-value analysis equations (e.g., see equations (3.1)-(3.4) in [8]) that if $r \in \overline{R(L)}$ and $J_r \ge 1$ then

$$\lambda_{r,1}(\underset{\sim}{K}^0,\underset{\sim}{J}^1) = J_r/\left[\sum_{\ell=1}^{L-1}(\theta_{r,\ell}\tau_{r,\ell}/\theta_{r,1}) + \sum_{\ell=1}^{L_1}(\theta_{r,\ell}\tau_{r,\ell}/\theta_{r,1})N_{,\ell}(\underset{\sim}{K}^0,\underset{\sim}{J}^1-\underset{\sim}{e}_r)\right] \tag{7.6}$$

and

$$N_{r,\ell}(\underset{\sim}{K}^0,\underset{\sim}{J}^1) = \lambda_{r,1}(\underset{\sim}{K}^0,\underset{\sim}{J}^1)(\theta_{r,\ell}\tau_{r,\ell}/\theta_{r,1})[1+N_{,\ell}(\underset{\sim}{K}^0,\underset{\sim}{J}^1-\underset{\sim}{e}_r)] \qquad \ell = 1,\ldots,L_1 \tag{7.7}$$

If $r \in \overline{R(L)}$ and $J_r = 0$ then of course $N_{r,\ell}(\underset{\sim}{K}^0,\underset{\sim}{J}^1) = 0$, $\ell = 1,\ldots,L_1$. For $k = 0,1,\ldots,|K^1|$ let

$$\mathcal{J}(k) = \{\underset{\sim}{J}^1\colon \underset{\sim}{0} \le \underset{\sim}{J}^1 \le \underset{\sim}{K}^1; |\underset{\sim}{J}^1| = k\} \tag{7.8}$$

where $|\underset{\sim}{V}|$ denotes the sum of the components of a vector $\underset{\sim}{V}$. The starting point in the recursive procedure is $\hat{N}_{r,\ell}(\underset{\sim}{K}^0,\underset{\sim}{0}) = 0, r \in \overline{R(L)}$, $\ell = 1,\ldots,L_1$. At the k-th step in the procedure, $k = 1,\ldots,|\underset{\sim}{K}^1|$, $\hat{N}_{r,\ell}(\underset{\sim}{K}^0,\underset{\sim}{J}^1)$ is computed for all $\underset{\sim}{J}^1 \in \mathcal{J}(k)$, $r \in \overline{R(L)}$, $\ell = 1,\ldots,L_1$ from $\hat{N}_{r,\ell}(\underset{\sim}{K}^0,\underset{\sim}{J}^1)$ for all $\underset{\sim}{J}^1 \in \mathcal{J}(k-1)$, $r \in \overline{R(L)}$, $\ell = 1,\ldots,L_1$ as follows:

For $\underset{\sim}{J}^1 \in \mathcal{J}(k)$ and $r \in \overline{R(L)}$ if $J_r = 0$ then

$$\hat{N}_{r,\ell}(\underset{\sim}{K}^0,\underset{\sim}{J}^1) = 0 \qquad \ell = 1,\ldots,L_1 \tag{7.9}$$

and if $J_r \ge 1$ then

$$\hat{N}^1_{,\ell}(\underset{\sim}{K}^0,\underset{\sim}{J}^1-\underset{\sim}{e}_r) = \sum_{s\in\overline{R(L)}} \hat{N}_{s,\ell}(\underset{\sim}{K}^0,\underset{\sim}{J}^1-\underset{\sim}{e}_r) \qquad \ell = 1,\ldots,L_1 \tag{7.10}$$

$$\hat{N}_{s,\ell}(\underset{\sim}{K}^0,\underset{\sim}{J}^1-\underset{\sim}{e}_r) = \frac{\rho_{s,\ell}[1-x_s(K_s,\hat{\underset{\sim}{N}}^1(\underset{\sim}{K}^0,\underset{\sim}{J}^1-\underset{\sim}{e}_r))][1+\hat{N}^1_{,\ell}(\underset{\sim}{K}^0,\underset{\sim}{J}^1-\underset{\sim}{e}_r)]}{1-\rho^0_\ell[1-x_s(K_s,\hat{\underset{\sim}{N}}^1(\underset{\sim}{K}^0,\underset{\sim}{J}^1-\underset{\sim}{e}_r))]}$$

$$s \in R(L), \quad \ell = 1,\ldots,L_1 \qquad (7.11)$$

$$\hat{N}_{,\ell}(\underset{\sim}{K}^0,\underset{\sim}{J}^1-\underset{\sim}{e}_r) = \left[\sum_{s\in R(L)} \hat{N}_{s,\ell}(\underset{\sim}{K}^0,\underset{\sim}{J}^1-\underset{\sim}{e}_r)\right] + \hat{N}^1_{,\ell}(\underset{\sim}{K}^0,\underset{\sim}{J}^1-\underset{\sim}{e}_r)$$

$$\ell = 1,\ldots,L_1 \qquad (7.12)$$

and $\hat{N}_{r,\ell}(\underset{\sim}{K}^0,\underset{\sim}{J}^1)$, $\ell = 1,\ldots,L_1$, is obtained by substituting $\hat{N}_{,\ell}(\underset{\sim}{K}^0,\underset{\sim}{J}^1-\underset{\sim}{e}_r)$ for $N_{,\ell}(\underset{\sim}{K}^0,\underset{\sim}{J}^1-\underset{\sim}{e}_r)$, $\ell = 1,\ldots,L_1$, in (7.6) and (7.7). Note that (7.11) is a consequence of (7.2) and (7.5). At any step of the procedure $\hat{\lambda}_r(\underset{\sim}{K}^0,\underset{\sim}{J}^1)$, $r \in R(L)$, can be obtained based on (4.3), $\hat{\lambda}_{r,1}(\underset{\sim}{K}^0,\underset{\sim}{J}^1)$, $r \in \overline{R(L)}$, can be obtained based on (7.6) and for any r

$$\hat{N}_{r,\ell}(\underset{\sim}{K}^0,\underset{\sim}{J}^1) = \hat{\lambda}_{r,\ell}(\underset{\sim}{K}^0,\underset{\sim}{J}^1)\tau_{r,\ell} \qquad \ell = L_1+1,\ldots,L-1. \qquad (7.13)$$

Finally, $\hat{T}_r(\underset{\sim}{K}^0,\underset{\sim}{J}^1)$, $r \in R(L)$, can be obtained based on (4.1). The computational complexity of the approximation procedure is proportional to the product of the components of $\underset{\sim}{K}^1$, whereas the computational complexity of the exact mean-value analysis procedure is proportional to the product of the components of $\underset{\sim}{K} = (\underset{\sim}{K}^0,\underset{\sim}{K}^1)$. Thus, the approximations should be much less expensive to compute than are the exact values when K is large.

We now prove that as $K \to \infty$ the approximations converge to the exact asymptotic values. We denote an asymptotic performance measure by replacing $\underset{\sim}{K}^0$ by $\underset{\sim}{\infty}$. Equation (4.7) holds for $r \in R(L)$ while (7.6) and (7.7) hold for $r \in \overline{R(L)}$. Taking the limit as $K \to \infty$ in these equations, then provided $\rho^0_\ell < 1$, $\ell = 1,\ldots,L_1$, it follows that if $r \in R(L)$ then

$$N_{r,\ell}(\underset{\sim}{\infty},\underset{\sim}{J}^1) = \rho_{r,\ell}[1 + N_{,\ell}(\underset{\sim}{\infty},\underset{\sim}{J}^1)] \qquad \ell = 1,\ldots,L_1 \qquad (7.14)$$

and if $r \in \overline{R(L)}$ then (7.6) and (7.7) hold with $\underset{\sim}{K}^0$ replaced by $\underset{\sim}{\infty}$. It follows from (7.14) that

$$N_{,\ell}(\underset{\sim}{\infty},\underset{\sim}{J}^1) = (N^1_{,\ell}(\underset{\sim}{\infty},\underset{\sim}{J}^1) + \rho^0_\ell)/(1-\rho^0_\ell) \qquad \ell = 1,\ldots,L_1. \tag{7.15}$$

Equations (7.6) and (7.7) with $\underset{\sim}{K}^0$ replaced by $\underset{\sim}{\infty}$ and (7.15) can be used to recursively compute the exact asymptotic mean queue sizes $N_{r,\ell}(\underset{\sim}{\infty},\underset{\sim}{K}^1)$, $r \in \overline{R(L)}$, $\ell = 1,\ldots,L_1$, where the recursion is over all $\underset{\sim}{J}^1$ such that $\underset{\sim}{0} \le \underset{\sim}{J}^1 \le \underset{\sim}{K}^1$. It follows from (7.14) and (7.15) that if $r \in R(L)$ then

$$N_{r,\ell}(\underset{\sim}{\infty},\underset{\sim}{J}^1) = \rho_{r,\ell}[1 + N^1_{,\ell}(\underset{\sim}{\infty},\underset{\sim}{J}^1)]/(1-\rho^0_\ell) \qquad \ell = 1,\ldots,L_1. \tag{7.16}$$

We wish to show that for all $\underset{\sim}{J}^1 \in \mathcal{J}(k)$, $k = 0,\ldots,|\underset{\sim}{K}^1|$,

$$\lim_{K\to\infty} N_{r,\ell}(\underset{\sim}{K}^0,\underset{\sim}{J}^1) = N_{r,\ell}(\underset{\sim}{\infty},\underset{\sim}{J}^1) \qquad r \in \overline{R(L)}, \quad \ell = 1,\ldots,L_1. \tag{7.17}$$

Clearly (7.17) holds for $k = 0$ since $\mathcal{J}(0) = \underset{\sim}{0}$ and $\hat{N}_{r,\ell}(\underset{\sim}{K}^0,\underset{\sim}{0}) = 0$, $r \in \overline{R(L)}$. We now show that if (7.17) holds for all $\underset{\sim}{J}^1 \in \mathcal{J}(k-1)$ then it holds for all $\underset{\sim}{J}^1 \in \mathcal{J}(k)$, $k = 1,\ldots,|\underset{\sim}{K}^1|$. Consider any $\underset{\sim}{J}^1 \in \mathcal{J}(k)$. Since $\underset{\sim}{J}^1 - \underset{\sim}{e}_r \in \mathcal{J}(k-1)$, $r \in \overline{R(L)}$, it can be shown that if $r \in \overline{R(L)}$ and $s \in R(L)$ then

$$\lim_{K\to\infty} x_s(K_s,\underset{\sim}{\hat{N}}^1(\underset{\sim}{K}^0,\underset{\sim}{J}^1-\underset{\sim}{e}_r)) = 0. \tag{7.18}$$

Hence, from (7.11), (7.17) and (7.18), if $r \in \overline{R(L)}$ and $s \in R(L)$ then

$$\lim_{K\to\infty} \hat{N}_{x,\ell}(\underset{\sim}{K}^0,\underset{\sim}{J}^1-\underset{\sim}{e}_r) = \rho_{s,\ell}[1 + N^1_{,\ell}(\underset{\sim}{\infty},\underset{\sim}{J}^1-\underset{\sim}{e}_r)]/(1-\rho^0_\ell)$$

$$= N_{s,\ell}(\underset{\sim}{\infty},\underset{\sim}{J}^1-\underset{\sim}{e}_r) \qquad \ell = 1,\ldots,L_1. \tag{7.19}$$

Therefore, for all $\underset{\sim}{J}^1 \in \mathcal{J}(k)$ if $r \in \overline{R(L)}$ then

$$\lim_{K\to\infty} \hat{N}_{,\ell}(\underset{\sim}{K}^0,\underset{\sim}{J}^1-\underset{\sim}{e}_r) = N_{,\ell}(\underset{\sim}{\infty},\underset{\sim}{J}^1-\underset{\sim}{e}_r), \qquad \ell = 1,\ldots,L_1. \tag{7.20}$$

Since $\hat{N}_{r,\ell}(\underset{\sim}{K}^0,\underset{\sim}{J}^1)$, $r \in \overline{R(L)}$, $\ell = 1,\ldots,L_1$, is obtained by substituting $\hat{N}_{,\ell}(\underset{\sim}{K}^0,\underset{\sim}{J}^1-\underset{\sim}{e}_r)$ for $N_{,\ell}(\underset{\sim}{K}^0,\underset{\sim}{J}^1-\underset{\sim}{e}_r)$ in (7.6) and (7.7) it follows from

(7.20) that (7.17) holds for all $\underset{\sim}{J}^1 \in \mathcal{J}(k)$. It is now trivial to show that as $K \to \infty$ the approximate throughputs, the approximate mean queue sizes at service centers $L_1+1,\ldots,L-1$, and the approximate mean response times also converge to the exact asymptotic values.

8. References

[1] Bard, Y., "Some Extensions to Multiclass Queueing Network Analysis," in Performance of Computer Systems, Arato, M., Butrimenko, A. and Gelenbe, E. (editors), North Holland, Amsterdam (1979).

[2] Baskett, F., Chandy, K. M., Muntz, R. R., and Palacios, F. G., "Open, Closed and Mixed Networks of Queues with Different Classes of Customers," J. Assoc. Comput. Mach., 22, 248-260 (1975).

[3] Chandy, K. M. and Neuse, D., "Fast Accurate Heuristic Algorithms for Queueing Network Models of Computing Systems," TR-157, Dept. of Computer Sciences, University of Texas at Austin (1980).

[4] Graham, G. S. (Editor), Special Issue: Queueing Network Models of Computer System Performance, ACM Computing Surveys, 10 (1978).

[5] Kelly, F. P., "Networks of Queues with Customers of Different Types," J. Appl. Probab., 12, 542-554 (1975).

[6] Pittel, B., "Closed Exponential Networks of Queues with Saturation: The Jackson-Type Stationary Distribution and Its Asymptotic Analysis," Math. Oper. Res., 4, 357-378 (1979).

[7] Reiser, M. and Kobayashi, H., "Queueing Networks with Multiple Closed Chains: Theory and Computational Algorithms," IBM J. Res. Develop., 19, 283-294 (1975).

[8] Reiser, M. and Lavenberg, S. S., "Mean-Value Analysis of Closed Multichain Queueing Networks," J. Assoc. Comput. Mach., 27, 313-322 (1980).

[9] Sauer, C. H. and Chandy, K. M., Computer Systems Performance Modeling, Prentice-Hall, Englewood Cliffs, New Jersey (1981).

[10] Schweitzer, P., "Approximate Analysis of Multiclass Closed Networks of Queues," (Abstract), Int. Conference on Stochastic Control and Optimization, Amsterdam (1979).

IBM Thomas J. Watson Research Center, Yorktown Heights, New York 10598.

Discussant's Report on
"On Closed Multichain Product Form Queueing Networks
With Large Population Sizes,"
by Stephen Lavenberg

Dr. Lavenberg has written a very clear and interesting paper. The queueing networks considered are closed multichain (multiclass) models of the BCMP type where at least one station is an infinite server (IS) center. Briefly, the paper establishes two results. First, Lavenberg proves that in the limit as the population size of such networks approaches infinity, the IS station behaves as a Poisson source (with a chain dependent rate) feeding jobs into the remainder of the network. This result is what one would expect. (His analysis also treats the case in which not all chains visit the IS station.)

The second result was the approximation of the asymptotic performance measures (throughputs, response times, and mean queue lengths) as the chain population sizes increase. (The computational complexity of the well established convolution and mean value analysis algorithms to determine these measures is proportional to the product of the population sizes for each chain.) The approximations are derived in a simple, straightforward manner from the (exact) mean value analysis expression for computing the mean queue length by chain at a station.

So often in papers presenting approximation techniques, no analysis is done to provide an upper bound on the error of the approximation. These, in general, are difficult problems. It is interesting to note that the paper by Schweitzer entitled "Bottleneck Determination in Networks of Queues" in this Proceedings is similar in nature to Lavenberg's paper.

The paper (as well as others) points out the importance of mean value analysis in the exact and approximate solution of queueing networks. It is the beauty of this technique that such conceptually clear algorithms as Lavenberg's can be found.

Dr. Lavenberg should be commended for his thorough and rigorous analysis. The total lack of typos attests to the care with which the ideas in this paper were derived and explained. This paper is a model of clarity once the notational conventions have been assimilated.

Discussant: Dr. Steven C. Bruell, Computer Science Department, University of Minnesota, Minneapolis, Minnesota 55455.

THE SIGNIFICANCE OF THE DECOMPOSITION AND THE ARRIVAL THEOREMS FOR THE EVALUATION OF CLOSED QUEUEING NETWORKS

M. Reiser

In this paper, we show how the convolution algorithm can be based on the decomposition theorem, and we introduce mean-value analysis as an application of the arrival theorem. In the queue-dependent case, the same recursive formula is at the heart of both algorithms. This formula relates the marginal probabilities, $p_i(k,K) \sim p_i(k-1, K-1)$, where i denotes a server, k is the queue size, and K denotes the population of the closed network. Thus, it is this recursion which relates seemingly very different computation methods. We also discuss properties of the two methods and disucss how they were extended heuristically to give approximate results for networks more general than the product-form class.

1. Introduction

Closed queueing-network models with product-form solution [1-3] have found widespread applications in performance analysis of computer systems and data communication networks.

Efficient numerical methods were first reported by Buzen [4] for a single class problem, and independently by Reiser and Kobayashi [5-7] who generalized it to multiclass (or multichain) problems.

It is the purpose of this paper to survey existing algorithms. Each algorithm is reduced to a fundamental property of the queueing network model such as conditioned system equivalent, the arrival theorem and full decomposability. We show how, once the underlying probabilistic arguments are exposed, heuristic extensions can be motivated.

The earliest algorithm is known as the convolution algorithm. Reiser gave a probabilistic interpretation in terms of an open queueing-network problem conditioned to contain a given number of customers [7]. This interpretation is based on the observation:

> *An open queueing system Q conditioned to contain exactly K customers is equivalent to a certain closed queueing network Q(K) with exactly K customers.*

The parameters of the closed network are easily obtained from the open one. However, many different open systems may have the same conditioned solution. The viewpoint of a conditioned open system is actually more representative or practical modeling problems. For example, a computer system conditioned to a certain multiprogramming level is such an example. If $p(\underline{k})$ is the probability distribution of state $\underline{k}$ in Q then the law of conditional probability yields

$$p(\underline{k},K) = Pr\{\underline{k}|\Sigma k_i=K\} = p(\underline{k})/g(K), \quad (1)$$

where k_i is the queue size at server i, and g(K) is the probability for the event: Q contains exactly K customers. Since in exponential systems queues are stochastically independent, we have

$$g(K) = Pr\{\Sigma k_i=K\} = \{p_1\}*\{p_2\}\ldots*\{p_N\} \mid \text{at } K, \quad (2)$$

where $\{p_i\}$ is the queue distribution of server i, and * denotes the convolution operator. Equation (2) allows an efficient calculation of the normalization constant. Note that a naive term-by-term summation of the defining equation for g(K) is infeasible for all but the smallest problems due to the large state space of Q(K).

Recently, a new approach to product-form networks called *mean-value analysis* was described by Reiser and Lavenberg [8-9]. Mean-value analysis is based on the arrival theorem [10-13] which states:

> *In a closed queueing network the (stationary) state probabilities at customer arrival epochs are identical to those of the same network in long-term equilibrium with one customer removed.*

Mean-value analysis is a program to analyze networks by means of the arrival theorem augmented by Little's formula. In the important practical case of closed networks with fixed-rate and time-delay servers, mean-value analysis allows a simple and numerically stable solution entirely in terms of mean queue size, mean waiting time and throughput. This contrasts with the earlier convolution method, which only yielded a joint probability distribution, and hence computation of performance measures had to be done separately.

In this paper, we shall give another interpretation of the convolution algorithm in terms of *complete decomposability* of closed queueing networks with product-form solution. The decomposition theorem states:

> *In a closed queueing network with product-form solution, the marginal probabilities of a given queue i are the same as those produced from an equivalent problem, where all queues other than i (called i-complement) are replaced by a single queue-dependent server. The rate of this equivalent server is the same as the throughput computed from the i-complement.*

The theorem is due to Chandy, Herzog, and Woo [14] (who termed it Norton's theorem) and who gave an algebraic proof. A probabilistic interpretation was given by Courtois [15]. The decomposition theorem allows the solution of a network with N queues by stepwise addition of one queue after another, thus evaluating a series of networks comprised of queues {1}, {1, 2}, {1, 2, 3} ... {1, 2, ..., N}. At each stage, the intermediate system {1, 2, ..., i - 1}, whose results (including throughput) are known for populations k = 1, 2, ... K, is viewed as the i-complement of the system {1, 2, ..., i}. A two-server cyclic problem has to be solved (see Figure 1). Its solution can be case in the form of the convolutions performed in the convolution algorithm.

For simplicity of notation, we restrict the paper to the single-chain case.

2. Definition of the Problem and Notation

We consider closed single-chain queueing networks with product-form solution [1-3]. We denote by Q(K) such a network with K customers and N queues (or service facilities). Q(K) is defined by the following parameters:

$\Theta_i = E[v_i]$: Expectation of v_i, the number of visits a customer makes to queue i between successive visits to i*, an arbitrarily chosen queue.

$\tau_i = E[s_i]$: Expectation of s_i the work brought into queue i at a given arrival epoch (measured in number of instructions to be executed, for example).

$\mu_i(k)$: The work rate of server i as a function of the number of customers queued at a given point in time (measured in instructions executed per second, for example).

If the routing rule is first-order stochastic [1-3] then the vector $\underline{\Theta} = (\Theta_1, \Theta_2, \ldots, \Theta_N)$ is the left-hand-side eigenvector of the eigenvalue one of the routing matrix, normalized such that $\Theta_{i*} = 1$

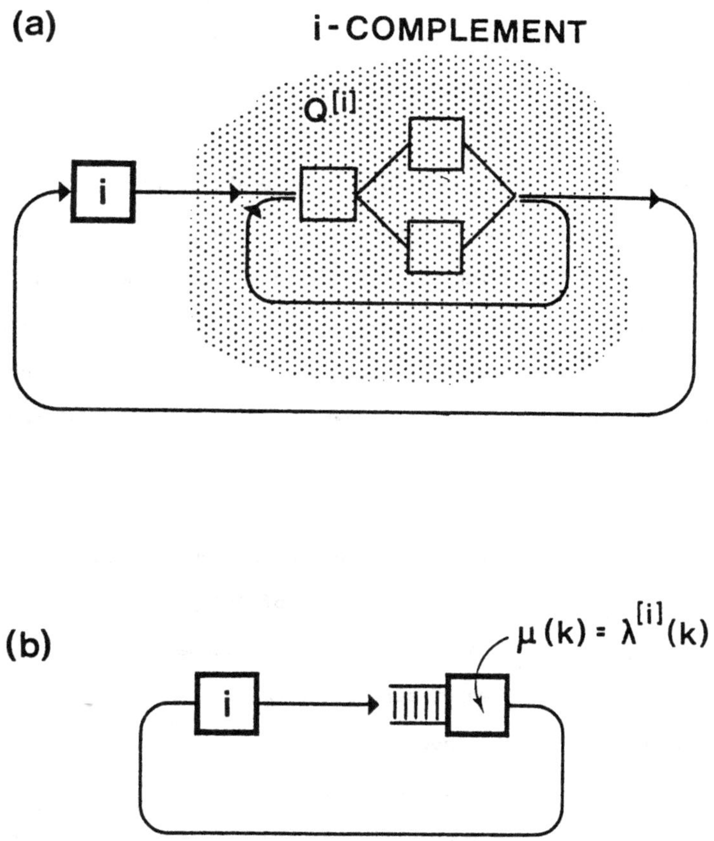

Figure 1. (a). i-complement $Q^{[i]}(K)$ of a queueing system Q(K). (b) Equivalent cyclic system. i-complement replaced by queue-dependent server whose rate equals the throughput $\lambda^{[i]}(k)$ obtained from the isolated i-complement system. (After [18].)

(where $i^* \in \{1, 2, \ldots N\}$ is an arbitrarily labeled queue). More general routing rules compatible with Q(K) are discussed in [16]. Without loss of generality, we choose units such that $\mu_i(1) = 1$ and to define $\mu_i(0) = 1$.

In order for a queueing network to have a product-form solution, assumptions on the service mechanism must be made. Service times s_i are assumed i.i.d. random variables of a Coxian distribution [17] if the queue is governed by the processor-sharing (PS), no-queueing (D) or last-come/first-served (LCFS) discipline. In the case of a first-come/first-served discipline, the extra restriction of exponential service times must be imposed.

We shall also introduce the notations:

k_i: Number of customers at queue i at a given instant in time,

$\underline{k} = (k_1, k_2, \ldots, k_N)$: state vector of Q(K),

$n_i = E[k_i]$: mean number of customers of queue i,

t_i: mean waiting time of server i (including service),

λ: throughput of the labeled queue i*,

$\lambda_i = \Theta_i \lambda$: throughput of queue i,

$p_i(k)$: steady-state probability of the event that there are k customers at queue i,

$w_i = \Theta_i \tau_i$: expectation of the work brought into queue i by a customer between successive visits to the labeled queue i*,

$W_i = \Theta_i t_i$: expected waiting time in queue i between successive visits to the labeled queue i*.

Where needed, we shall use an argument K to denote that the named quantities are for the system Q(K) with K customers, for example, $n_i(K)$, $t_i(K)$, $\lambda(K)$ or $p_i(k,K)$ for the marginal probability distribution of queue i.

We shall also need the notion of the *i-complement of* Q(K) denoted by $Q^{[i]}(K)$ which is defined as the queueing system without the queue i (see Figure 1). Superscripts [i] indicate that the named variables are for the i-complement (e.g., $\lambda^{[i]}(K)$ is the throughput of $Q^{[i]}(K)$, etc.).

3. Solution of Q(K)

The stationary state probabilities $p(\underline{k})$ of a queueing network with product-form solution are

$$p(\underline{k}) = \pi_1(k_1)\pi_2(k_2) \ldots \pi_N(k_N)/g(K), \tag{3}$$

where

$$\pi_i(k) = \frac{w_i^{\,k}}{\mu_i(1)\mu_i(2)\ldots\mu_i(k)} \tag{4}$$

$\pi_i(0) = 1$, and g(K) is a normalization constant defined by the sum $\Sigma\, \pi_i(k_1)\pi_2(k_2) \ldots \pi_N(k_N)$ which extends over the set $F(K) = \{\underline{k}\colon \underline{k} \geq 0$ and $k_1 + k_2 + \cdots + k_N = K\}$.

From [4] and [5] we know that the following relations hold

$$p_i(k,K) = \pi_i(k)g^{[i]}(K-k)/g(K), \tag{5}$$

$$\lambda(K) = g(K-1)/g(K). \tag{6}$$

Their derivation from the product-form solution is straightforward. The marginal distribution of queue i in Q(K), $p_i(k,K)$, is defined

$$p_i(k,K) = \sum_{\substack{F(K)\\ k_i=k}} p(\underline{k},K) = \sum_{\substack{F(K)\\ k_i=k}} \pi_1(k_1)\pi_2(k_2)\ldots\pi_N(k_N)/g(K). \tag{7}$$

Set without loss of generality i = N. Then

$$p_N(k,K) = \pi_N(k) \sum_{F^{[N]}(K-k)} \pi_1(k_1)\pi_2(k_2)\ldots\pi_{N-1}(k_{N-1})/g(K), \tag{8}$$

where $F^{[N]}(j) = \{(k_1, k_2, \ldots k_{N-1})\colon\ k_\ell \geq 0,\ \ell = 1, \ldots N-1,\ \sum_{\ell=1}^{N-1} k_\ell = j\}$. The sum in (8) is the definition of $g^{[N]}(K-k)$. This proves (5). Now assume that the labeled queue $i^* = N$. Then the throughput $\lambda(K)$ is defined by

$$\lambda(K) = \sum_{j=1}^{K} p_N(j,K)\mu_N(j)/w_N = \sum_{j=1}^{K} \frac{g^{[N]}(K-j)}{g(K)} \frac{\mu_N(j)}{w_N} \pi_N(j)$$

$$= \sum_{j=1}^{K} \frac{g^{[N]}(K-j)}{g(K)} \frac{\mu_N(j)}{w_N} \frac{w_N^j}{\mu_N(1)\mu_N(2)\ldots\mu_N(j)}$$

$$= \frac{1}{g(K)} \sum_{j=0}^{K-1} g^{[N]}(K-1-j)\pi_N(j) = \frac{g(K-1)}{g(K)} \tag{9}$$

which proves (6).

The key for mean-value analysis and for the new interpretation of the convolution algorithm is the following recursion, first reported in [9]

$$p_i(k,K) = \frac{w_i\lambda(k)}{\mu_i(k)} p_i(k-1, K-1). \tag{10}$$

We call (10) *mean-value recursion*. It is derived simply from (5) and (6), viz.,

$$p_i(k,K) = \frac{w_i^{\,k}}{\mu_i(1)\mu_i(2)\ldots\mu_i(k)} \frac{g^{[i]}(K-k)}{g(K)}$$

$$= \frac{w_i}{\mu_i(k)} \pi_i(k-1) \frac{g^{[i]}(K-1-[k-1])}{g(K-1)} \frac{g(K-1)}{g(K)}$$

$$= \frac{w_i}{\mu_i(k)} p_i(k-1, K-1)\, \lambda(K), \tag{11}$$

where we utilized (6) in the third step.

Figure 2 shows how the mean-value recursion can be used to calculate all values $p_i(k,K)$, $k = 1, 2, \ldots K$ from the boundary values $p_i(0,K-k)$. Thus, from the empty queue probability $p_i(0,K)$ $K = 0, 1 \ldots K_{max}$, all the marginal distributions for the systems $Q(K)$ $K = 0,1, \ldots K_{max}$ can be derived.

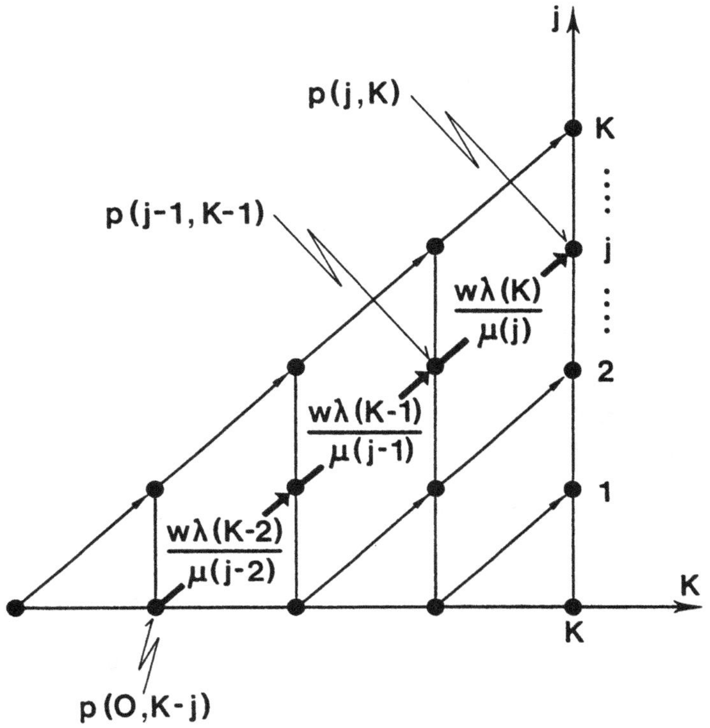

Figure 2. Recursive computation scheme for marginal queue-size probabilities. The quantities w, p(j,k) and μ(j) are understood to be for queue i (subscript i dropped for clarity). (After [18].)

4. Mean-Value Analysis

If Q(K) consists entirely of single-server queues of fixed service rates normalized to $\mu_i = 1$ (i = 1, 2, ...N) then the expectation of the waiting time, t_i, of an arriving customer is given as

$$t_i = \tau_i + \tau_i \cdot \{\text{expected number of customers at arrival epoch}\}. \quad (12)$$

By the arrival theorem, Equation (12) becomes

$$t_i(K) = \tau_i + \tau_i \, n_i(K-1). \quad (13)$$

Little's formula for the set of queues {1, 2, ...N yields}

$\lambda(K)$ = K/{expected time a customer spends in the network between successive arrivals to i*}

$$= K/\sum_{i=1}^{N} \Theta_i t_i(K). \tag{14}$$

Set $W_i(k) = \Theta_i t_i(k)$, then

$$W_i(K) = w_i[1+n_i(K-1)] \tag{15}$$

$$\lambda(K) = K/\sum_{i=1}^{N} W_i(K), \tag{16}$$

$$n_i(K) = \lambda(K)W_i(K), \tag{17}$$

where the last equation is Little's formula for queue i. Starting with $n_i(0) = 0$, Equations (15) to (17) allow a simple recursive calculation of the mean values of the system Q(K). If queue i is of type D (time delay of "infinite servers") then (15) is replaced by $W_i(K) = w_i$.

If queue i has a queue-dependent service rate, then the mean waiting time is no longer determined by what is encountered by an arriving customer, since later arrivals may change the service rate. It is therefore necessary to calculate marginal probabilities in addition to mean values.

Using Little's formula, we may write for the expected waiting time

$$t_i(K) = \frac{n_i(K)}{\lambda_i(K)} = \sum_{j=1}^{K} j\,\frac{p_i(j,K)}{\lambda_i(K)}$$

$$= \sum_{j=1}^{K} \frac{j\tau_i}{\mu_i(j)}\, p_i(j-1,\ K-1), \tag{18}$$

where we utilized (10) to obtain the last equality. The mean-value recursion now becomes

$$p_i^*(k,K) = \frac{w_i}{\mu_i(k)}\, p_i(k-1,\ K-1) \tag{19}$$

$$W_i(K) = \sum_{j=1}^{K} j p_i^*(j, K) \tag{20}$$

$$\lambda(K) = K / \sum_{i=1}^{N} W_i, \tag{21}$$

$$p_i(k, K) = \lambda(K) p_i^*(k, K) \tag{22}$$

$$p_i(0,K) = 1 - \sum_{j=1}^{K} p_i(j,K), \tag{23}$$

where indices i range over i = 1, 2, ...N, indices k range over k = 1, 2, ...K, and p(0,0) = 0 starts the recursion. A diagram of this computation is given in Figure 3.

Through normalization, $p_i(0)$ is obtained for each new value K. If $p_i(0)$ is very small (order of the floating-point accuracy), then large numerical errors may be generated through the subtraction in (24). This numerical instability was reported in [13]. However, the difference in (24) can be avoided if the i-complement system $Q^{[i]}(K)$ is also evaluated. Then $p_i(0)$ follows from (3) and (5) as

$$p_i(0,K) = \frac{g^{[i]}(K)}{g(K)} = \frac{g^{[i]}(K-1)}{g(K-1)} \frac{g^{[i]}(K)}{g^{[i]}(K-1)} \frac{g(K-1)}{g(K)}$$

$$= p_i(0,K-1)\lambda(K)/\lambda^{[i]}(K). \tag{24}$$

Thus at the expense of evaluating the i-complement system, we obtain a stable version of MVA if (24) is used instead of (23). Of course, if there are fixed-rate servers or delay queues in Q(K) then the calculation of marginal probabilities in MVA is optional, and (13) is used to obtain W_i.

5. Convolution Algorithm

The convolution algorithm evaluates the normalization constant g(K) by a sequence of convolutions. One probabilistic interpretation

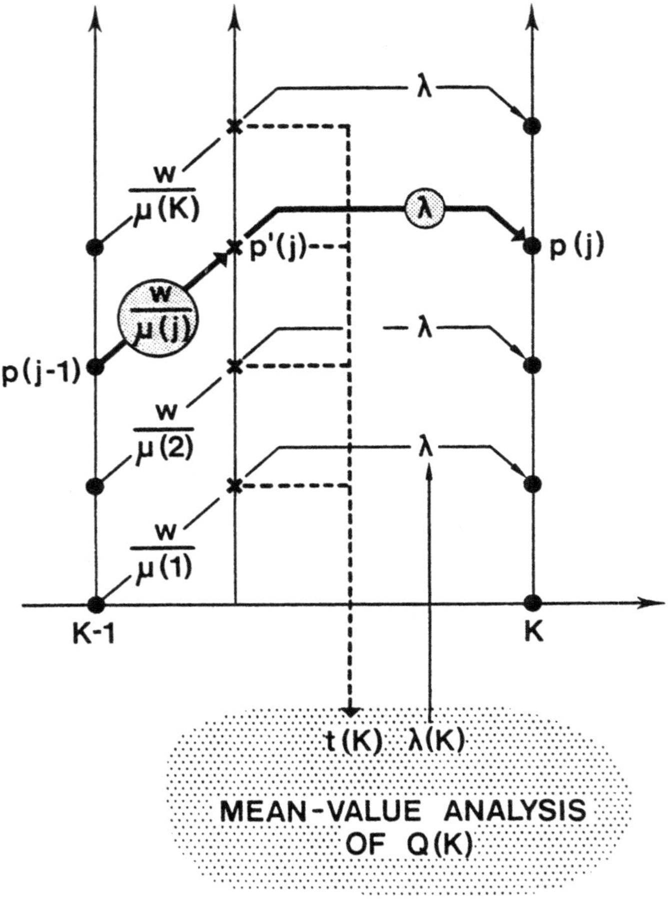

Figure 3. Computation diagram of the mean-value algorithm MVA. (After [18].)

is in terms of an open system with population constraint as was elaborated in [7]. Here we wish to give another interpretation which remains in the domain of closed networks. We view the convolutions as solving a series of networks with queues {1}, {1, 2}, {1, 2, 3}, ... , {1, 2, ... N}. At each stage, the intermediate result is viewed as the complement of a system with one more server. For notational convenience, let us focus on the last step. In this case, the

intermediate results are just those for the N-complement system $Q^{[N]}$.
From (5), we have

$$\sum_{j=0}^{K} p_i(j,K) = 1 = \frac{1}{g(K)} \sum_{j=0}^{K} \pi_i(j) g^{[N]}(K-j)$$

or

$$g(K) = \sum_{j=0}^{K} \pi_N(j) g^{[N]}(K-j). \tag{25}$$

Equation (25) is called convolution algorithm.

We may interpret this successive addition of servers by means of the decomposition theorem. Assume that we have evaluated the system $Q^{[N]}$ of queues $\{1, 2, \ldots, N-1\}$. We represent it by an equivalent queue with variable rates given by the throughput of $Q^{[N]}$. Now, we add queue N. Thus, we need to solve the cyclic system shown in Figure 1(b). The normalization constant of this system is

$$g(K) = \sum_{j=0}^{K} \pi_N(j) \frac{1}{\lambda^{[N]}(1)\lambda^{[N]}(2)\ldots\lambda^{[N]}(K-j)}$$

$$= \sum_{j=0}^{K} \pi_N(j) g^{[N]}(K-j), \tag{26}$$

where we made repetitive use of (6). Equation (26) is identical to (25). Using the same argument as in the proof of (11), we obtain

$$g(K) = g^{[N]}(K) + \sum_{j=1}^{K} \frac{w_N}{\mu_N(j)} \pi_N(j-1) g^{[N]}[K-1-(j-1)]$$

$$= g^{[N]}(K) + \sum_{j=1}^{K} \frac{w_N}{\mu_N(j)} p'_N(j-1, K-1), \tag{27}$$

where $p'_i(j)$ is an unnormalized marginal probability, i.e., $p_i(j) = p'_i(j)/g(K)$. Based on (27), we compute the normalization constants $g(K)$, $k = 1, 2, \ldots$ by the scheme

$$p'_N(0,K) = g^{[N]}(K). \tag{28}$$

$$p_N'(j,K) = p_N'(j-1, K-1)w_N/\mu_N(j) \text{ for } j = 1, 2, \ldots, K, \quad (29)$$

$$g(K) = \sum_{j=0}^{K} p_N'(j). \quad (30)$$

A diagram for the computations (28) to (30) is shown in Figure 4. The similarity with MVA is obvious. The normalization constants g(K) can become very large (especially if delay queues with long delay times are present). This may produce floating-point overflows. The original solution, proposed by Reiser, is scaling [7].

No scaling is needed if we normalize each step in the computations (28) to (30), by g(K-1) which yields

$$p_N''(0) = p_N(0,K-1)/\lambda^{[N]}(K), \quad (31)$$

$$p_N''(j) = p_N(j-1, K-1)w_N/\mu_N(j) \text{ for } j = 1, 2, \ldots K, \quad (32)$$

$$\lambda(K) = 1/\sum_{j=0}^{K} p_N''(j), \quad (33)$$

$$p_N(j,K) = p_N''(j)\lambda(K) \text{ for } j = 0, 1, \ldots K. \quad (34)$$

Note that throughputs $\lambda(K)$ now replace the normalization constrants g(K). We call the computations (31) to (34) normalized convolution algorithm (NCA).

Comparison of Figure 3 with Figure 4 shows the basic similarity of MVA and NCA. In the first case, results for all queues are computed *in parallel* for populations $K = 1, 2, \ldots K_{max}$. In the second case, results for all populations are computed for a sequence of networks with consecutively more servers. A discussion of numerical properties is found in [18].

6. Extensions

It is an interesting observation that all three of the properties of closed queueing networks listed in the introduction were used as starting points for heuristic extensions.

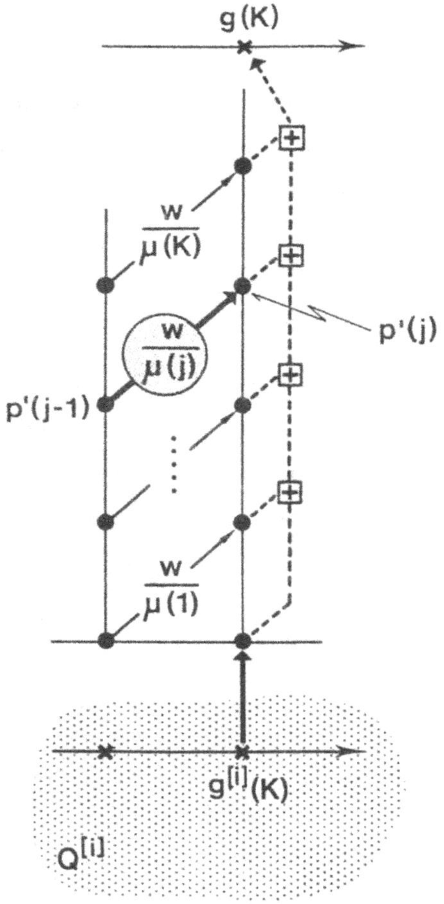

Figure 4. Computation diagram of convolution algorithm in the form of Equations (28) to (30). (After [18].)

Shum used the view of a conditioned open system as basis for her extended product-form technique (EPF) [19]. She considered closed networks with general service-time distributions and probabilistic routing. In an open product-form network, each queue has a queue-size distribution identical to the one of the server, isolated from the network, and subjected to Poisson arrivals. Shum postulated this property for general networks. This allows her to express the marginal distributions through known M/G/1 solutions. Then, assuming independence, the closed-network solution is obtained following (1) - (2). Note that the conditioned solution of (1) is invariant with respect to the arrival rate of the underlying open network. The EPF algorithm, however, does not have this invariance, and Shum discusses the choice of arrival rates which yield good results.

Chandy, Herzog and Woo use the decomposability of closed networks as starting points [14]. Their method is best illustrated assuming that the closed queueing network has just one server with general service-time distribution. The decomposition theorem is now postulated. This allows lumping all exponential servers into an equivalent queue with queue-dependent service rates. Thus, there remains a cyclic queue problem with a general server and a queue-dependent server (as in Figure 1). Such systems are amenable to numerical solutions. If there are several general servers, Chandy, Herzog and Woo give a technique to iteratively isolate one server at a time and lump the other ones.

Both EPF and the iterative technique are reported to yield good results in cases where exponential servers mix with queues having hyperexponential and Erlangian service-time distributions.

In the sequel, we wish to illustrate how MVA motivates heuristic methods.

6.1 Multiserver Queues

Besides servers with fixed rates and pure time delays ("infinite servers"), multiserver stations often occur in practice. A station with m servers each of rate μ may be treated as a queue-dependent server with rate function

$$\mu(k) = \min(k,m)\mu. \tag{35}$$

For notational convenience, in (35) and in the following we omit subscripts referring to queues. Unfortunately, a queue-dependent server requires the complicated general algorithm with its high storage demand and operations count. Using (35), we can rewrite the waiting-time equation (18) as follows (see [9])

$$t(K) = \tau + \frac{\tau}{m} q(K-1) + \frac{\tau}{m} B(K-1), \tag{36}$$

where we assumed without loss of generality $\mu = 1$ and where

$q(K)$: Mean waiting-line size of the multiserver (excluding customers in service) given a population of size K,

$B(K)$: Probability that all m servers are busy given a population of size K.

The mean waiting-line size relates to the mean queue size by

$$q(K) = n(K) - \lambda(K)\tau. \tag{37}$$

Unfortunately, to calculate B(K) we need the marginal probabilities. To avoid the expense of the evaluation, let us estimate B(K) from the known quantities of Q(K). A very simple estimate can be obtained from the well-known M/M/m solution with arrival rate $\lambda(K)$ which is explicitly available in the MVA computation. Using such an estimtae, we obtain an algorithm almost as simple as MVA for fixed-rate servers. The error vanishes for very light and very heavy loads of the multiserver. It is largest if the mean queue size is around m. For a large set of examples, the error was always found below 5%, clearly good enough in practice. The benefit is a very simple and fast algorithm. This approximation shows how the underlying fundamental property guides the modeler to efficient solution. (This example is not part of the earlier works, on which this survey is based.)

6.2 General Servers

In this case, we postulate the arrival theorem. The waiting-time equation may be written as

$$t(K) = \tau + \tau q(K-1) + \{\text{mean residual service time upon arrival}\}. \qquad (38)$$

We estimate the last term by a renewal argument, assuming that arrivals are at random epochs with respect to service intervals. This yields

$$t(K) = \tau + \tau q(K-1) + u(K-1)\,\frac{m_2}{2\tau}, \qquad (39)$$

where u(K-1) is the utilization of the server and where m_2 denotes the second moment of the service-time distribution. For an exponential distribution, (39) yields $t(K) = \tau + \tau n(K-1)$, the correct mean waiting equation. In the case of constant service times, (39) specializes to

$$t(K) = \tau + \tau[n(K-1) - \frac{1}{2}\,u(K-1)]. \qquad (40)$$

We have not enough experience to comment on the accuracy of this heuristic. In [20], Reiser discusses multichain solutions with exponential and constant servers in the context of communication networks. Useful accuracy was reported in this instance (less than 10% error).

6.3 Multichain Case

Multichain networks are valuable models for computer systems with different application subsystems [21] as well as for communication networks with window flow control [20]. The mean-value algorithm as well as the convolution algorithm generalize easily to the multichain case (see [9]). The operation count, however, is

$$\prod_{r=1}^{R} K^r, \qquad (41)$$

where R is the number of closed chains, and K^r is the population of chain r. This operation count is large and makes feasible only solutions of problems with few chains (less than ten). Mean-value analysis leads the way to a heuristic algorithm of low computational complexity. We motivate the algorithm for a single-chain network (where it is not needed of course). Assume that we can estimate the solution of Q(K-1) from those values arising in Q(K). Then we may rewrite (15) to (17) as follows:

$$n_i(K-1) = f(\lambda(K), n_i(K), i = 1, 2, \ldots N), \tag{42}$$

$$W_i(K) = W_i[1 + n_i(K-1)], \tag{43}$$

$$\lambda(K) = K/\Sigma W_i(K), \tag{44}$$

$$n_i(K) = \lambda(K)W_i(K). \tag{45}$$

Thus, we replaced the recursions (15) to (17) by a set of nonlinear equations. It can be solved by standard methods.

In general, the evaluation of (42) is as complex as the original problem. Not only is nothing gained, quite on the contrary, we iterate where we could solve the problem exactly in one step. Thus, an efficient estimation of the function f in (42) is called for.

We may write

$$n_i(K-1) = n_i(K) - \varepsilon_i(K), \tag{46}$$

where $0 < \varepsilon_i(K) < 1$ and $\Sigma\varepsilon_i(K) = 1$. Thus, the quantities ε_1 measure how the one missing customer in Q(K-1) is distributed over the individual queues. P. Schweitzer suggests [22], that

$$\varepsilon_i(K) \approx \frac{t_i}{T} = \frac{K-1}{K}, \tag{47}$$

where T is the average time between successive visits to the labeled queue i*. With the Schweitzer approximation, sequential iterations through (42) to (45) prove to be an adequate solution method of the system of nonlinear equations. The Schweitzer approximation is simple but yields solutions with less pronounced bottleneck behavior compared to exact solutions.

In [20], Reiser proposes a better approximation to the multichain case which estimates the quantities $\{\varepsilon_i\}$ through a substitute single-chain problem with aggregated parameters. This heuristic is more successful in the representation of bottleneck behavior. It yields exact solutions for the single-chain case.

Again, extensive validation results are still missing. Bard [21] and Reiser [20] report encouraging initial results. It is interesting to observe that the iterative heuristic correctly identifies bottlenecks, as the total population $K = K^1 + K^2 + \ldots + K^R$ tends to infinity such that $K^r/K = \alpha^r$ = const. (see discussion in [9]). The assurance that bottlehecks will be identified accurately is very important, since bottleneck identification is nontrivial in the case of multichain networks. In practice, the location of bottlenecks is in fact valuable even if queue sizes should be less accurate. As a consequence of the robustness with respect to bottlenecks, we also expect throughput results to be more accurate than delays, for example. This expectation seems to be borne out by the data reported in [20].

7. Summary

For product-form networks with fixed and delay servers (no general queue-dependent rates), mean-value analysis is clearly the simplest and most stable evaluation method. If there are such queue-dependent servers, the older convolution algorithm may be as useful. A good approach then seems to solve the fixed rate/delay subnetwork using MVA and then add the queue-dependent serving using NCA. Note that through (6), normalization constants can always be obtained from throughputs.

The area of closed-network heuristics requires further study. Methods should be compared, and regimes of applicability identified.

8. References

[1] J. R. Jackson, "Jobshop-Like Queueing Systems," Management Sci., 10, 1973, pp. 131-142.

[2] W. T. Gordon and G. F. Newell, "Closed Queueing Systems with Exponential Servers," Oper. Res., 15, 1967, pp. 252-265.

[3] F. Baskett, K. M. Chandy, R. R. Muntz, and F. G. Palacios, "Open, Closed and Mixed Networks of Queues with Different Classes of Customers," J. Assoc. Comput. Mach., 22, 1975, pp. 248-260.

[4] J. P. Buzen, "Computational Algorithms for Closed Queueing Networks with Exponential Servers," Comm. ACM, 16, 1973, pp. 527-531.

[5] M. Reiser and H. Kobayashi, "Queueing Networks with Multiple Closed Chains: Theory and Computational Algorithms," IBM J. Res. Develop., 19, 1975, pp. 283-294.

[6] M. Reiser and H. Kobayashi, "Horner's Rule for the Evaluation of General Closed Queueing Networks," Comm. ACM, 18, 1975, pp. 592-593.

[7] M. Reiser, "Numerical Methods in Separable Queueing Networks," Studies in the Management Sci., 7, 1977, pp. 113-142.

[8] M. Reiser, "Mean Value Analysis of Queueing Networks, A New Look at an Old Problem," Proc. 4th Intl. Symp. on Modeling and Perf. Eval. of Computer Systems, Vienna, 1979.

[9] M. Reiser and S. S. Lavenberg, "Mean Value Analysis of Closed Multi-chain Queueing Networks," J. Assoc. Comput. Mach., 22, 1980, pp. 313-322.

[10] S. S. Lavenberg and M. Reiser, "Stationary State Probabilities of Arrival Instants for Closed Queueing Networks with Multiple Types of Customers," J. Appl. Probab., to appear, 1980.

[11] K. C. Sevcik and I. Mitrani, "The Distribution of Queueing Network States at Input and Output Instants," Proc. 4th Intl. Symp. on Modeling and Perf. Eval. of Computer Syst., Vienna, 1979.

[12] J. P. Buzen and P. J. Denning, "Operational Treatment of Queue Distributions and Mean Value Analysis," Technical Report, Purdue University, 1979.

[13] K. M. Chandy and C. H. Sauer, "Computational Algorithms for Product Form Queueing Networks," IBM Research Report RC7950, IBM Thomas J. Watson Research Center, Yorktown Heights, New York, U.S.A., 1980 (also in Proceedings Supplement, Performance 80, ACM Press, Toronto, Ontario, Canada).

[14] K. M. Chandy, U. Herzog and L. Woo, "Parametric Analysis of Queueing Networks," IBM J. Res. Develop., 19, 1975, pp. 36-42.

[15] P. F. Courtois, Decomposability: Queueing and Computer System Application, (ACM Monograph Series) Academic Press, New York, 1977.

[16] H. Kobayashi and M. Reiser, "On Generalization of Job Routing Behavior in a Queueing Network Model," IBM Research Report RC5252, IBM Thomas J. Watson Research Center, Yorktown Heights, New York, U.S.A. 1975.

[17] D. R. Cox, "A Use of Complex Probabilities in the Theory of Stochastic Processes," Proc. Cambridge Phil. Soc., 51, 1955. pp. 131-139.

[18] M. Reiser, "Mean Value Analysis and Convolution Method for Queue-Dependent Servers in Closed Queueing Networks," Performance Evaluation, 1, 1981, pp. 7-18.

[19] A. W. Shum, "Queueing Models for Computer Systems with General Service Time Distribution," Ph.D. Thesis, Harvard University, 1976.

[20] M. Reiser, "A Queueing Network Analysis of Computer Communication Networks with Window Flow Control," IEEE Trans. Comm., COM-27, 1979, pp. 1199-1209.

[21] Y. Bard, "Some Extensions to Multiclass Queueing Network Analysis," Proc. 4th Intl. Symp. on Modeling and Perf. Eval. of Computer Systems, IFIP "6 7.3, 1979, North Holland Publ. Co.

[22] P. Schweitzer, private communication.

IBM Zurich Research Laboratory, 8803 Rüschlikon, Switzerland.

ON COMPUTING THE STATIONARY PROBABILITY VECTOR OF A NETWORK OF TWO COXIAN SERVERS

William J. Stewart

Abstract

This paper presents an efficient direct method for the solution of a queueing network consisting of two Coxian servers. The structure of the transition rate matrix associated with this network is examined in some detail, and it is shown how advantage may be taken of this structure to derive the required stationary probability vector in a number of operations which is proportional to N, the number of customers in the network. The direct method which is employed, is based on a single step of inverse iteration. The method is shown to be stable.

1. Introduction

In recent years a considerable research effort has been spent in investigating approximate methods for the solution of almost general queueing networks. This research has been spurred by the emergence of a consensus that significant advances in analytical techniques for solving general queueing networks are still some way off, and by a demand for methods which compute solutions to models of rather complex systems but to an accuracy commensurate only with that of the modelling process itself.

One class of approximate solution method which has attracted much attention is that termed "iterative". These methods were initially developed by Chandy, Herzog and Woo [CHW 75] and later extended by Marie [MARI 78]. Fundamental to these methods is the concept of a "complementary" server. Each station of the network is, in turn, isolated and the remainder of the network replaced by a single exponential server called the complementary server. The service rate of this server is computed from an "auxilary" network which is topologically equivalent to the original network, has the same number of customers, but in which all servers are exponential. The complementary server is supposed to capture the interactions between the selected station and the rest of the network. In other words, when the "reduced" network consisting of the chosen station and the complementary exponential server is analyzed, the probability measures obtained for the chosen station should (in the best of worlds) be identical to the corresponding measures for that station in the original network. When each of the stations in the network has been so analyzed, certain network parameters are calculated and if prespecified tolerance criteria are satisfied, the process halts; otherwise the auxilary network is altered to reflect the information obtained in the previous iteration and the process gone through once again. It has been proved by Stewart and Zeiszler [STEW 80] that an exponential server (or even a load dependent exponential server) cannot, in general, exactly model the interactions between a Coxian server and the remainder of a network. If better accuracy is required a more general composite server must be used. One possibility would be to use a composite Coxian server, but this would involve solving a reduced network consisting of two Coxian servers. The purpose of the present paper is to show how this might be achieved.

Since at least one two-server Coxian network must be solved in each iteration of the global iterative process, it is essential that the solution technique be highly efficient and stable. In this paper we present such an algorithm for numerically solving the global balance equations. We show that the complexity of the algorithm is O(N) where N is the number of customers in the network; in fact, the number of operations required when each of the Coxian servers has two fictitious service stages is given by 82N. We note [MARI 78] that in cases where only the first two moments of the service time distribution are of importance, a two stage Coxian representation is sufficient to model any server for which the service time distribution has a rational Laplace transform and coefficient of variation $\geq$ 0.5.

A further area in which the efficient solution of a network consisting of two Coxian servers will be of value is related to the work of Zahorjan [ZARO 77]. His approximation technique consists of decomposing networks of M stations into (M-1) multiple components consisting of two stations each; obtaining the solution for each of these components and then combining these solutions to obtain an approximate overall solution. The stationary probability vector corresponding to each component is computed by numerically solving the global balance equations. This paper provides an efficient way of performing this step of his algorithm.

Before proceeding further, attention should be called to the work of Marcel F. Neuts, [NEUT 81]. Neuts and his co-workers have for some time now used numerical techniques to obtain elegant solutions to a variety of queueing systems, although, in general, the type of system he considers consists only of a single server. However, instead of representing the service time distribution at a server by a law of Cox as we do in this paper, Neuts chooses to use phase type distributions. Both may be used to approximate any distribution arbitrarily closely. In this paper, however, it is implicitly assumed that the number of stages in the representation of a Coxian server is finite and this is undoubtedly a limitation. In practice however this will not be a major restriction since it is possible with the algorithm presented in this paper, to choose quite a large number of stages and still require only a modest amount of computation steps and computer storage.

We point out that it may well be possible to solve the network discussed here by the methodology of Neuts. Additionally, it may also be possible, by considering various embedded processes (for example, at epochs of departure of server 1 or 2) to reduce the dimensionality of the problem. Although neither of these approaches were examined in this paper, it is likely that they will provide fruitful areas for future research. Finally, the interested reader may find the analysis employed by Wong, Giffin and Disney, [WONG 77], to obtain the steady state solution for two finite M/M/1 queues in tandem, to be of value.

It is easy to verify that the stationary probability vector P of a general queueing network may be obtained as the solution of the homogeneous system of linear equations $S^T P = 0$, where the matrix S is of order equal to the number of states in the network and whose elements s_{ij}, $i \neq j$, denote the rate of transition from state i to state j. Note that $s_{ij} \geq 0$ for all i, j with $i \neq j$. It is usual to define the diagonal element in any row to be equal to the negated sum of the off-diagonal elements in that row,

i.e., $s_{ii} = -\sum_{j \neq i}^{n} s_{ij}$

Alternatively, by letting $W^T = S^T \Delta t + I$, the problem may be formulated as an eigenvalue problem in which the stationary probability vector P is obtained as the right eigenvector corresponding to a dominant unit eigenvalue; i.e., $(S^T \Delta t + I)P = P$. It is usual to choose Δt so that W^T is stochastic and has only one eigenvalue of unit modulus. This formulation is due to Wallace and is discussed in the Recursive Queue Analyzer [WALL 66].

In the next section we consider the structure of the transition rate matrix of a network consisting of two Coxian servers of order r_1 and r_2 respectively. Of particular interest is its block tridiagonal structure and the fact that most of the blocks are identical. This feature leads to considerable saving both in computation time and memory requirements. Some theoretical aspects of the method employed for the computation of the stationary probability vector are discussed in section 3, while in section 4, details concerning its actual implementation are discussed and a complexity argument is presented. Finally, in section 5, some conclusions are drawn.

2. The Transition Rate Matrix

Consider a closed queueing network consisting of two single server stations. We assume that the service time distribution at each server has a rational Laplace transform so that each may be modelled as a Coxian server. The queueing discipline is first-come first-served and a fixed number, N, of customers circulates within the network. This is illustrated in figure 1 below.

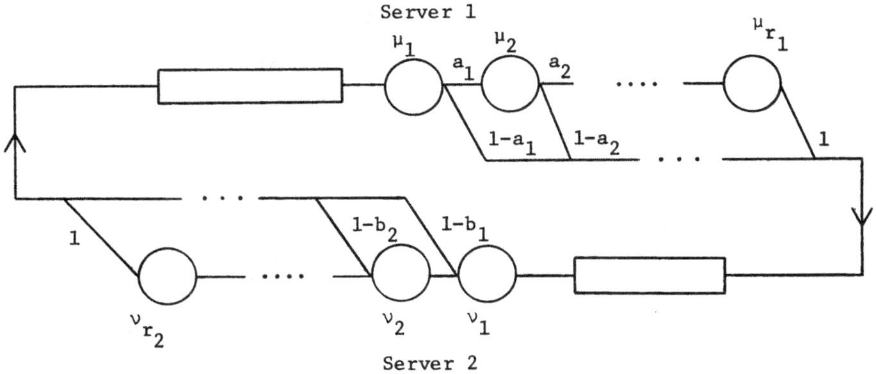

Figure 1. Network of Two Coxian Servers

Server 1 (respectively, server 2) is represented by a law of Cox of order r_1, (r_2), in which the mean service rate at stage i is μ_i, (ν_i). Upon completion of service at stage i, the customer proceeds to stage i+1 with probability a_i, (b_i), or leaves the server with probability $1-a_i$, $(1-b_i)$. By convention a_{r_1} and b_{r_2} are chosen to be zero.

We note that the state of the system at any time t is completely specified by the quadruple (n_1, i, n_2, j). Here n_1 and n_2 denote respectively the number of customers at server 1 and server 2 and since the total number of customers in the network is N, we must have $n_1 + n_2 = N$. The parameters i and j denote the current phase of service at server 1 and server 2 respectively. Obviously $0 \leq i \leq r_1$ and $0 \leq j \leq r_2$ and are equal to zero only if the corresponding server is idle.

It is convenient to arrange the states of the system, first in decreasing order of the number of customers at server 1, second according to the service phase of server 1 and finally according to the service phase of server 2. This is illustrated in figure 2.

There are a total of (N-1) blocks of states with n $(0 < n < N)$ customers at the first station and it may be observed from figure 2 that each of these blocks contains a total of $r_1 r_2$ states. In addition, there are r_1 states in which all customers are at the first station $(n = N)$ and r_2 states in which all N are at the second station $(n = 0)$. Consequently, the total number of states of the system is given by

$$m = (N-1)r_1 r_2 + r_1 + r_2.$$

In particular, if both servers are second order Coxians, the number of states is given by 4N. The position of any arbitrary state (n, i, N-n, j) in this list of states may be obtained from the relation:

$$\text{posn}(n,i,N-n,j) = \begin{cases} i, & \text{if } n = N; \\ r_1 + (N-n-1)r_1r_2 + (i-1)r_2 + j & \text{if } 0 < n < N; \\ r_1 + (N-1)r_1r_2 + j, & \text{if } n = 0. \end{cases}$$

$(n_1$	i	n_2	$j)$
N	1	0	0
$\vdots$	$\vdots$	$\vdots$	$\vdots$
$n+1$	r_1	$N-n-1$	r_2-1
$n+1$	r_1	$N-n-1$	r_2
n	1	$N-n$	1
n	1	$N-n$	2
$\vdots$	$\vdots$	$\vdots$	$\vdots$
n	1	$N-n$	r_2
n	2	$N-n$	1
n	2	$N-n$	2
$\vdots$	$\vdots$	$\vdots$	$\vdots$
n	2	$N-n$	r_2
$\vdots$	$\vdots$	$\vdots$	$\vdots$
n	r_1	$N-n$	r_2-1
n	r_1	$N-n$	r_2
$n-1$	1	$N-n+1$	1
$n-1$	1	$N-n+1$	2
$\vdots$	$\vdots$	$\vdots$	$\vdots$
0	0	N	r_2

Figure 2. The Ordering of the States of the Network

From an arbitrary state of the system, the following transitions can occur:

Transition	Rate of Transition.
1) A customer leaves stage i, $(i \neq r_1)$ of server 1 and proceeds to state i+1.	$\mu_i a_i$
2) A customer leaves station 1 from stage i.	$\mu_i(1-a_i)$
3) A customer leaves stage j, $(j \neq r_2)$ of server 2 and proceeds to stage j+1.	$\nu_j b_j$
4) A customer leaves station 2 from stage j.	$\nu_j(1-b_j)$

This information is sufficient to generate any row of the matrix S^T. We point out that the (i-j) element of S^T denotes the rate of transition to state i from state j, so that when generating row i of S^T we need to determine all states from which a single step transition leads to state i. It may be shown that the transition rate matrix has the block tridiagonal structure

$$S^T = \begin{bmatrix} D_0 & U_0 & & & & & \\ L_0 & D & U & & \underline{\text{Zero}} & & \\ & L & D & U & & & \\ & & \cdot & \cdot & \cdot & & \\ & & & L & D & U & \\ & \underline{\text{Zero}} & & & L & D & U_N \\ & & & & & L_N & D_N \end{bmatrix}$$

The submatrices are dimensioned as follows: L, D, and U are of order $r_1 r_2$.

$$D_0 \in \mathbb{R}^{r_1 x r_1}; \quad L_0 \in \mathbb{R}^{r_1 r_2 x r_1}; \quad U_0 \in \mathbb{R}^{r_1 x r_1 r_2}$$

$$D_N \in \mathbb{R}^{r_2 x r_2}; \quad L_N \in \mathbb{R}^{r_2 x r_1 r_2}; \quad U_N \in \mathbb{R}^{r_1 r_2 x r_2}$$

Blocks which are not specifically labelled must contain only zero elements since the contrary would imply that two transitions from one station to the second could occur in a single time interval. Also the interior blocks are identical since the rates of transition are independent of the number of customers in the queue. The structure of this matrix is the single most critical factor which effects the efficiency of the proposed numerical algorithm and consequently it is important to study the tridiagonal blocks of S^T in some detail. We will consider the blocks L, D and U, first in the particular case where $r_1 = 4$ and $r_2 = 3$, and then in the general case.

<u>The diagonal blocks, D:</u> $r_1 = 4,\ r_2 = 3.$

	$(n_1\ 1\ n_2\ 1)$	. 1 . 2	. 1 . 3	. 2 . 1	. 2 . 2	. 2 . 3	. 3 . 1	. 3 . 2	. 3 . 3	. 4 . 1	. 4 . 2	. 4 . 3
$(n_1\ 1\ n_2\ 1)$	*											
. 1 . 2	$b_1\nu_1$	*										
. 1 . 3		$b_2\nu_2$	*									
. 2 . 1	$a_1\mu_1$			*								
. 2 . 2		$a_1\mu_1$		$b_1\nu_1$	*							
. 2 . 3			$a_1\mu_1$		$b_2\nu_2$	*						
. 3 . 1				$a_2\mu_2$			*					
. 3 . 2					$a_2\mu_2$		$b_1\nu_1$	*				
. 3 . 3						$a_2\mu_2$		$b_2\nu_2$	*			
. 4 . 1							$a_3\mu_3$			*		
. 4 . 2								$a_3\mu_3$		$b_1\nu_1$	*	
. 4 . 3									$a_3\mu_3$		$b_2\nu_2$	*

More generally, we have

$$D = \begin{bmatrix} D_{11} & & & & \\ D_{21} & D_{22} & & \underline{\text{Zero}} & \\ & D_{32} & D_{33} & & \\ & & \cdot & \cdot & \\ & \underline{\text{Zero}} & & & \\ & & & D_{r_1 r_1 - 1} & D_{r_1 r_1} \end{bmatrix} \in \mathbb{R}^{r_1 r_2 \times r_1 r_2}$$

in which all the subblocks are of order r_2.

The diagonal subblocks D_{ii} ($i = 1, 2, \ldots r_1$) contain non-zero elements only along the diagonal and subdiagonal. Each diagonal element of D_{ii} is also the diagonal element of the overall transition rate matrix and equals the negated sum of the off-diagonal elements (of S^T) in the column in which it occurs. In general, the diagonal elements of D will differ. The non-zero elements of the subblocks D_{ii} represent possible transitions among states which differ only in the service phase of the second server. Naturally the only transitions possible among these states are those which take server 2 from one service phase to the next. Consequently, the subdiagonal elements (k, k-1) of any of the subblocks, D_{ii}, are given by $b_{k-1}\nu_{k-1}$, $k = 2, 3, \ldots r_2$. Non-zero elements of subdiagonal blocks, $D_{i\ i-1}$, represent transitions from states in which server 1 is in phase i-1 to states in which it is in phase i. Therefore, taking into account the ordering which is imposed on the states, it follows that the subdiagonal blocks, $D_{i\ i-1}$ of D, must be defined by

$$D_{i\ i-1} = a_{i-1}\mu_{i-1}\ I, \quad i = 2, 3, \ldots r_1.$$

<u>The subdiagonal blocks, L:</u> $r_1 = 4,\ r_2 = 3.$

	$(n_1{+}1\ 1\ n_2{-}1\ 1)$	. 1 . 2	. 1 . 3	. 2 . 1	. 2 . 2	. 2 . 3	. 3 . 1	. 3 . 2	. 3 . 3	. 4 . 1	. 4 . 2	. 4 . 3
$(n_1\ 1\ n_2\ 1)$	$(1-a_1)\mu_1$			$(1-a_2)\mu_2$			$(1-a_3)\mu_3$			$(1-a_4)\mu_4$		
. 1 . 2		$(1-a_1)\mu_1$			$(1-a_2)\mu_2$			$(1-a_3)\mu_3$			$(1-a_4)\mu_4$	
. 1 . 3			$(1-a_1)\mu_1$			$(1-a_2)\mu_2$			$(1-a_3)\mu_3$			$(1-a_4)\mu_4$
. 2 . 1												
. 2 . 2												
. 2 . 3												
. 3 . 1												
. 3 . 2						<u>ZERO</u>						
. 3 . 3												
. 4 . 1												
. 4 . 2												
. 4 . 3												

More generally, we have

$$L = \begin{bmatrix} L_{11} & L_{12} & \cdots & L_{1r_1} \\ & & & \\ & \underline{\text{Zero}} & & \\ & & & \end{bmatrix} \in \mathbb{R}^{r_1 r_2 x r_1 r_2}$$

in which all subblocks are of order r_2.

The non-zero elements of L denote transitions which occur when a customer finishes service at any phase of server 1 and joins the queue at server 2. The next customer at server 1 immediately enters service phase 1 and hence all subblocks L_{ij} for $i > 1$ must contain only zero elements. In addition, the arrival of a customer to server 2 does not alter the service phase of that server and thus only the diagonal elements of the blocks L_{ij} are non-zero. These blocks are given by

$$L_{ij} = (1-a_j)\mu_j\, I, \quad j = 1, 2, \ldots r_1.$$

<u>The superdiagonal blocks, U:</u> $r_1 = 4,\ r_2 = 3.$

$$
\begin{array}{c|ccc|ccc|ccc|ccc|}
 & (n_1{-}1\ 1\ n_2{+}1\ 1) & .\ 1\ .\ 2 & .\ 1\ .\ 3 & .\ 2\ .\ 1 & .\ 2\ .\ 2 & .\ 2\ .\ 3 & .\ 3\ .\ 1 & .\ 3\ .\ 2 & .\ 3\ .\ 3 & .\ 4\ .\ 1 & .\ 4\ .\ 2 & .\ 4\ .\ 3 \\
\hline
(n_1\ 1\ n_2\ 1) & \alpha_1 & \alpha_2 & \alpha_3 & & & & & & & & & \\
.\ 1\ .\ 2 & & & & & & & & & & & & \\
.\ 1\ .\ 3 & & & & & & & & & & \text{ZERO} & & \\
\hline
.\ 2\ .\ 1 & & & & \alpha_1 & \alpha_2 & \alpha_3 & & & & & & \\
.\ 2\ .\ 2 & & & & & & & & & & & & \\
.\ 2\ .\ 3 & & & & & & & & & & & & \\
\hline
.\ 3\ .\ 1 & & & & & & & \alpha_1 & \alpha_2 & \alpha_3 & & & \\
.\ 3\ .\ 2 & & & & & & & & & & & & \\
.\ 3\ .\ 3 & & & \text{ZERO} & & & & & & & & & \\
\hline
.\ 4\ .\ 1 & & & & & & & & & & \alpha_1 & \alpha_2 & \alpha_3 \\
.\ 4\ .\ 2 & & & & & & & & & & & & \\
.\ 4\ .\ 3 & & & & & & & & & & & & \\
\hline
\end{array}
$$

where

$$\alpha_1 = (1-b_1)\nu_1$$

$$\alpha_2 = (1-b_2)\nu_2$$

$$\alpha_3 = (1-b_3)\nu_3 = \nu_3$$

More generally, we have

$$U = \begin{bmatrix} U' & & & & \underline{Zero} \\ & U' & & & \\ & & \cdot & & \\ & & & U' & \\ \underline{Zero} & & & & U' \end{bmatrix} \quad \varepsilon\ \mathbb{R}^{r_1 r_2 x r_1 r_2}$$

in which all subblocks are of order r_2. The (i-j)th element of each of the diagonal submatrices is given by

$$U'_{ij} = (1-b_j)\nu_j\delta_{i1} \text{ where } \delta_{ij} \text{ is the kronecker delta.}$$

This structure is a result of the fact that non-zero elements of U indicate transitions that occur when a customer finishes service at any phase of server 2 and proceeds to join the queue at server 1. The particular pattern of the non-zero elements arises since server 2 must recommence at service phase 1 while the service phase of server 1 remains unchanged.

In conclusion, the transition rate matrix for the two station network has a unique block tridiagonal structure, is easy to generate and requires very little storage. Three arrays of declared dimensions (r_1r_2 x r_1r_2) may be used to hold blocks L_0, D_0 and U_0 initially; L, D and U throughout the central portion of the algorithm, and finally blocks L_N, D_N, and U_N. Alternatively, the algorithm may be written in a form that requires only the non-zero elements of the blocks to be stored, in which case L, D and U, for example, require only

$r_1r_2 + 2(r_1 + r_2 - 1)$ storage positions. This is discussed in more detail in the following section where it becomes evident that the algorithm proceeds by generating a row of the transition rate matrix, operating on this row, and then moving onto the next row. All that is required then is the ability to determine efficiently and in sequence, the rows of S^T.

3. The Numerical Method

When solving the homogeneous system of linear equations $S^TP = 0$ or the equivalent stochastic eigenvalue problem $W^TP = P$, $(W^T = S^T\Delta t + I)$, it is usual to employ numerical iterative methods. A number of factors favour an iterative approach over a direct approach, viz: advantage may be taken of good approximations to the solution vector, storage requirements may be minimized by the use of compact storage schemes, and the iterative procedure may be halted once a prescribed tolerance criteria has been met. However, iterative methods commonly suffer from the failing that the number of iterations required to achieve even a small number of decimal places of accuracy is very large. On the other hand, direct methods not only require large amounts of time in the general network case, but also considerable amounts of memory due largely to "fill-in"--the process by which zero elements become non-zero due to the addition of rows in the reduction stage. In the special case of a two station network however, the structure of the matrix is such that fill-in is restricted to a narrow band along the diagonal and therefore memory requirements need not be excessive. Also the operation count is considerably reduced, so that direct methods have much to recommend them.

For a non-trivial solution to the set of homogeneous linear equations

$$S^T P = 0$$

the matrix S must be singular and hence ill-conditioned as regards equation solving. However, in general the Markovian model will be ergodic (i.e., the associated stochastic matrix will be irreducible and consequently the matrix S will possess a unique zero eigenvalue), so that it is sufficient to replace one of the rows of S^T with the vector $\underset{\sim}{1} = (1, 1, \ldots, 1)$ of length m, equal to the order of the matrix S^T, and to set the corresponding element of the right-hand side also equal to unity. This is equivalent to normalizing the solution vector. It is usual to replace the last row of S^T in this fashion since this will not cause any additional fill-in which would later require to be eliminated, and also because the right-hand side may be ignored until the back substitution is initiated.

An alternative approach is the method of inverse iteration [WILK 65]. Consider an iterative scheme based on the relationship

$$z^{(k)} = (S^T - \mu I)^{-1} z^{(k-1)}.$$

$z^{(0)}$ is arbitrary and may be written in the form:

$$z^{(0)} = \sum_{i=1}^{m} \alpha_i q_i$$

where the q_i are the right eigenvectors of the matrix S^T corresponding to eigenvalues λ_i. Then

$$z^{(k)} = (S^T - \mu I)^{-k} z^{(0)}$$

$$= \sum_{i=1}^{m} \alpha_i (\lambda_i - \mu)^{-k} q_i$$

$$= (\lambda_r - \mu)^{-k} \{\alpha_r q_r + \sum_{i \neq r}^{m} \alpha_i (\lambda_r - \mu)^k (\lambda_i - \mu)^{-k} q_i\}$$

Consequently, if for all $i \neq r$, $|\lambda_r - \mu| \ll |\lambda_i - \mu|$ convergence to the eigenvector q_r is rapid. If $\mu = \lambda_r$ then the summation term equals zero and the vector q_r will be obtained to full machine precision in a single iteration.

It is usually recommended that instead of forming the inverse of the shifted matrix and then postmultiplying it with the trial vector as indicated in the recurrence formula, inverse iteration be conducted by solving the set of linear equations

$$(S^T - \mu I)\, z^{(k)} = z^{(k-1)}$$

If $\mu = \lambda_r$ then it is simply sufficient to replace the zero pivot which arises due to the singularity of the matrix by a small value ε. This should be chosen to be the smallest number for which $1 + \varepsilon > 1$ on the particular computer being used. This results in a very inaccurate solution to the set of equations but a rigorous error analysis [WILK 65] will show that since the elements of the solution vector possess errors in the same ratio, normalizing this vector will yield a very accurate eigenvector.

This approach has an advantage over the first method in that it is possible to observe the build-up of rounding error. Theoretically, it is known that we should obtain a zero pivot during the reduction of the final row. However, due to rounding error, this will hardly ever be exactly zero; its nonzero value will yeild an indication of the rounding error build up.

Inverse iteration has another advantage in that it requires fewer arithmetic operations. Replacing the last row of S^T by $\underset{\sim}{1}$ requires that the first (m-1) elements of this vector be eliminated, each elimination requiring a certain number of multiplications, additions, and a division. The number of elements to be eliminated in the final row using inverse iteration will be substantially less than (m-1). Further, the normalization requirement in inverse iteration requires only (m-1) additions and usually a much smaller number of divisions. Also note that the reduction of all (m-1) elements in the first method entails a further disadvantage in that this elimination requires access to be available to all (m-1) previous rows of the matrix.

From a practical point of view, the only difference between the Gaussian elimination method and the method of inverse iteration for this particular type of application, is in the handling of the normalization. In the next section we consider some of the practical details in the implementation of inverse iteration in the context of the matrix described in section 2.

4. Implementation Considerations and Operation Count

When applying a direct equation solving method such as Gaussian elimination, it is usually assumed that the complete set of linear equations has already been derived and that the entire coefficient matrix is stored in the computer memory, possibly in compact form. The reduction phase begins by using the first equation to eliminate all non-zero elements in the first column of the coefficient matrix from column position 2 through m. More generally, during the i-th reduction step, the i-th equation is used to eliminate all non-zero elements in the i-th column from positions (i+1) through m. Partial pivoting is almost invariably used to ensure that a stable reduction results.

However, since we are responsible for both the initial generation of the system of equations and for its solution, it is possible to envisage an alternative approach which has several advantages over the traditional method outlined above. In section 2 it was shown that consecutive rows of the transition rate matrix may be produced with relatively little effort. Thus immediately after the second row has been obtained it is possible to eliminate its sub-diagonal element in position (2,1) by adding a multiple of the first row to it. This process may be continued recursively so that when the i-th row of the coefficient matrix is generated, rows 1 through (i-1) will already have been derived and reduced to upper triangular form. The first (i-1) rows may therefore be used to eliminate all non-zero elements in row i from column positions (i,1) through (i,i-1), thus putting it into the desired triangular form.

It is only possible to implement a scheme of this nature when pivoting is not necessary; for example, when the coefficient matrix is diagonally dominant. This is precisely the case with the matrix of transition rates, since by definition, the diagonal element in any column is equal to the negated sum of the off-diagonal elements in that column and all off-diagonal elements are greater than or equal to zero. Thus the above procedure automatically implements a pivoting strategy and this together with the sparsity of the matrix is sufficient to guarantee stability.

When a row of S^T has been generated and reduced in this fashion it must then be stored so that it can participate in the reduction of succeeding rows and also in the back substitution. The structure of the matrix dictates that a two dimensional array is most suitable. The number of rows should obviously be equal to the number of states in the queueing network ($m = (N-1)r_1r_2 + r_1 + r_2$) while the number of columns should be sufficiently large to allow the storage of any elements to the right of the diagonal element which might become non-zero. From the structure of the matrix this is equal to $r_1r_2 + r_2$. The diagonal element of a reduced row will always be the first element stored in the corresponding row of the array. It is very likely that all elements within the band will be non-zero as a result of fill-in during the reduction.

As described in the previous section, once the matrix has been completely reduced, the final pivotal element (which is theoretically equal to zero) should be replaced by machine epsilon and the backsubstitution initiated. For convenience the right hand side of the equation may be set equal to the vector I_m, each component of which is zero except the last, which is equal to unity. No additional storage is required for the solution vector since the components, obtained in reverse order, may simply overwrite corresponding components in one column of the array used to store the reduced matrix.

5. Operation Count

Figure 3 is useful in determining the number of operations involved in reducing to triangular form those rows which correspond to states in which k customers are at server 1. It is assumed that all rows prior to (k, 1, N-k, 1) have already been reduced. It is apparent that the first r_2 rows will require considerably more operations than the remaining $r_2(r_1-1)$ rows. It is also apparent that once the reduction in any row is commenced, all zero elements between the element being reduced and the diagonal element will experience fill-in. Consequently in each of the first r_2 rows, $r_1 r_2$ elements will eventually have to be reduced before the row conforms to the upper triangular form. The reduction of each of these elements requires 1 division and at most $(r_1 r_2 + r_2 - 1)$ multiplications and additions. Therefore a total of $r_2 x r_1 r_2 x (r_1 r_2 + r_2)$ multiplications and divisions are required to bring the first r_2 rows to upper triangular form. In a similar manner it may be shown that $r_2 x (r_1 r_2 + r_2)$ multiplications and divisions are required for each of the remaining $r_1 r_2 - r_2$ rows.

The total number of multiplications and divisions for section k of the reduction is then strictly less than:

$$r_2 x r_1 r_2 x (r_1 r_2 + r_2) + (r_1 r_2 - r_2) x r_2 x (r_1 r_2 + r_2).$$

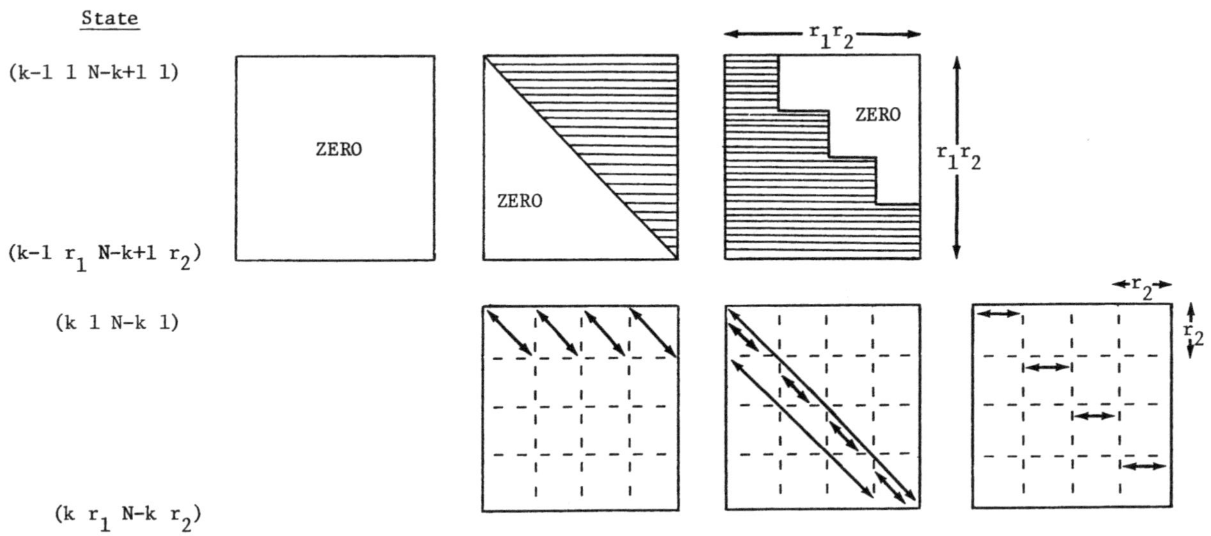

Figure 3. Matrix structure prior to reduction of section k. Non-zero elements are indicated by solid lines with arrowheads at both ends.

In the back substitution stage each element of the solution vector may be obtained with a maximum of $(r_1 r_2 + r_2 - 1)$ multiplications and additions and 1 division. A total of $r_1 r_2 (r_1 r_2 + r_2)$ multiplications and divisions is therefore sufficient to obtain all components of the solution vector corresponding to section k. An upper bound for the operation count for the entire algorithm is then given by

$$(N+1)\ \{2r_1^2 r_2^3 + r_1^2 r_2^2 + r_1 r_2^3 + r_1 r_2^2 - r_2^3\}$$

In the special case when both servers are Coxians of order 2 $(r_1 = r_2 = 2)$, the operation count is strictly less than $96(N+1)$.

6. Conclusions

In this paper we presented a method for the numerical solution of a two-server Coxian queueing network. It was shown how advantage could be taken of the unique structure of the transition rate matrix to produce a stable numerical algorithm based on inverse iteration which is efficient in terms of both memory requirements and computation time. It is envisaged that this method will prove useful in ongoing research aimed at obtaining better approximate methods for general networks.

7. References

[CHW 75] Chandy, K. M., Herzog, V. and Woo, L., "Approximate Analysis of General Queueing Networks," IBM J. Res. Develop., 19, January 1, 1975.

[MARI 78] Marie, R., "Modelization par Réseaux de Files d'Attente," Thesis: Docteur-es-Sciences Mathématiques. Université de Rennes, France, November, 1978.

[NEUT 81] Neuts, M. F., Matrix - Geometric Solutions in Stochastic Models - An Algorithmic Approach, The John Hopkins University Press, 1981.

[STEW 80] Stewart, W. J., and Zeiszler, G. A., "On the Existence of Composite Flow Equivalent Markovian Servers". The 7th IFIP W.G.7.3 International Symposium on Computer Performance Modelling, Measurement and Evaluation. May 28-30, 1980, Toronto, Canada.

[WALL 66] Wallace, V. L. and Rosenberg, R. S., "RQA-1: The Recursive Queue Analyzer," Dept. of Electrical Engineering, System Engineering Laboratory, Technical Report No. 2, The University of Michigan, Ann Arbor, (1966).

[WILK 65] Wilkinson, J. H., The Algebraic Eigenvalue Problem, Oxford University Press, London, 1965.

[WONG 77] Wong, B., Giffin, W., and Disney, R. L., "Two Finite M/M/1 Queues in Tandem: A Matrix Solution for the Steady State," OPSEARCH, Vol. 14, 1977.

[ZAHO 77] Zahorjan, J., "Iterative Aggregation with Global Balance," Project SAM Notes, Computer Systems Research Group, University of Toronto, Canada, February 1977.

This research was supported in part by NSF grant MCS80-04345.

Department of Computer Science, North Carolina State University, Raleigh, N.C. 27650.

PERFORMANCE AND RELIABILITY
Donald Gross, Chairman

M. L. Shooman & R. W. Schmidt
J. P. Lehoczky & D. P. Gaver

FITTING OF SOFTWARE ERROR AND RELIABILITY MODELS TO FIELD FAILURE DATA

Professor M. L. Shooman

and

Captain R. W. Schmidt

Abstract

The quantitative prediction and measurement of software reliability is of vital importance in the development of high quality cost effective software. Many software reliability models have been postulated in the literature [11], however few have been applied to field data. A model based upon the assumption that the failure rate of the software is proportional to the number of residual software errors leads to a constant failure rate and an exponential reliability function, [1]. The model contains two constants: the proportionality constant K and the initial (total) number of errors E_T.

The constants K and E_T can be estimated during early design by comparison of the present project with historical data. During the integration test phase, a more accurate determination of the model parameters can be obtained by using simulator test data as if it were operational failure data. The simulator data is collected at two different points in the integration test phase and the two parameters can be determined from moment estimator formulas [9]. The more powerful maximum likelihood method can also be employed to obtain point and interval estimates [3]. It is also possible to use least squares methods to obtain parameter estimates which is the simplest method and provides insight into the analysis of the data [12].

This paper utilizes a set of development and file data taken on 16 different software developments [10] as a vehicle to study the ease of calculation and the correspondence of the three methods of parameter estimation. The sensitivity of the reliability predictions to parameter changes are studied and compared with field results. Additional theoretical studies plus comparisons between estimates and field measurements are needed to determine the optimum method of parameter estimation.

The results show that if data is carefully collected, software reliability models are practical and yield useful results. These can serve as one measure to help in choosing among competitive designs and as a gauge of when to terminate the integration test phase.

1. Introduction

Software presently represents the highest cost item in the development of computer systems. There is a paucity of quantitative measures to judge the quality of the final software and use as a measure of progress during the test and debugging phase. The reliability and meantime to failure (MTTF) of the software is a most useful metric for both of the above purposes.

An important class of software reliability models [1], [8] make the assumption that the operational software failure rate is proportional to the remaining number of errors. Thus the failure rate is dependent on development time τ, but not on operating time t. This leads to a constant hazard and exponential reliability model, with two unknown parameters K and E_T.

A major focus of this paper is to investigate and provide insight into a number of issues related to the estimation of these two parameters:

1. The accuracy one can obtain by using historical (handbook-type) data to determine K and E_T.
2. The relative advantages of using field test, simulators, and development test data to determine K and E_T.
3. A comparison of the accuracy obtained using three different methods of parameter estimation, the maximum likelihood, moments, least squares.
4. A comparison of predicted values of MTTF and observed MTTF values from field failure data.

The conclusions reached in this study clearly indicate that parameter estimation during system development is a highly practical tool which can be used to successfully predict software MTTF and the debugging time required to achieve that goal. Additional theoretical studies plus comparisons between estimates and field measurements are needed to determine the optimum method of parameter estimation.

2. Development of Error and Reliability Models

2.1 An Error Removal Model

The reliability model used in this paper has been described in detail in a number of references [1], [2], [3]. In brief, the model assumes that the program enters the integration test phase with E_T total errors remaining in the software. As integration testing proceeds, all detected errors are promptly corrected, and at any point in the development cycle (after τ months of development time[1]), a total of $E_c(\tau)$ errors have been corrected, and the remaining number of errors is

$$E_r(\tau) = E_T - E_c(\tau) \tag{1}$$

In a more advanced model [4], it is assumed that new errors are generated during development. One can often normalize the above equation through division by the number of object code instructions, I_T

$$\varepsilon_r(\tau) = \frac{E_T}{I_T} - \varepsilon_c(\tau) \tag{2}$$

where

$$\varepsilon_r = E_r/I_T \qquad \varepsilon_c = E_c/I_T$$

Basically the error model used in this paper assumes that the total number of errors in the program are fixed and if we record the cumulative number of errors corrected during debugging, then the difference represents the remaining errors. Sometimes greater insight is obtained if we deal with the rate of error removal

$$\frac{dE_r(\tau)}{d\tau} \equiv r_r(\tau) \tag{3}$$

which can be normalized to yield

$$\frac{1}{I_T} \cdot \frac{dE_r(\tau)}{d\tau} = \rho_r(\tau) \tag{4}$$

where

$$\rho_r(\tau) = \text{errors removed/total number of instructions/test hours} \tag{5}$$

$$\varepsilon_r(\tau) = \int_0^\tau \rho(x)dx = \text{cumulative errors/total number of instructions.} \tag{6}$$

In Reference [5] error data are reported for seven large supervisory programs and applications programs. In Figure 1 the normalized error rate $\rho(\tau)$ calculated from this data is plotted as a function of τ the number of months of debugging after release for three of the seven systems. Although several curve shapes might be fitted to this data, one characteristic is common for all curves. The normalized error rate decreases over the entire curve or at least over the latter two-thirds or half of the curve; whereas initial behavior of $\rho(\tau)$ differs from example to example. A curve of the cumulative error data for the supervisory system A of Figure 1 is shown in Figure 2. If similar curves for $\varepsilon(\tau)$ were drawn for the other examples of Figure 1 all would build up initially with a constant or increasing slope and then exhibit a decreasing curvature appearing to become asymptotic. The smoothing nature of integration makes all the $\varepsilon(\tau)$ curves look more alike than the $\rho(\tau)$ curves do. Both $\varepsilon(\tau)$ and $\rho(\tau)$ curves are needed for a detailed study.

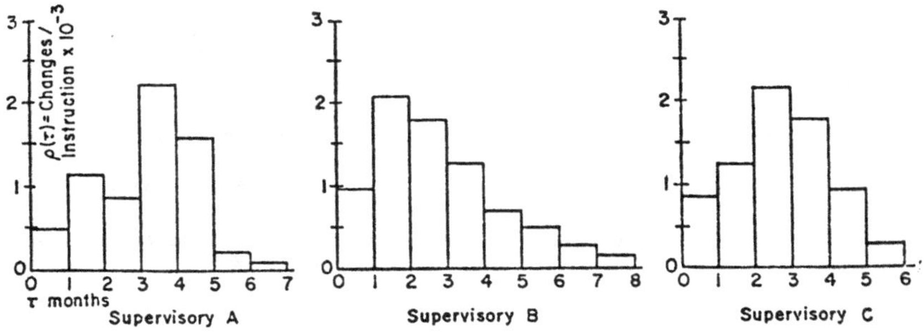

Figure 1. Normalized Error Rate vs Debugging Time for Three Supervisory Programs

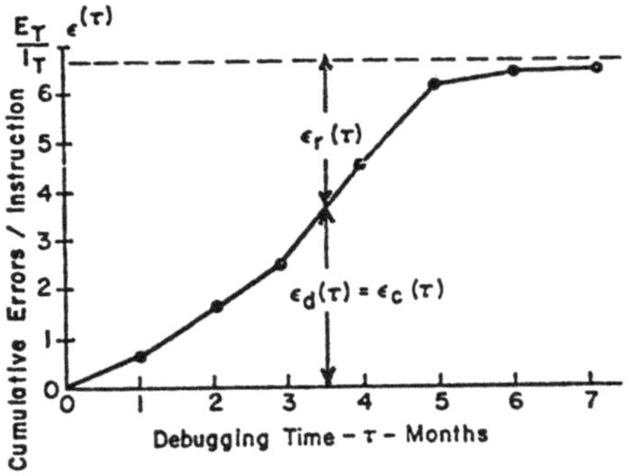

Figure 2. Cumulative Error Curve for Supervisory System A Given in Figure 1

If we assume that the total number of errors in the program E_T is constant and that the program contains I_T instructions, then the asymptote which the $\varepsilon(\tau)$ curves approach is E_T/I_T.

The error model developed above will be used in the following section to formulate a reliability model.

2.2 Development of the Exponential Reliability Model

We assume that operational software errors occur due to the occasional traversing of a portion of the program in which a hidden software error is lurking. We begin by writing an expression for the probability that an error is encountered in the time interval Δt after t successful hours of operation. We make the assumption that this probability is proportional to the remaining number of errors. (See [6] for substantiating data.)

From a study of basic probability and reliability theory we learn that the probability of failure in time interval t to t + Δt given that no failures have occurred up till time Δt is proportional to the failure rate (hazard function) Z(t).

$$P(t < t_f \leq t + \Delta t \mid t_f > t) = Z(t)\Delta t = K\varepsilon_r(\tau)\Delta t \quad (7)$$

where $t_f \equiv$ time to failure, (occurrence of a software error)

$P(t < t_f \leq t + \Delta_t \mid t_f > t \equiv$ probability of failure in interval Δt, given no previous failure.

K = an arbitrary constant.

From reliability theory we can show that the probability of no system failures in the interval 0 to t is the reliability function which is related to the hazard, Z(t), by:

$$R(t) = e^{-\int_0^t Z(x)dx} \quad (8)$$

If we substitute our expression for Z(t) from Equation 7 into Equation 8 and assume K, and $\varepsilon_r(\tau)$, are independent of <u>operating</u> time, we obtain

$$R(t) = e^{-[K\varepsilon_r(\tau)]t} = e^{-\gamma t} \tag{9}$$

Basically the above equation states that the probability of successful operation without software errors is an exponential function of operating time. When the system is first turned on, $t = 0$ and $R(0) = 1$. As operating time increases the reliability monotonically decreases as shown in Figure 3. We depict the reliability function for three values of debugging time, $\tau_0 < \tau_1 < \tau_2$. From this curve we may make various predictions about the system reliability. For example, looking along the vertical line $t = 1/\gamma$ we may state:

1. If we spend τ_0 hours of debugging then $R(1/\gamma) \approx 0.35$
2. If we spend τ_1 hours of debugging then $R(1/\gamma) \approx 0.50$
3. If we spend τ_2 hours of debugging then $R(1/\gamma) \approx 0.75$.

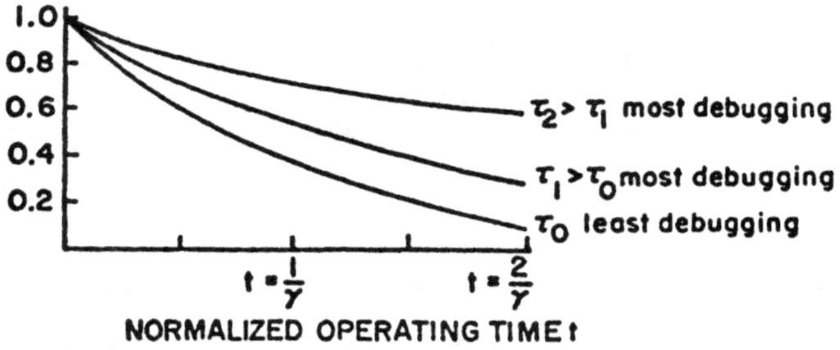

Figure 3. Variation of Reliability Function R(t) with Debugging Time τ

2.3 Meantime to Software Failure

A simpler way to summarize the results of the reliability model is to compute the mean time to (software) failure, MTTF, for the system.

$$\text{MTTF} = \int_0^\infty R(t)\,dt. \tag{10}$$

For purposes of illustrations we let $\rho(\tau)$ be modeled by a constant rate of error correction ρ_0 (see [5] for other models) then solutions of Equations 10 and 2 yields

$$\text{MTTF} = \frac{1}{K\frac{E_T}{I_T} - \rho_0\tau} = \frac{1}{\beta(1-\alpha\tau)} \tag{11}$$

where

$$\beta = \frac{E_T}{I_T} K \quad \text{and} \quad \alpha = \frac{\rho_0 I_T}{E_T}$$

In Figure 4, β x MTTF is plotted vs. $\alpha\tau$. We see that the most improvement in MTTF occurs during the last 1/4 of the debugging.

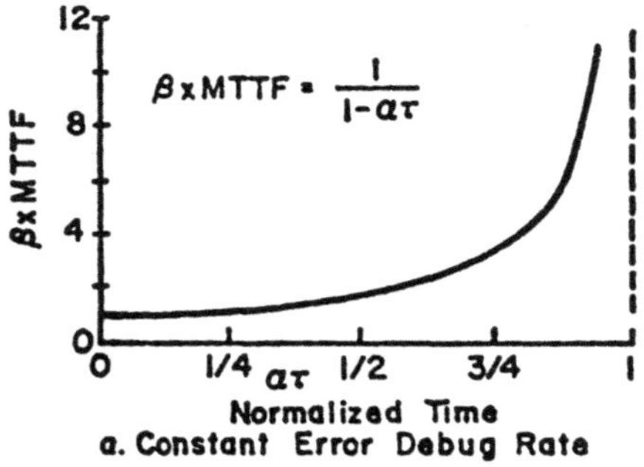

Figure 4. Comparison of MTTF vs Debug Time for The Error Model

The rapid rise in MTTF toward the end of the integration phase is of considerable importance in Project Management. Field data by others (see Figures 5 and 6) confirms this basic shape. If a manager is pressed to release a system to the field at an early date, he may accept the current reliability and deliver the software. However, if we believe Figures 4, 5, and 6 then a few more weeks may yield a big improvement. The model predicts such behavior, however, if one had only test data for $\alpha\tau \leq 1/2$ as shown in Figure 4, then it would be difficult to predict the sharp rise near $\alpha\tau = 1$. Thus, the model is of great use in managing a project and setting its release date.

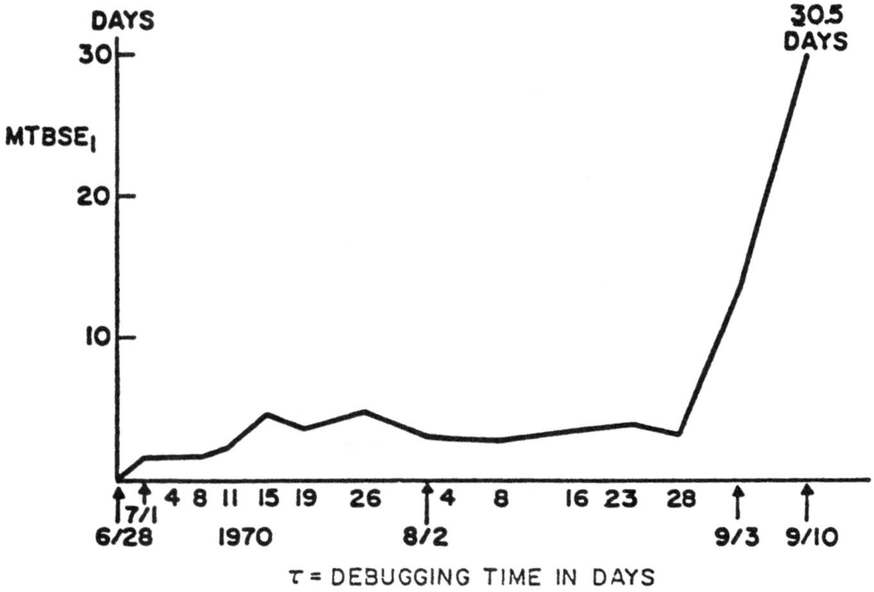

Figure 5. Growth Curve of Software Reliability Mean Time Between Software Errors (from I. Miyamoto, ref. 13)

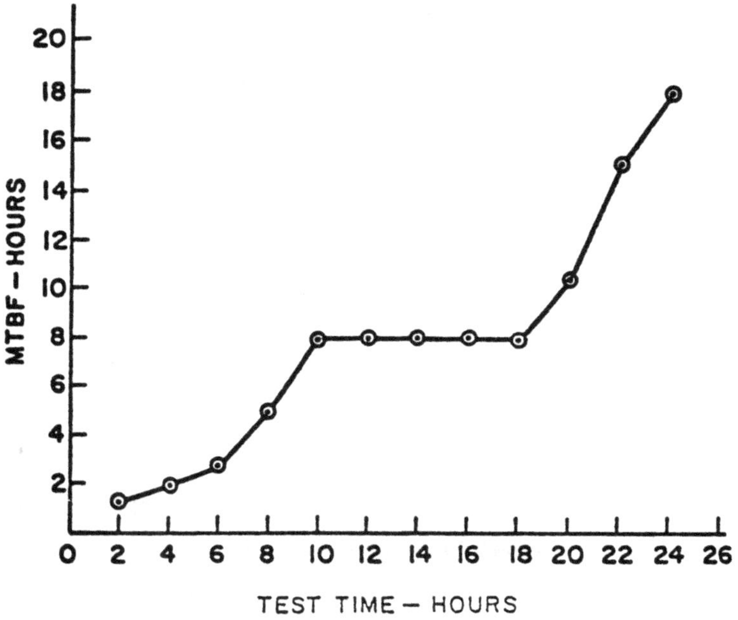

Figure 6. MTBF vs. Test Time for Project 1 (replotted from the data of Figure 3, J. Musa, ref. 8)

2.4 Musa's Model

Musa has developed a model similar to that given in Section 2.2. However, instead of basing his model on Development time as the resourse measure during integration, he utilized actual CPU time. His model is given by [8]

$$R(\tau') = \exp\ (-\tau'/T) \tag{12}$$

$$T = T_0 \exp\ (\frac{C\tau}{M_0 T_0}) \tag{13}$$

where

τ' = hours of program operation = t

T = mean time to failure in operating hours = MTTF

T_0 = MTTF at the start of test ($\tau = 0$)

C = ratio of equivalent operating time/test time

M_0 = number of failures which must occur to uncover all errors = E_T

τ = the CPU time in hours during testing.

The similarities and differences between the two models is explored in Section 5.7.4 [3]. Since we will be using Musa's data and some of his results in Section 5.0 of this paper we must carefully account for the different definitions of time in Musa's model and the exponential model.

3. Estimation of Model Parameters

3.1 Introduction

The exponential reliability model given in Equation 9 contains the parameters K and E_T which must be estimated. In many cases one wishes to use such a model to roughly predict MTTF during a proposal phase or early design of the project. In such a case the only available technique for determining values of the parameters is to use historical data. Presently Rome Air Development Center is developing a handbook and data base on Software reliability for just such a purpose. At present the data in the literature is scattered.

3.2 Moment Estimate

In [9], a method is discussed for measurement of the two needed parameters based on simulation testing of the software. A program simulating the field environment is generally available for all real time computer programs. It is necessary that this program be run following τ_1 months of development and the r_1 failures out of H_1 total hours be recorded. Similarly, after τ_2 months of testing the simulator is run and H_2 hours and r_2 failures are obtained. The MTTF for the data is given by the familiar values H_2/r_2 and is equated to the MTTF expression obtained by substituting equation 9 into equation 10. (Note, the normalizing factor I_T has been absorbed in the constant K.) The two equations (for τ_1 and τ_2) allow us to solve for the constants K and E_T

$$\frac{H_1}{r_1} = \frac{1}{K[E_T - E_c(\tau_1)]} \tag{14}$$

$$\frac{H_2}{r_2} = \frac{1}{K[E_T - E_c(\tau_2)]} \tag{15}$$

Simultaneous solution of equations 14 and 15 yields the desired values of K and E_T. Note that in deriving the above results we have assumed that the failure rate is constant and using the common notation for constant failure rates

$$z(t) = \lambda = r/H = 1/\text{MTTF}$$

3.3 Least Square Estimates

Another method of estimating model parameters is to rewrite equations 14-15 in the form

$$\lambda_i = K(E_T - E_c(\tau_i)) \tag{16}$$

where λ = failure rate to yield

$$E_c(\tau_i) = E_T - \frac{1}{K}\lambda_i \tag{17}$$

This equation represents a straight line whose parameters can be determined from the slope and intercept of a least squares fit of the data, where

$$K = \frac{-1}{\text{slope}} \tag{18}$$

$$E_T = \text{intercept} \tag{19}$$

Naturally, the larger the data set and the more accurate the error removal data, the more precisely can the model parameters be predicted.

3.4 Maximum Likelihood Estimates

Another method which can also be used to estimate the values of $\hat{K}$ and $\hat{E}_t$ is known as the Maximum Likelihood Estimation technique, (MLE). The likelihood function, L is the joint probability of occurrence for the observed set of test values. If during simulation testing we observe r_1 failure times $(t_1, t_2, \cdots t_{r_1})$ and n_1-r_1 successful runs testing $(T_1, T_2, \cdots, T_{n_1-r_1})$ then the likelihood function for a single test after $E_c(\tau_1)$ errors have been removed is given by

$$L(K,E_T) = f(t_1)f(t_2)\cdots\cdot f(t_{r_1})R(t_1)R(T_2)\cdots\cdot R(T_{m-n_1})$$

where

$$f(t_i) = \text{the density function } KE_r(\tau_1)e^{-K\varepsilon_r(\tau_1)t_i}$$

$$R(T_t) = \text{the reliability function } e^{-K\varepsilon_r(\tau_1)T_i}$$

To maximize the likelihood function we take partial derivitives with respect to K and E_T and set them equal to zero. To solve for the two parameters we need a second equation obtained from another likelihood equation based on a second set of test data at time τ_2. Applying MLE theory to two tests with r_1 and r_2 failures over H_1 and H_2 total hours, we obtain (Appendix A, [3])

$$\hat{K} = \frac{r_1 + r_2}{H_1[E_T - E_c(\tau_1)] + H_2[E_T - E_c(\tau_2)]} \tag{20}$$

$$\hat{K} = \frac{1}{H_1 + H_2}\left[\frac{r_1}{E_T - E_c(\tau_1)} + \frac{r_2}{E_T - E_c(\tau_2)}\right] \tag{21}$$

As is often the case with MLE, the above equations require numerical solution, however, most statisticians believe them to yield superior results to moment estimates. An iterative computer solution of equations 20 and 21 is easily implemented, yet a graphical solution using a simple calculator suffices in most cases. The first step is to obtain starting values for E_T and K using some other method, such as Method of Moments. Values of E_T above and below the starting value are substituted into equations 20, 21 and the curves of K vs E_T plotted on the same axes. Their intersection determines the value of $\hat{E}_T$ and $\hat{K}$.

4. Musa's Data

4.1 Introduction

The software reliability data used in this paper was compiled by John D. Musa [10] over a period of time on a variety of large software systems, ranging from tens to hundreds of thousands of object code instructions. His objective was "to present in detail a substantial body of data that has been gathered in the application of the execution time theory of software reliability"; the end product is ideally suited for the purpose of this paper, as it presents a wealth of precise software failure data obtained under carefully controlled circumstances.

4.2 Description of Raw Data

Software failure interval data was presented in the following format:

failure number failure interval day of failure

Failure interval was measured in seconds, and represents either running clock time, (operating time on the computer), or, in one case, actual CPU time. 'Day of Failure' is the working day counted from the start of project on which the failure occurred.

For the purpose of this paper, failures occurring on the same working day were summed together to form one statistical data point, and were tabulated under the following format:

S	WD	E	Ec	z	T	Tc

where:

S = sequential serial number assigned to each statistical data point

WD = working day on which failure(s) occurred

E = number of errors occurring that working day

Ec = cumulative errors to date

T = total operating time (failure interval time) for that working day

Tc = cumulative failure interval time to date

z = failure rate (E/T) for that working day

Presentation of Musa's data in this format had a two-fold purpose:

a) reduce the sheer bulk of the raw data without affecting its statistical significance by grouping the occurrence of software failures by working day; and,

b) tabulate the data in a format more suitable to subsequent calculations.

4.3 System Characteristics

Systems 1-4 studied by Musa are real time command and control software packages consisting of 21,700 to 33,500 object instructions and a failure sample size of 38 to 136.

System 5 is a real time commercial application, consisting of 2,445,000 changes, involving 21% of the source code, which were introduced after approximately 30% of total testing time had elapsed.

5. Estimation of Model Constants

In the following section, we will be estimating the model constants $\hat{K}$ and $\hat{E}_T$ of several systems, using system #3 as a working example to demonstrate the calculations used for least squares, method of moments, and maximum likelihood estimates.

5.1 Method of Moments

The data reported by Musa for system #3 has been processed and grouped by working day as described in Section 4.2 (see Table 1). Using 2 data points for the 9th and 18th group of failures, {(Ec 9, Tc 9), (Ec 18, Tc 18)}, we obtain the average failure rate for the interval 0-9

$$\lambda_1 = \frac{Ec9}{Tc9} = \frac{25 \text{ failures}}{14{,}260 \text{ sec.}} = .00175 \text{ failures/sec} = 6.3 \text{ failures/hr.}$$

and similarly the average failure rate for the interval 10-18

$$\lambda_2 = \frac{Ec18-Ec9}{Tc18-Tc9} = \frac{38-25}{67{,}362-14{,}260}$$

$$= 2.448 \times 10^{-4} \text{ failures/sec}$$

$$= .881 \text{ failures/hr.}$$

$$\lambda_2/\lambda_1 = .1399$$

$$\therefore \hat{E}_T = \frac{(\lambda_2/\lambda_1 \; Ec9)-Ec18}{(\lambda_2/\lambda_1 - 1)} = \frac{(.1399 \cdot 25)-38}{(.1399 - 1)}$$

$$= 40.114$$

Table 1. System # 3 Failure Rate Data

S	WD	E	Ec	Z	T	Tc
1	1	2	2	.01739	115	115
2	3	10	12	.00523	1,911	2,026
3	6	1	13	.00165	606	2,632
4	8	2	15	.00163	1,229	3,861
5	18	4	19	.00236	1,697	5,558
6	26	3	22	.00302	994	6,552
7	27	1	23	.00054	1,863	8,415
8	30	1	24	.00075	1,337	9,752
9	36	1	25	.00022	4,508	14,260
10	38	1	26	.00119	834	15,094
11	40	2	28	.00059	3,406	18,500
12	42	1	29	.00022	4,561	23,061
13	44	1	30	.00031	3,186	26,247
14	47	3	33	.00022	13,904	40,151
15	48	1	34	.00153	652	40,803
16	50	1	35	.00018	5,593	46,396
17	54	2	37	.00011	18,420	64,816
18	55	1	38	.00039	2,546	67,362

$$K' = \frac{\lambda_1}{\hat{E}_T - Ec9} = \frac{.00175}{40.114-25} = 1.158 \times 10^{-4}$$

Decreasing the sample interval (using more data points from the same system) and averaging over them often yields more accurate results. For example, if 3 data points are used (P6, P12, P18)

$\hat{E}_{T1} = 30.01$ using P6, P12

$\hat{K}_1 = 4.192 \times 10^{-4}$

$\hat{E}_{T2} = 39.73$ using P12, P18

$\hat{K}_2 = 1.171 \times 10^{-4}$

$$\therefore \overline{\hat{E}_T} = \frac{\hat{E}_{T1} + \hat{E}_{T2}}{2} = 34.87$$

$$\overline{\hat{K}} = 2.682 \times 10^{-4}$$

See Table 5-2 for a summary of all system

5.2 Least Squares Linear Regression

In this method, we plot the failure rate Z vs cumulative errors Ec. Theoretically the best fit y-intercept yields $\hat{E}_T$ and the negative reciprocal of the slope equals $\hat{K}$.

The primary equations used are:

$$\text{slope } M = \frac{\Sigma xy - \frac{\Sigma x \Sigma y}{N}}{\Sigma x^2 - \frac{(\Sigma x)^2}{N}}$$

$$y' = \frac{\Sigma y - m\Sigma x}{N}$$

$$y = mx + y'$$

Here again, we can use a varying number of intervals of the system to obtain different results. Referring back to Table 5-1,

$E_{c9} = 25$ $E_{c18} = 38$

$T_{c9} = 14{,}260$ $T_{c18} = 67{,}362$

We plot

Table 2. Method of Moments Estimation of $\hat{E}_T$, $\hat{K}$

System					
1	#Pts	2	4	8	16
	$\hat{E}_T$	151.6	140	109.2	88.5
	$\hat{K}$ *	5.514	6.566	19.19	34.56
2	#Pts	2	4	8	12
	$\hat{E}_T$	61.6	49.2	41.0	35.2
	$\hat{K}$ *	3.909	7.660	13.77	17.91
3	#Pts	2	3	6	9
	$\hat{E}_T$	40.1	34.87	31.9	30.2
	$\hat{K}$ *	11.60	26.813	39.47	64.77
4	#Pts	2	3	6	19
	$\hat{E}_T$	54.9	54.4	131.2	37.5
	$\hat{K}$ *	19.88	16.64	34.51	67.37
5	#Pts	2	4	8	16
	$\hat{E}_T$	2390	1369	1247	506
	$\hat{K}$ *	8.088	33.23	66.16	111.2

*N.B. $\hat{K}$ values x 10^{-5}

Table 3. Least Squares Regression Estimates of $\hat{E}_T$, $\hat{K}$

System							
1	#Pts	2	4*	8	47		
	$\hat{E}_T$	151.6	140.1	103.3	74.6		
	$\hat{K}$	5.515	5.245	17.609	543.48		
2	#Pts	2	4	8	12	24	
	$\hat{E}_T$	61.6	51.7	45.5	44.9	42.9	
	$\hat{K}$	3.909	6.253	8.333	9.395	10.279	
3	#Pts	2	3	6	9	18	
	$\hat{E}_T$	40.1	34.9	32.8	30.6	28.4	
	$\hat{K}$	11.603*	25.490	23.928	27.942	56.593	
4	#Pts	2	4	10	19		
	$\hat{E}_T$	54.9*	56.8	49.8	42.4		
	$\hat{K}$	19.885	14.029	20.400	40.601		
5	#Pts	2	4	8	16	32	290
	$\hat{E}_T$	2390	998.9	805.2	742.9	597.6	481.1
	$\hat{K}$	8.089	30.675	44.982	51.475	94.518	290.13

*N.B. $\hat{K}$ values x 10^{-5}

Table 4. Maximum Likelihood Estimates of $\hat{E}_T$, $\hat{K}$

System	$\hat{E}_T$	$\hat{K}$
1	151.59	5.141×10^{-5}
2	61.61	3.909×10^{-5}
3	40.11	1.160×10^{-4}
4	54.80	2.249×10^{-4}
5	2390.29	8.088×10^{-5}

Table 5. Comparison of Model Predictions with Field Experience

Syst. No.	$Kx10^{-5}$	E_T	E_c	$MTBF_T$	C	$MTBF_0$ ($CxMTBF_T$)	$MTBF_F$	$\%\Delta_1$	$MTBF_0$ (MUSA)	$\%\Delta_2$	
1	6.566	140	136	1.06	15.1	16.0	14.6	9.4	20.4	39.7	MOM
	5.245	140		1.29		19.5		33.5			LS
	5.141	152		.346		5.2		64.2			MLE
2	3.909	62	54	.935	13.6	12.7	31.4	59.5	43.5	38.5	
	3.909	62		.935		12.7		59.5			
	3.909	62		.934		12.7		59.6			
3	11.60	40	38	1.14	13.2	15.0	30.3	50.4	30.4	0.33	
	11.603	40		1.14		15.0		50.3			
	11.60	40		1.13		15.0		50.6			
4	19.88	55	53	.735	13.1	9.63	9.17	5.1	14.5	58.1	
	19.89	55		.735		9.63		5.0			
	22.49	55		.686		8.98		2.0			

LEGEND

$MTBF_T$ - Predicted MTBF based on testing time

$MTBF_0$ - Predicted operational (field) MTBF

$MTBF_F$ - Actual operational (field) MTBF

$\%\Delta_1$ - % difference between actual and predicted MTBF (exponential model)

$\%\Delta_2$ - % difference between actual and predicted MTBF (MUSA'S model)

MOM - Method of Moments

LS - Least Squares

MLE - Maximum Likelihood Estimate

$$x_1 = Z_1 = \frac{25}{14{,}260} \qquad y_1 = E_{c9} = 25$$

$$x_2 = Z_2 = \frac{38-25}{67{,}362-14{,}260} \qquad y_2 = E_{c18} = 38$$

To yield

$$\hat{E}_T = 40.1 \qquad \hat{K} = 1.1603 \times 10^{-4}$$

Using 3 data points P6, P12, P18 yields

$$\hat{E}_T = 34.9 \qquad \hat{K} = 2.5490 \times 10^{-4}$$

The system plot using 2, 9, and 18 data points is presented in Figures 5-1. All system model constants are summarized in Tables 5-3.

5.3 Maximum Likelihood Estimates (MLE)

The primary equations used (see Section 3.4) are:

$$\hat{K} = \frac{r_1 + r_2}{H_1(E_T-E_{c1})+H_2(E_T-E_{c2})}$$

$$\hat{K}_2 = \frac{1}{H_1+H_2} \cdot \left[\frac{r_1}{E_T-E_{c1}} + \frac{r_2}{E_T-E_{c2}}\right]$$

Iterative Numerical Methods can be used to obtain a solution, however a simpler approach is to plot both equations on the same axis vs a given value of $\hat{E}_T$.

The intersection of both curves yields an approximation for $\hat{K}$ and $\hat{E}_T$.

A more aesthetic approach is to use a starting value of $\hat{E}_T$ which has been obtained by other methods, such as least squares or method of moments. Once an approximate value is obtained from the curve's intersection, iterative methods can be used to obtain the required degree of accuracy. This is illustrated in Figures 5-2.

6. Comparison of Model Predictions with Field Experience

6.1 Introduction

Predicting final system performance or required debugging time to achieve the desired level of performance is a field in which many theories abound. Many software designers are content if their estimate is of the same order of magnitude as the final 'evidence'. High quality failure data, logical reasoning, and a sound knowledge of mathematics can hone the 'guessing game' to an accurate science.

6.2 Model Predictions

In the preceding section, we presented the model predictions of K and E_T (Tables 2,3,4) using three methods of calculation: Method of Moments, Least Squares, and Maximum Likelihood. These estimates were tabulated using a varying number of data-size groupings for the same system for illustrative purposes. In general, as the number of sample points increased, E_T decreased while K increased for the same system. Obviously, since E_T must be larger than E_c, only data points which reflected this fact were used to calculate system MTBF.

Table 5 summarizes the predictions obtained using the three methods of parameter estimation. The predictions are compared with field measurements of MTBF for the four systems studied.

We can evaluate the accuracy of these software reliability perdictions by studing typical error levels obtained in the well known hardware reliability prediction process. In predicting the reliability of a system, one makes simplifying assumptions and utilizes handbook failure data, often from small populations. In addition, data is seldom available for all specified components and generic data for particular component types must often be used. Thus, although the mathematics is precise, the pressures of engineering practice make estimation an approximate process. We will illustrate how an analyist deals with these matters by an example.

Suppose a system specification called for an 800 hour MTBF, and the initial estimate was 1000 hours. A prudent engineer would still undertake some modest redesign to raise his estimate to prehaps 1200 hours. Proper design includes such conservatism to insure that the product will meet specifications even if some of the assumptions prove to be optimistic.

Using the above discussion as a basis, we can draw the following conclusions:

1. The predicted MTBF compared favorably with field experience. The predictions varied above and below the measured values from 2% to 64%, with an average (absolute value) error of 37%.
2. All three estimation techniques, method of moments (MOM), least squares (LS), and maximum likelihood estimate (MLE) yield similar results in systems 2,3, and 4. In system 1, MOM produced a superior estimate.
3. Further studies are needed to determine why MLE, with its many theoretically superior properties did not uniformly excell.
4. Lacking further studies of estimate accuracy, we would recommend that the LS method be used for the following reasons: (a) it is well known by engineers in the field, (b) LS programs are radily available on all computer systems, (c) the graph upon which the LS method is based provides an immediate assessment of the data variability (correlation coefficient).

Musa reported software reliability estimates for 12 other systems, however, they were not studied in detail, since no coroborative field experience was available. It is hoped that the heightened interest in the field of software reliability will lead to additional data gathering,

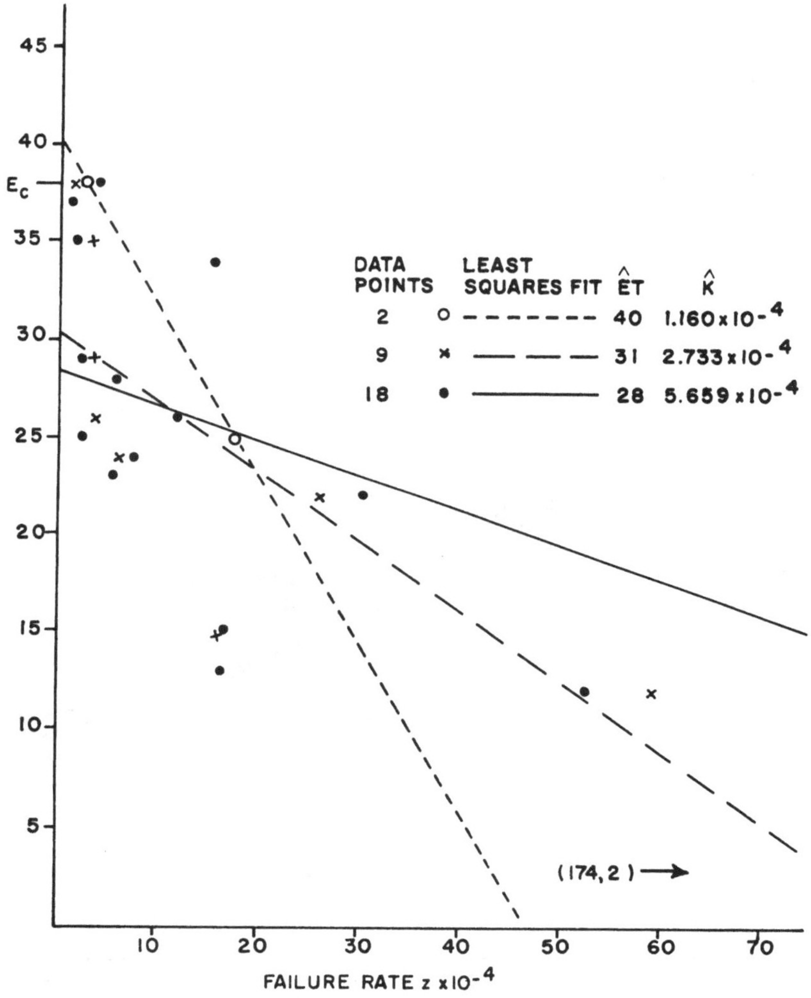

Figure 7. Cumulative Errors vs Failure Rate System 3

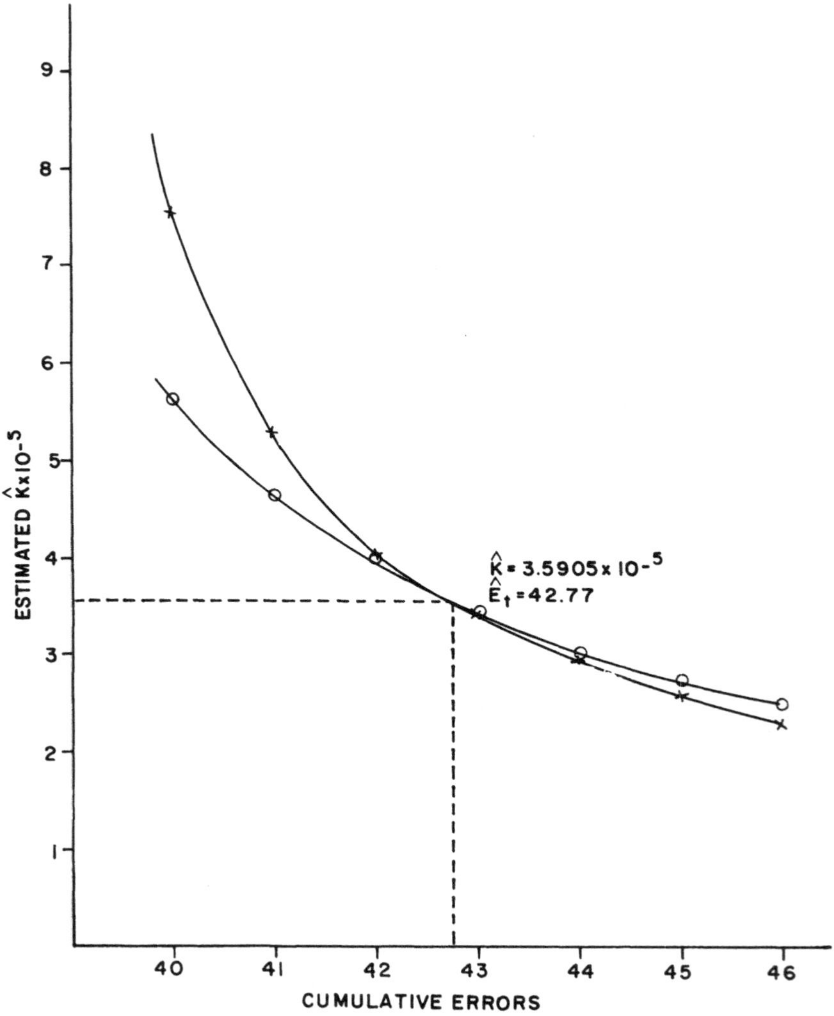

Figure 8. Estimated $\hat{K}$ vs Cumulative Errors System 3

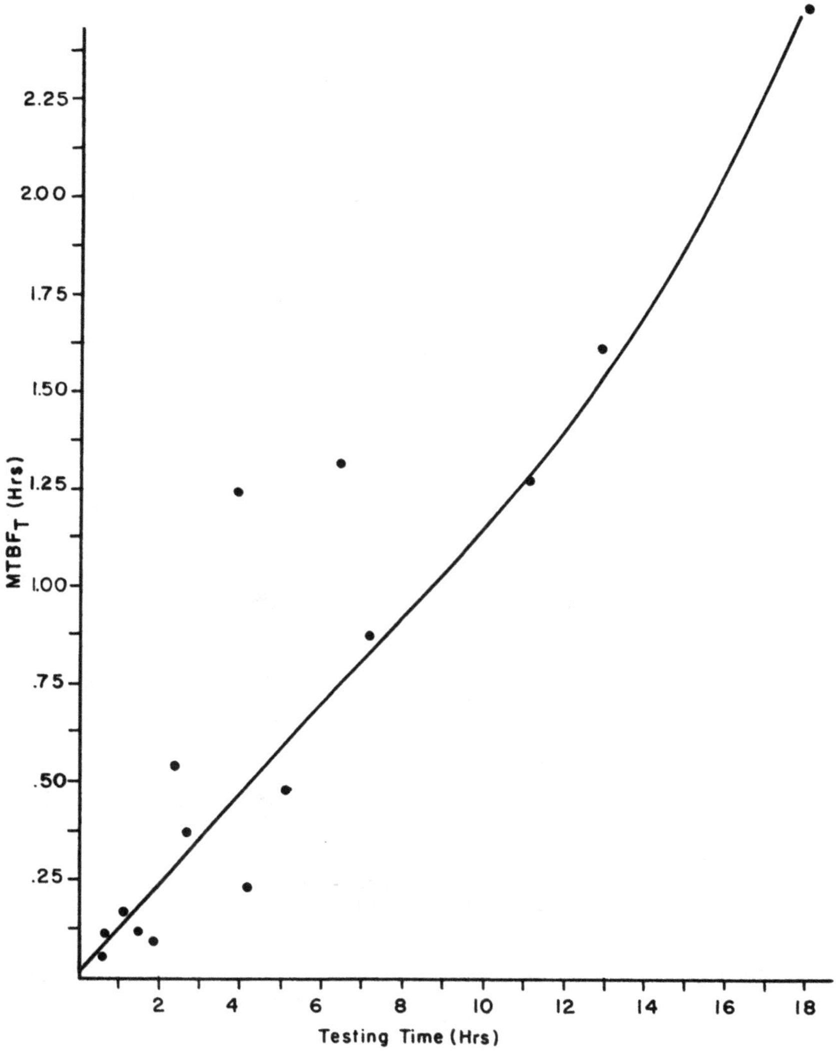

Figure 9. $MTBF_T$ vs Testing Time System 3

allowing a larger comparative study in the future.

7. Conclusion

The primary aim of this paper was to present a model of software system parameter estimation, and to subsequently compare predicted system characteristics with actual field data.

As indicated by the results tabulated in Sections 5 and 6, use of the exponential model is a viable approach which can readily be used by software engineers and system supervisors to quantitatively predict software MTTF based on failure data gathered during project development.

8. References

[1] Martin L. Shooman, "Probabilistic Models for Software Reliability Prediction", in Statistical Computer Performance Evaluation, Walter Freiberger editor, Academic Press, New York, 1972, pp. 485-497.

[2] Martin L. Shooman, "Software Reliability", in Computing Systems Reliability, T. Anderson and B. Randell editors, Cambridge University Press, New York, 1979.

[3] Martin L. Shooman, Software Engineering: Design, Reliability, Management, McGraw-Hill Book Co., New York, 1982.

[4] M. L. Shooman and S. Natarajan, "Effect of Manpower Deployment and Bug Generation on Software Error Models", Proceedings of the Polytechnic Symposium on Software Engineering, Polytechnic Press, Brooklyn, New York, 1976.

[5] J. Dickson, J. Hesse, A. Kientz, M. Shooman, "Quantitative Analysis of Software Reliability," 1972 Annual Reliability Symposium Proceedings, IEEE, Jan. 1972.

[6] John D. Musa, "Validity of Execution-Time Theory of Software Reliability," IEEE Transactions on Reliability, Special Issue on Software Reliability, 1979, Vol. R-28, pp. 181-191.

[7] Martin L. Shooman, Probabilistic Reliability: An Engineering Approach, McGraw-Hill Book Co., New York, 1968.

[8] John D. Musa, "A Theory of Software Reliability and Its Application," IEEE Trans. Software Engrg., Vol. SE-1, 1975, pp. 312-327.

[9] M. L. Shooman, "Operational Testing and Software Reliability Estimation During Program Development," Record 1973, IEEE Symposium on Computer Software Reliability, April 30-May 2, 1973, IEEE, New York, pp. 51-57.

[10] John D. Musa, "Software Reliability Data," Bell Telephone Laboratories, 1979 available through Data and Analysis Center for Software, Rome Air Development Center, Grifiss Air Force Base, New York 13441.

[11] "Software Engineering Research Review-Quantitative Software Models," Data and Analysis Center for Software, Rome Air Development Center, Grifiss Air Force Base, New York, 13441, March 1979.

[12] M. L. Shooman, "Software Reliability Data Analysis and Model Fitting," Proceedings of Workshop on Quantitative Software Models, IEEE, New York, Oct. 1979, pp. 182-188.

[13] I. Miyamoto, "Software Reliability in Online Real Time Environment," Proceedings, International Conference on Reliable Software, April 1975, IEEE Cat. No. 75CH0940-7CSR.

9. Endnotes

[1] In some cases the actual number of test hours is estimated and is used as the development time variable rather than the cruder calendar days.

This work was sponsored in part under Contract No. N00014-75-C-0858 with the Operations Research Program of the Office of Naval Research, Arlington, Virginia.

Department of Electrical Engineering and Computer Science, Polytechnic Institute of New York, Long Island Center, Route 110, Farmingdale, New York 11735.

PERFORMANCE EVALUATION OF VOICE/DATA QUEUEING SYSTEMS

John P. Lehoczky[1*]

and

D. P. Gaver[**]

Abstract

The behavior of voice/data queueing systems is presented. Several methods of analysis are reviewed including matrix geometric methods of Neuts, fluid flow approximations, and diffusion approximations. The methods are applied to the case of a TASI system for voice. The diffusion approximation is also applied to a system with a non-exponential service distribution.

1. Introduction

The concept of an integrated voice and data communication system like the SENET network was introduced originally by Coviello and Vena (1975) and Barbacci and Oakley (1976). There are numerous papers in the literature giving a probabilistic analysis of such systems. These include early works by Halfin and Segal (1972) and Halfin (1972). Recent work includes Fischer (1977), Fischer and Harris (1976), Bhat and Fischer (1976), Chang (1977), Weinstein, Malpass, and Fischer (1980), Arthurs and Stuck (1979), Feldman and Claybaugh (1980), and Lehoczky and Gaver (1979, 1981). These papers generally study the system performance under Markovian conditions. Voice and data traffic are independent Poisson processes with parameters λ and δ respectively.

The service time for voice calls and data packets are exponential with means μ^{-1} and η^{-1} respectively. Voice operates as a loss system with priority over data. Lehoczky and Gaver (1979) noted that extremely large data queues are possible, even when the traffic intensities are not great. This surprising phenomenon results from the fact that voice holding times are generally orders of magnitude longer than data packet lengths. Situations in which voice occupies all or nearly all of the channels, though infrequent, tend to persist for long periods and thus create long data queues.

Several design solutions to overcome these long queue lengths are possible. First, one might allocate some of the channels for exclusive use of data. This increases the voice blocking probability but can drastically reduce the data queue lengths. Second, one might treat the data itself as a loss system. This corresponds to finite buffer space and is called a finite waiting room problem. Third, one might increase the number of voice channels by decreasing the voice bandwidth. Such a solution improves both voice and data queueing performance at the expense of voice quality. Finally, one might introduce a TASI system for the voice to increase channel capacity for data.

The normal telephone conversation consists of an alternating sequence of talkspurts and silent periods (Brady 1968). With TASI (Time Assignment Speech Interpolation), it is possible to detect the silent periods and use that channel for other traffic (voice or data) during these periods. If the talkspurts have length the same order of magnitude as data packets, then the data queue lengths will be far shorter under such a system. Fischer (1978, 1980) presented an analysis of TASI and TASI in a voice-data context. The latter was restricted to a single channel.

This paper deals with several models. First the TASI voice-data system is introduced and analyzed using a variety of techniques. These techniques include the matrix geometric methods of Neuts (1978) and Feldman and Claybaugh (1980). In addition fluid flow and diffusion approximations are presented. Finally a simple voice-data model with non-exponential service times is studied.

2. Problem Formulation

We assume a queueing system having a total of N channels shared between voice and data traffic. Voice calls and data packets arrive according to independent Poisson processes with parameters λ and δ respectively. Data packets have independent exponential service times with mean μ^{-1}. Voice calls begin in a talkspurt and have alternate talkspurt and silent periods. Talkspurts are assumed to be of exponential length with mean α^{-1}, while silent periods have an exponential distribution with mean β^{-1}. The voice calls are of independent exponential total length with mean μ^{-1}. This means that voice calls may end in either a talkspurt or a silent period. There are several ways to introduce a TASI assumption. First, one might allow more than N callers into the system at one time. The silent periods are used to transmit talkspurts from other callers. Limited talkspurt queueing is possible. Second, and the approach followed here, is to allow at most N callers into the system and to use silent periods for data packet transmissions. Define $V(t)$ = number of voice customers (in a talkspurt or silent period) at time t. Also let $A(t)$ = number of voice calls having talkspurts at time t and $Q(t)$ = number of data packets in the system at time t. The voice will behave as an M/M/N/N queueing system with traffic intensity $\rho_v = \lambda/\mu$. The behavior of the data stream at any time depends on the number of channels available at that time, thus at time t data behaves as an M/M/N-A(t) queueing system with traffic intensity ρ_d. The triple $\{(Q(t), V(t), A(t))\}$, $t \geq 0$ is therefore a Markov chain with transition probabilities given by

Time t	Time t+dt	probability
(i,j,k)	(i,j+1,k+1)	$\lambda dt+o(dt)$, $o \leq j < N$
	(i,j-1,k)	$(j-k)\mu dt+o(dt)$
	(i,j-1,k-1)	$k\mu dt+o(dt)$
	(i,j,k-1)	$k\alpha dt+o(dt)$
	(i,j,k+1)	$(j-k)\beta dt+o(dt)$
	(i+1,j,k)	$\delta dt+o(dt)$
	(i-1,j,k)	$\eta \min(i,N-k)dt+o(dt)$

(2.1)

It is clear that $\{(V(t), A(t)), t \geq 0\}$ is also a Markov chain since voice has priority over data. The state space for this marginal process has $(N+1) \cdot (N+2)/2$ states. It is convenient to relabel these states in some fashion $1,2,\ldots,(N+1)(N+2)/2$. For the special case N=1 there are 3 states and the transition matrix is given by

$$\underset{\sim}{M} = \begin{pmatrix} -\lambda & \lambda & o \\ \mu & -(\mu+\alpha) & \alpha \\ \mu & \beta & -(\mu+\beta) \end{pmatrix} \tag{2.2}$$

with states 1, 2 and 3 meaning no voice users present, a voice user in a talkspurt, and a voice user in a silent period respectively. We let $\underset{\sim}{\pi}_N$ denote the equilibrium distribution associated with the voice process for N channels. $\underset{\sim}{\pi}_1$ is easily derived from (2.2) and is given by

$$\begin{aligned} \pi_{11} &= 1/(1+\rho_v) \\ \pi_{12} &= \rho_v(\mu+\beta)/(1+\rho_v)(\mu+\alpha+\beta)) \\ \pi_{13} &= \rho_v\alpha/((1+\rho_v)(\mu+\alpha+\beta)). \end{aligned} \tag{2.3}$$

2.1 Matrix Geometric Methods

Once one has relabeled the voice states $1,\ldots,(N+1)(N+2)/2$, the transition probability matrix for the voice-data process takes on a tridiagonal form: for N=1

$$\begin{pmatrix} \underset{\sim}{C}_1 & \underset{\sim}{B}_0 & & & 0 & \\ \underset{\sim}{B}_2 & \underset{\sim}{B}_1 & \underset{\sim}{B}_0 & & & \\ & \underset{\sim}{B}_2 & \underset{\sim}{B}_1 & \underset{\sim}{B}_0 & & \\ & \cdot & \cdot & \cdot & & \\ 0 & & \cdot & \cdot & \cdot & \\ & & & \cdot & \cdot & \cdot \end{pmatrix} \tag{2.4}$$

with $\underset{\sim}{B}_0 = \delta \underset{\sim}{I}_3$

$$\underset{\sim}{B}_1 = \begin{pmatrix} -(\delta+\eta+\lambda) & \lambda & 0 \\ \mu & -(\delta+\mu+\alpha) & \alpha \\ \mu & \beta & -(\delta+\eta+\mu+\beta) \end{pmatrix}$$

$$\underset{\sim}{B}_2 = \begin{pmatrix} \eta & 0 & 0 \\ 0 & 0 & 0 \\ 0 & 0 & \eta \end{pmatrix}$$

$$\underset{\sim}{C}_1 = \begin{pmatrix} -(\delta+\lambda) & \lambda & 0 \\ \mu & -(\delta+\mu+\alpha) & \alpha \\ \mu & \beta & -(\delta+\mu+\beta) \end{pmatrix} .$$

For N>1, the form of (2.4) remains, but the first N rows are modified to account for the min(i, N-k) = i case in (2.1). Neuts (1978) has studied such stochastic systems and proved that the equilibrium distribution has a matrix geometric form:

$$\begin{aligned} \underset{\sim}{\pi}_N &= (\underset{\sim}{\pi}_{N0}, \underset{\sim}{\pi}_{N1}, \ldots) \quad \text{and} \\ \underset{\sim}{\pi}_{Nk} &= \underset{\sim}{\pi}_{N0}\, \underset{\sim}{R}^{k-1} \quad k>1 \end{aligned} \tag{2.5}$$

where $\underset{\sim}{R}$ is a solution of $\underset{\sim}{0} = \underset{\sim}{B}_0 + \underset{\sim}{R}\underset{\sim}{B}_1 + \underset{\sim}{R}^2\underset{\sim}{B}_2$. Neuts also studies numerical methods for computing $\underset{\sim}{R}$ as well as other queueing characteristics. Feldman and Claybaugh (1980) also applied these methods to the voice-data problem without TASI.

This method has the advantage of providing insight into the nature of the equilibrium distribution and offering computational methods. Research on queueing with matrix geometric invariant vectors has been very active. There has been substantial work on numerical evaluation of these distributions and associated performance quantities, so this provides a complete equilibrium solution. There are, however, some limitations. First, $\underset{\sim}{R}$ cannot be found in closed form, since it requires solving simultaneous general second degree equations. Furthermore, as N increases the solution becomes more complicated and the geometric structure less useful. Finally, the solution is entirely an equilibrium one and the method doesn't seem to apply with non-Markovian assumptions. It is important to gain some insight into the transient behavior especially when η/μ is very large. For these reasons we outline in the next section several alternative methods of analysis based on approximations of various types.

3. Fluid Flow Approximations

One approximation method explored in Lehoczky and Gaver (1979) is a fluid flow approximation. In this method, the data queueing process is replaced by a deterministic process with the same mean. Thus, if at time t voice is in state (j,k), then data has N-k channels available. Rather than treating data as an M/M/N-k system, we assume its trajectory to be linear with slope $\delta - \eta(N-k) = \eta(\rho_d+k-N)$. Each voice state has a rate associated with it called r_ℓ, $1 \leq \ell \leq (N+1)(N+2)/2$; r_ℓ represents the net rate of data increase when voice is in state ℓ. Some r_ℓ's are >0 indicating a growth in data queue length. These are referred to as "up" states. The other r_ℓ's are <0 and are "down" states. For the case N=1, $r_1 = r_3 = \eta(\rho_d - 1)$, and $r_2 = \eta\rho_d$. A natural traffic intensity parameter is given by

$$\rho = \rho_d + \frac{\rho_v}{1+\rho_v} \frac{\mu+\beta}{\mu+\alpha+\beta}$$

which must be less than 1 for stability. Thus states 1 and 3 are down, while state 2 is up. We may think of the voice as furnishing a "random environment" for data service.

The fluid flow approximation replaces the discrete state space for data by a continuous state space. Motion on that data space is influenced only by the randomness of the voice environment. One can find the equilibrium distribution for the data queue length from the forward equations. Let $f(x, \ell, t)$ be the density that the data queue length is x and the voice state is ℓ at time t. A straightforward argument for N=1 and $x > 0$ gives

$$\begin{aligned} f(x, 1, t+dt) &= f(x-r_1dt, 1, t)(1-\lambda dt) + f(x-r_2dt, 2, t)\mu dt \\ &\quad + f(x-r_3dt, 3, t)\mu dt + o(dt), \\ f(x, 2, t+dt) &= f(x-r_2dt, 2, t)(1-(\mu+\alpha)dt) + f(x-r_1dt, 1, t)\lambda dt \\ &\quad + f(x-r_3dt, 3, t)\beta dt + o(dt), \qquad (3.1) \\ f(x, 3, t+dt) &= f(x-r_3dt, 3, t)(1-(\mu+\beta)dt) + f(x-r_2dt, 2, t)\alpha dt \\ &\quad + o(dt). \end{aligned}$$

Expanding in a Taylor series and letting $dt \to 0$, one finds

$$\begin{aligned} f_t(x,1,t) + r_1 f_x(x,1,t) &= -\lambda f(x,1,t) + \mu f(x,2,t) + \mu f(x,3,t) \\ f_t(x,2,t) + r_2 f_x(x,2,t) &= -(\mu+\alpha) f(x,2,t) + \lambda f(x,1,t) + \beta f(x,3,t) \qquad (3.2) \\ f_t(x,3,t) + r_3 f_x(x,3,t) &= -(\mu+\beta) f(x,3,t) + \alpha f(x,2,t). \end{aligned}$$

where

$$f_t(x,i,t) = \frac{\partial}{\partial t} f(x,i,t)$$

and

$$f_x(x,i,t) = \frac{\partial}{\partial x} f(x,i,t).$$

To find an equilibrium distribution, one lets $t \to +\infty$ and assumes $f_t(x,i,t) \to 0$. Defining $\lim_{t\to\infty} f(x,i,t) = f_i(x)$ we find

$$\underset{\sim}{f}'(x) = \underset{\sim}{S}^{-1}\underset{\sim}{M}^T\underset{\sim}{f}(x) \qquad x > 0 \tag{3.3}$$

where

$$\underset{\sim}{S} = \begin{pmatrix} r_1 & & 0 \\ & r_2 & \\ 0 & & r_3 \end{pmatrix}$$

and $\underset{\sim}{M}$ is given by (2.2).

The derivation is actually quite general. For $N>1$, (3.3) still holds with $\underset{\sim}{S}$ the diagonal matrix of rates and $\underset{\sim}{M}$ the generator of the voice process. A condition for stability is $\sum_i \pi_{Ni} r_i < 0$.

Equation (3.3) has a simple exponential solution for $x>0$. The solution method is presented in Gaver and Lehoczky (1979). It remains to calculate the boundary probabilities. Under fluid flow the boundary probability will be 0 for an up state and will be positive for a down state. Finally, the integral over x of $f_i(x)$ must be π_{Ni} minus the boundary probability, if any. The equilibrium distribution is expressed as a mixture of exponentials. This is not surprising as it is the continuous analogue of the matrix geometric distribution.

The fluid flow method allows one to gain insight into the transient behavior of the system. One interesting quantity is the time it takes to reduce a backlog of x data packets with voice in a particular state. One can define

$t_\ell(x)$ = E(time for data to reach 0|initially in state (x,ℓ))

$a_\ell(x)$ = E(area under data sample path until it reaches 0|(x,ℓ)).

The second quantity is the total accumulated waiting time over this part of the busy period. A backward equation approach yields for N=1

$$t_1(x) = dt+t_1(x+r_1dt)(1-\lambda dt) + t_2(x+r_1dt)\lambda dt + o(dt)$$

$$t_2(x) = dt+t_2(x+r_2dt)(1-(\mu+\alpha)dt) + t_1(x+r_2dt)\mu dt + t_3(x+r_2dt)\alpha dt + o(dt) \quad (3.4)$$

$$t_3(x) = dt+t_3(x+r_3dt)(1-(\mu+\beta)dt) + t_1(x+r_3dt)\mu dt + t_2(x+r_2dt)\beta dt + o(dt).$$

Standard manipulations result in

$$-\underset{\sim}{t}'(x) = \underset{\sim}{S}^{-1}\underset{\sim}{M}\underset{\sim}{t}(x) + \underset{\sim}{S}^{-1}\underset{\sim}{1}$$
$$\text{with } \underset{\sim}{1} = (1,1,1)^T \text{ and } \underset{\sim}{t}(x) = (t_1(x),\ t_2(x),\ t_3(x))^T. \quad (3.5)$$

This derivation is also quite general, and (3.5) holds for any $N \geq 1$ where the elements of the equation are given the obvious interpretations for $N \geq 1$. Equation (3.5) has a straightforward matrix exponential solution. The exact solution is rather complicated; however, for reasonably large values of x, $t_1(x)$ is approximately linear, that is proportional to x.

The deviation of a(x) is very similar to that of (3.5). One finds

$$-\underset{\sim}{a}'(x) = \underset{\sim}{S}^{-1}\underset{\sim}{M}\underset{\sim}{a}(x) + x\underset{\sim}{S}^{-1}\underset{\sim}{1}. \quad (3.6)$$

This again yields an exact solution which is quite detailed, but which behaves quadratically for large values of x. The reader is referred to Gaver and Lehoczky (1979) for a derivation of these facts and a discussion of boundary conditions. Not only do these functions give insight into first passage probabilities and hence transient behavior, they can be used to characterize the busy periods of the system.

The form of the matrix $\underset{\sim}{M}$, whether for N=1 or N>1 is important and indicates that the TASI system will have great advantages over the ordinary voice-data system. In Gaver and Lehoczky (1979), it is shown that the matrix $\underset{\sim}{S}^{-1}\underset{\sim}{M} = \frac{\mu}{\eta}\underset{\sim}{U}$ where $\underset{\sim}{U}$ is some matrix. This results in μ/η being a scale parameter for the data queue length distribution, and the mean data queue length is scaled by η/μ which can be as large as 10^4. The introduction of TASI changes everything. Now for N=1,

$$\underset{\sim}{S}^{-1}\underset{\sim}{M} = \frac{\mu}{\eta}\begin{pmatrix} -\frac{\rho_v}{\rho_d-1} & \frac{\rho_v}{\rho_d-1} & 0 \\ \frac{1}{\rho_d} & \frac{-(1+\frac{\alpha}{\mu})}{\rho_d} & \frac{\alpha}{\mu\rho_d} \\ \frac{1}{\rho_d-1} & \frac{\beta}{\mu(\rho_d-1)} & \frac{-(1+\frac{\beta}{\mu})}{\rho_d-1} \end{pmatrix} \tag{3.7}$$

If the talkspurt and silent periods have a length the same order of magnitude as δ, then the factors α/μ and β/μ will be very large, order of 10^4. An eigenvalues of $\underset{\sim}{S}^{-1}\underset{\sim}{M}$ will be order η/μ and the resulting queue length distribution will be far smaller than that of a voice-data system. A TASI system with short talkspurts provides a real solution to the problem.

4. Diffusion Approximation

A second method available for studying the dynamic behavior of a TASI system is the diffusion approximation. This method replaces the actual queueing process with a Wiener process. The parameters of the Wiener process are fit by moments. The methods for developing such an approximation are based on the work of Burman (1979); the reader should consult Burman or Gaver and Lehoczky (1980) for a full description of the method including a discussion of the behavior at the boundary.

We assume N=1 and find the generator of the TASI Markov chain. Recall that the voice process has 3 states: 1, 2, and 3. Introduce a smooth function f(i,j) where i corresponds to the data system size, while j gives the voice state. Using (2.1) one finds the generator to be

$$Af(i,j) = \begin{cases} \lambda(f(i,2)-f(i,1)) + \delta(f(i+1,1)-f(i,1)) \\ \qquad + \eta \min(i,1)(f(i-1,1) - f(i,1)) \quad j=1 \\ \mu(f(i,1)-f(i,2) + \alpha(f(i,3)-f(i,2)) \\ \qquad + \delta(f(i+1,2)-f(i,2)) \quad j=2 \\ \mu(f(i,1)-f(i,3)) + \beta(f(i,2)-f(i,3)) + \delta(f(i+1,3)-f(i,3)) \\ \qquad - \eta \min(i,1)(f(i-1,3) - f(i,3)) \quad j=3. \end{cases} \tag{4.1}$$

One now scales time by n^{-1} and scales the data queue length by $n^{-1/2}$. As $n \to +\infty$, the data process becomes continuous. One calculates the generator of the scaled process and applies it to a sequence of functions

$$f_n(x,j) = g(x) + \frac{1}{\sqrt{n}} u(x,j) + \frac{1}{n} v(x,j).$$

The result is

$$A_n f_n = \begin{cases} \sqrt{n}\ (\lambda(u(x,2)-u(x,1)) + (\delta-\eta)g'(x)) + (\lambda(v(x,2)-v(x,1)) \\ \quad + (\delta-\eta)u_x(x,1) + \frac{1}{2}g''(x)(\delta+\eta)) + 0(n^{-1/2}) \quad j=1 \\ \sqrt{n}\ (\mu(u(x,1)-u(x,2)) + \alpha(u(x,3)-u(x,2)) + \delta g'(x)) \\ \quad + (\mu(v(x,2)-v(x,1)) + \alpha(v(x,3)-v(x,2)) + \delta u_x(x,2) \\ \quad + \frac{1}{2}\delta g''(x)) + 0(n^{-1/2}) \quad j=2 \\ \sqrt{n}\ (\mu(u(x,1)-u(x,3)) + \beta(u(x,2)-u(x,3)) + (\delta-\eta)g'(x)) \\ \quad + (\mu(v(x,1)-v(x,3)) + \beta(v(x,2)-v(x,3)) + (\delta-\eta)u_x(x,3) \\ \quad + \frac{1}{2}(\delta+\eta)g''(x)) + 0(n^{-1/2}) \quad j=3. \end{cases} \tag{4.2}$$

The functions u and v are to be selected to have (4.2) converge to a diffusion limit which is independent of j as $n \to +\infty$. Thus each of the three coefficients of $\sqrt{n}$ must equal some function S(x), while each of the coefficients of the 0(1) term must equal some T(x).

One can find

$$S(x) = g'(x) \sum_{i=1}^{3} r_i \pi_i = g'(x)\eta(\rho-1)$$

where

$$\rho = \rho_d + \frac{\rho_v}{1+\rho_v} \frac{\mu+\beta}{\alpha+\mu+\beta} \quad .$$

The functions u are determined up to a constant $k_1 g'(x)$ so

$$u_1(x) = k_1 g'(x)$$

$$u_2(x) = (k_1 + \frac{\eta}{\mu} \frac{(\rho-\rho_d)}{\rho_v}) g'(x) = (k_1+k_2)g'(x) \tag{4.3}$$

$$u_3(x) = u_2(x) + \frac{\eta}{\alpha} ((\frac{\rho-\rho_d}{\lambda} - \rho_d)(\frac{1+\rho_v}{\rho_v}) + \frac{\mu}{\lambda})g'(x) = (k_1+k_2+k_3)g'(x).$$

The function T(x) is found in the same way but is more complicated. It is given by

$$\begin{aligned} T(x) = {} & [\pi_1(\delta-\eta)k_1 + \tfrac{1}{2}(\delta+\eta)) + \pi_2(\delta(k_1+k_2) + \tfrac{1}{2}\delta) \\ & + \pi_3((\delta-\eta)(k_1+k_2+k_3) + \tfrac{1}{2}(\delta+\eta))]g''(x) = \sigma g''(x) \end{aligned} \tag{4.4}$$

The constant k_1 is arbitrary and contributes $k_1\eta(\rho-1)g''(x)$ to T(x). One finally finds

$$A_n f_n = \sqrt{n}\, S(x) + \sigma g''(x) + 0(n^{-1/2}) \qquad x > o \tag{4.5}$$

In order to have a limit as $n \to +\infty$, one must introduce a heavy traffic assumption, $\sqrt{n}\,\eta(\rho-1) \to \eta\theta < 0$ or $\rho = 1 - \frac{\theta}{\sqrt{n}}$. With this assumption $A_n f_n \to A_\infty g(x) = -\eta\theta g'(x) + T(x)$. This corresponds to a Wiener process with drift $\eta\theta$, scale σ, and reflection at 0. This characterization of the data queue length process is especially convenient. First, the equilibrium distribution will be exponential with parameter $\frac{\eta\theta}{2\sigma}$. Since this distribution is for the data queue scaled by $\sqrt{n}$, it suggests the distribution of the data queue length is exponential with parameter $\frac{\eta(1-\rho)}{2\sigma}$. The Wiener process characterization is also useful in a characterization of the dynamic behavior of the queue and first passage behavior.

It is important to note that this analysis given for N=1 generalizes to N>1. The drift term is again $\sum_i r_i \pi_i = N\eta(\rho-1)$. The calculation of the scale parameter is more complicated but still straightforward.

The numerical evaluation of queueing performance quantities using the diffusion approximation is discussed in Gaver and Lehoczky (1981).

5. Non-Markov Service Times

In this section we illustrate the flexibility of the diffusion method by applying it to a voice data system in which the voice service times are non-exponential. We present a simple special case, N=1 with no TASI, but the analysis can be generalized.

Specifically we assume N=1, data can transmit only when no voice users are present, and voice calls have a service distribution with distribution function G, density g, and hazard function h. It is no longer necessary to distinguish between talkspurts and silent periods. We again construct the generator of the joint process and take limits appropriately. The generator has three aspects: the size of the data queue, whether a voice customer is present (1) or not (0), and the elapsed service time if a customer is present. We introduce a function f(i,j,y) and find

$$
Af = \begin{cases} \lambda(f(x,1,0)-f(x,0,0)) + \delta(f(x+1,0,0)-f(x,0,0)) \\ \quad + \eta \min(x,1)(f(x-1,0,0) - f(x,0,0)) \qquad j=0 \\ \\ f_y(x,1,y) + h(y)(f(x,0,0)-f(x,1,y)) \\ \quad + \delta(f(x+1,1,y) - f(x,1,y)) \qquad j=1. \end{cases} \tag{5.1}
$$

Next time is scaled by n and the data queue length by $\frac{1}{\sqrt{n}}$. The function f is replaced by $f_n(x,j,y) = w(x) + \frac{1}{\sqrt{n}} u(x,j,y) + \frac{1}{n} v(x,j,y)$. The resulting generator is given by

$$
A_n f_n = \begin{cases} \sqrt{n}(\lambda(u(x,1,0)-u(x,0,0)) + (\delta-\eta)w'(x)) \\ \quad + (\lambda(v(x,1,0)-v(x,0,0)) + (\delta-\eta)u_x(x,0,0) \\ \quad + \frac{1}{2}(\delta+\eta)w''(x)) + O(n^{-1/2}) \qquad j=0 \\ \\ \sqrt{n}(u_y(x,1,y) + h(y)(u(x,0,0)-u(x,1,y)) + \delta w'(x)) \\ \quad + (v_y(x,1,y) + h(y)(v(x,0,0)-v(x,1,y)) + \delta u_x(x,1,y) \\ \quad + \frac{1}{2}\delta w''(x)) + O(n^{-1/2}) \qquad j=1 . \end{cases} \tag{5.2}
$$

Again the functions u and v are selected to have the generator be independent of j and y. Set both coefficients of $\sqrt{n}$ to S(x). It is straightforward to compute

$$
\begin{aligned} S(x) &= -\eta(1-\rho)w'(x) \quad \text{with} \quad \rho = \rho_d + \rho_v/(1+\rho_v) \\ u(x,1,y) &= u(x,0,0) + \frac{w'(x)\eta}{1+\rho_v} \cdot \int_y^\infty \frac{(1-G(z))dz}{1-G(y)} . \end{aligned} \tag{5.3}
$$

The term $u(x,0,0) = kw'(x)$ with k arbitrary. One requires u(x,1,y) to be bounded. An application of L'Hospital's rule shows that this will hold if and only if the hazard function is bounded away from 0.

The scale terms are handled in a similar fashion. Both are set to $T(x)w''(x)$.

One finds

$$T(x) = \frac{\eta}{2} (\rho_d+(1-\rho)(k(1+\rho_v)-1) + \frac{\eta}{\mu} \frac{\rho_d\rho_v}{(1+\rho_v)^2} I) \tag{5.4}$$

where

$$\mu^2 I = \int_0^\infty \int_y^\infty (1-G(z))dzdy = \frac{1}{2} \int_0^\infty y^2 dG(y),$$

the second moment of the service distribution. Furthermore

$$\begin{aligned} v(x,1,y) = v(x,0,0) + \frac{1}{1-G(y)} [&\delta(\delta+\eta(1-\rho))\int_y^\infty \int_z^\infty (1-G(u))dudz \\ &- (T(x)-\frac{1}{2}\delta-k\delta) \int_y^\infty (1-G(z))dz]. \end{aligned} \tag{5.5}$$

It is interesting to observe the way in which the assumption of a general G has modified the scale parameter. The crucial term in (5.4) is

$$\frac{\eta}{\mu} \frac{\rho_d\rho_v}{(1+\rho_v)^2} I$$

which involves the coefficient of variation of the service distribution. For exponentials I=1. Thus the data queue length distribution contains the (1+coefficient of variation) as a scale parameter.

The Wiener process approximation follows in a way similar to the results of section 4. The data queue length process is approximately a Wiener process with drift $-\eta(1-\rho)$ and scale given by (5.4). The equilibrium distribution is thus exponential and dynamic behavior can be addressed.

The preceding derivation should serve as an illustration of the importance of the contribution of Burman (1979). The methods developed by Burman have a wide range of applicability. They also offer tractable methods for complicated stochastic systems, even non-Markovian systems. Nevertheless, much work remains to be done. The initial numerical studies indicate that the Wiener process approximation so derived offers good accuracy only for very high traffic intensities, at least .95. This is simply not adequate for many practical situations. One possible approach is to investigate the importance of the arbitrary constant $u(x,0,0)$. It is multiplied by $(1-\rho)$ and is therefore asymptotically negligible; however, its contribution for moderate ρ may be important. A second approach is to fit the coefficients of the diffusion approximation differently. Rather than matching moments of infinitesimal rates as is done here, other methods may lead to greater accuracy. Finally, it may be overly optimistic to expect the marginal data process to be modelled by the marginal diffusion process.

It is clear that each of the techniques discussed in this paper has a role to play in the analysis of complicated communication system models. Far more work is needed by applied probabilists in refining these methods so that they can be applied to even more complex systems.

6. Summary

This paper has given methods for carrying out a performance evaluation of voice/data queueing systems. First, the problem was formulated in such a way that the equilibrium probability vector had matrix geometric form. This allows one to utilize the algorithmic work of Neuts (1978) for calculating system performance quantities and provides a complete solution for the Markovian formulation in equilibrium. Second, two approximation methods were given, a fluid flow approximation and a diffusion approximation. These approximation methods are especially useful for studying busy period problems and for dealing with non-exponential systems. The numerical implementation is discussed in Gaver and Lehoczky (1981). The approximations give insight into the dynamic behavior of the system; however, the numerical accuracy was acceptable only when the traffic intensity was very high. Possible improvements were mentioned.

In general, TASI was shown to be a control strategy of great promise. It leads to dramatic decreases in data queue lengths with no additional increase in voice blocking.

7. References

[1] Arthurs, E. and Stuck, B. W. (1979), "A Theoretical Traffic Performance Analysis of an Integrated Voice-data Virtual Circuit Packet Switch," IEEE Trans. Comm., COM-27, 1104-1111.

[2] Barbacci, M. R. and Oakley, J. D. (1976), "The Integration of Circuit and Packet Switching Networks Toward a SENET Implementation," 15th NBS-ACM Annual Technique Symposium.

[3] Bhat, U. N. and Fischer, M. J. (1976), "Multichannel Queueing Systems with Heterogeneous Classes of Arrivals," Naval Res. Logist. Quart., 23, 271-283.

[4] Brady, P. T. (1967), "A Statistical Analysis of On-Off Patterns in 16 Conversations," Bell System Tech. J., 73-91.

[5] Burman, D. Y. (1979), "An Analytic Approach to Diffusion Approximations in Queueing," Ph. D. Dissertation, New York University.

[6] Chang, Lih-Hsing (1977), "Analysis of Integrated Voice and Data Communication Networks," Ph. D. Dissertation, Department of Electrical Engineering, Carnegie-Mellon University, November.

[7] Coviello, G. and Vena, P. A. (1975), "Integration of Circuit/ Packet Switching in a SENET (Slotted Envelop NETwork) Concept," National Telecommunications Conference, New Orleans, December, pp. 42-12 to 42-17.

[8] Feldman, R. M. and Claybaugh, C. A. (1980), "A Computational Model for a Data/Voice Communication Queueing System," Technical Report, Industrial Engineering Department, Texas A&M University.

[9] Fischer, M. J. (1977a), "A Queueing Analysis of an Integrated Telecommunications System with Priorities," INFOR, 15, 277-288.

[10] Fischer, M. J. (1978), "Data Performance in a System where Data Packets are Transmitted during Voice Silent Periods - Single Channel Case," DCA Technical Report.

[11] Fischer, M. J. (1980), "Delay Analysis of TASI with Random Fluctuations in the Number of Voice Calls," IEEE Trans. Comm., COM-28, November, 1883-1889.

[12] Fischer, M. J. and Harris, T. C. (1976), "A Model for Evaluating the Performance of an Integrated Circuit- and Packet-Switched Multiplex Structure," IEEE Trans. Comm., COM-24, February.

[13] Gaver, D. P. and Lehoczky, J. P. (1981), "Channels that Cooperatively Service a Data Stream and Voice Messages," to appear IEEE Trans. Comm.

[14] Halfin, S. (1972), "Steady-state Distribution for the Buffer Content of an M/G/1 Queue with Varying Service Rate," SIAM J. Appl. Math., 23, 356-363.

[15] Halfin, S. and Segal, M. (1972), "A Priority Queueing Model for a Mixture of Two Types of Customers," SIAM J. Appl. Math., 23, 369-379.

[16] Lehoczky, J. P. and Gaver, D. P. (1979), "Channels that Cooperatively Service a Data Stream and Voice Messages," Carnegie-Mellon University Technical Report No. 171.

[17] Lehoczky, J. P. and Gaver, D. P. (1980), "Diffusion Approximations for the Cooperative Service of Voice and Data Messages," to appear J. Appl. Probab.

[18] Neuts, M. F. (1978), "Markov Chains with Applications in Queueing Theory, which have a Matrix-Geometric Invariant Probability Vector," Adv. in Appl. Probab., 10, 185-212.

[19] Weinstein, C. J., Malpass, M. L., and Fischer, M. J. (1980), "Performance of Data Traffic in an Integrated Circuit and Packet Switched Multiplex Structure," IEEE Trans. Comm., COM-28, 873-877.

[1]Research supported in part by a Grant from the National Science Foundation NSF-ENG-7905526.

*Carnegie-Mellon University, Pittsburgh, Pennsylvania 15213

**Naval Postgraduate School, Monterey, California 93940

PROBABILISTIC ASPECTS OF SIMULATION
Peter Lewis, Chairman

P. Heidelberger & P. D. Welch
A. J. Lawrance & P. A. W. Lewis
L. W. Schruben

ON A SPECTRAL APPROACH TO SIMULATION RUN LENGTH CONTROL

Philip Heidelberger
and
Peter D. Welch

This paper is concerned with the problems of generating confidence intervals for the steady state mean of an output sequence from a single run, discrete event simulation and using these confidence intervals to control the length of the simulation. It summarizes the results of two earlier papers, [5] and [6], and the reader is referred to those papers for a more detailed discussion.

We assume that the simulation generates a covariance stationary process $\{X(n), n \geq 1\}$ with mean $\mu = E[X(n)]$ and spectral density $p(f)$. Under general conditions the sample mean, $\bar{X}$, is, for large samples, approximately normally distributed with mean μ and variance $p(0)/N$ where N is the sample size. Confidence intervals on μ are generated through the estimation of $p(0)$. The problem of run length control is addressed by defining a sequential procedure which continues the simulation until a confidence interval of desired accuracy is obtained.

To avoid storing the entire output sequence, we work with a set of batch means which occupy a fixed amount of storage. The variance of $\bar{X}$ is unaffected by the batching procedure. Let B be the batch size and N_B be the number of batches ($N = BN_B$), then Variance $(\bar{X}) \approx p(0)/N = p_B(0)/N_B$ where $p_B(f)$ is the spectral density of the batch means. Thus to generate a confidence interval it is sufficient to estimate $p_B(0)$. This quantity is estimated by smoothing the logarithm of the averaged periodogram of the batch means.

Let $I(n/N_B)$ be the periodogram of the batch means defined by

$$I(n/N_B) = \left|\sum_{j=1}^{N_B} \bar{X}_B(j)e^{-2\pi i(j-1)n/N_B}\right|^2/N_B.$$

Let $f_n = (4n-1)/2N_B$ and define $J(f_n) = \log(\{I((2n-1)/N_B) + I(2n/N_B)\}/2)$. The quantity $J(f_1)$ is the logarithm of the average of $I(1/N_B)$ and $I(2/N_B)$, $J(f_2)$ is the logarithm of the average of $I(3/N_B)$ and $I(4/N_B)$, etc. If $0 < n,m < N_B/4$ then $J(f_n)$ has the following approximate properties (see [2] and [3])

$$E[J(f_n)] \approx \log(p_B(f_n)) - 0.270$$

$$\text{Variance}[J(f_n)] \approx 0.645$$

$$\text{Covariance}[J(f_n),J(f_m)] \approx 0.0 \quad n \neq m$$

We will estimate $p_B(0)$ by fitting a smooth function to $J(f_n)$. The sequence $J(f_n)$ is used because it has a constant variance and an approximately symmetric distribution.

In [5] $\log(p_B(f))$ was estimated by using ordinary least squares to fit a polynomial of degree d, $g(f_n) = a_0 + a_1 f_n + \ldots + a_d f_n^d$, to $J(f_n) + 0.270$ for $n = 1,\ldots,K$ where K is a fixed integer. From this fit, a variance estimate $\hat{p}_B(0)$ is obtained and a confidence interval for μ is generated based on a Student's t approximation to the distribution of $(\bar{X}-\mu)/(\hat{p}_B(0)/N_B)^{1/2}$. The number of degrees of freedom is determined by the coefficient of variation of $\hat{p}_B(0)$. This coefficient of variation is a decreasing function of d.

A major problem is to determine an appropriate value for d. For small samples a relatively large value of d is required to properly approximate $\log(p_B(f))$. However, in [5] we showed that $\lim_{B\to\infty} Bp_B(f) = p(0)$ so that for large batch sizes $p_B(f)$ is nearly flat and a large value of d is unnecessary. Thus the potential exists for increasing the stability of $\hat{p}_B(0)$ by reducing d as the shape of $p_B(f)$ changes.

In [5] the parameters K = 25 and d = 2 were recommended. This represented a compromise between small sample flexibility and large sample stability. In [6] adaptive smoothing techniques were considered. These adaptive methods attempt to obtain both a more robust small sample and a more stable large sample estimate of $p_B(0)$. Three types of adaptive methods were considered: polynomial regression with the degree selected by stepwise regression (see [4]), polynomial regression with the degree selected by cross validation (see [1] and [7]), and smoothing splines with the amount of smoothing determined by cross validation (see [8] and [9]).

Empirical tests on a variety of queueing models showed that the large sample behavior of the adaptive methods was only slightly better than that of the fixed degree quadratic method. Furthermore, when incorporated into a sequential run length control procedure, the adaptive methods produced significantly degraded confidence interval coverage. This severe degradation in coverage was not a problem for the quadratic method. As a result of these studies, the fixed degree quadratic method is recommended as an effective, simple and practical technique for simulation confidence interval generation and run length control.

1. References

[1] Allen, D. M. (1974). The relationship between variable selection and data augmentation and a method of prediction. Technometrics. 16, 125-127.

[2] Bartlett, M.S. and Kendall, D.G. (1946). The statistical analysis of variance hetereogeniety and the logarithmic transformation. J. Roy. Statist. Soc. (Suppl.), 8, 128-138.

[3] Brillinger, D.R. (1975). Time Series, Data Analysis and Theory. Holt, Rinehart and Winston, Inc., New York.

[4] Draper, N.R. and Smith, H. (1966). Applied Regression Analysis. John Wiley and Sons, Inc., New York.

[5] Heidelberger, P. and Welch, P.D. (1981). A spectral method for simulation confidence interval generation and run length control. Comm. ACM, 24, 233-245.

[6] Heidelberger, P. and Welch, P.D. (1981). On the statistical control of simulation run length. IBM J. Res. Develop., 25, 860-876.

[7] Stone, M. (1974). Cross-validatory choice and assessment of statistical predictions. J. Roy. Statist. Soc. Ser. B. 36, 111-147.

[8] Wahba, G. and Wold, S. (1975). A completely automatic French Curve: fitting spline functions by cross validation. Comm. Statist., 4, 1-17.

[9] Wahba, G. and Wold, S. (1975). Periodic splines for spectral density estimation: the use of cross validation for determining the degree of smoothing. Comm. Statist., 4, 125-141.

IBM Thomas J. Watson Research Center, Yorktown Heights, New York 10598.

GENERATION OF SOME FIRST-ORDER AUTOREGRESSIVE MARKOVIAN SEQUENCES OF POSITIVE RANDOM VARIABLES WITH GIVEN MARGINAL DISTRIBUTIONS

A. J. Lawrance†

and

P. A. W. Lewis††

Abstract

Methods for simulating dependent sequences of continuous positive-valued random variables with exponential, uniform, Gamma, and mixed exponential marginal distributions are given. In most cases the sequences are first-order, linear autoregressive, Markovian processes. A very broad two-parameter family of this type, GNEAR(1), with exponential marginals and both positive and negative correlation is defined and its transformation to a similar multiplicative process with uniform marginals is given. It is shown that for a subclass of this two-parameter family extension to mixed exponential marginals is possible, giving a model of broad applicability for analyzing data and modelling stochastic systems, although negative correlation is more difficult to obtain than in the exponential case. Finally, two schemes for autoregressive sequences with Gamma distributed marginals are outlined. Efficient simulation of some of these schemes is discussed.

1. Introduction

In a recent series of papers [1,2,3,4,5,6,7,8,9] some simple models have been derived for stationary dependent sequences of positive, continuous random variables with given first-order marginal distributions. In general the dependency structure, as measured by second-order joint moments (serial correlations) mimics that of the usual linear mixed autoregressive-moving average (ARMA) models which have been used for so long in time-series analysis. In the ARMA models, which are defined quite generally, there is in usage an implicit assumption of marginal normality of the random variables. This is clearly not the case if the random variables are positive, say the times between events in a series of events (Cox and Lewis [10]) or the successive response times at a computer terminal. Thus the new models are derived to accommodate situations in which the dependent random variables have, for instance, exponential, Gamma, uniform and mixed exponential marginal distributions. The exponential case is the most highly developed, with the nomenclature (Lawrance and Lewis [4]) EARMA (p,q) (exponential process with mixed moving average-autoregressive structures of orders p and q respectively) and NEARMA(p,q) (new EARMA(p,q)). A generalization to extend the range of attainable autocorrelations to negative values has been defined with the nomenclature GNEARMA(p,q).

The development of the probabilistic properties of these processes is given in the referenced papers, applications to queueing models and computer systems modelling by Lewis and Shedler [11] and Jacobs [12,13], while development of estimation and testing procedures has just begun.

The object of the present paper is to define and discuss the simulation of the processes on digital computers, though for the sake of brevity only the first-order Markovian, autoregressive case is considered. The simplicity of structure of these models--in general they are linear additive mixtures of random variables--makes them ideal for this purpose. However, stationarity conditions are sometimes difficult to derive analytically and in some cases it is not simple to generate the innovation random variables in the processes. A striking example of this is the case of the Gamma first-order autoregressive process, given in Section 4.1, for which an efficient means of simultion was reported by Lawrance [7] for some parametric values. This procedure carries over into another Gamma process, the Gamma-Beta process, which will be discussed in Section 4.2.

In Section 3 it is shown that a simple transformation of the exponential sequences gives a direct multiplicative method for generating dependent processes with uniform marginals. These could be the basis in simulations for many other types of dependent sequences.

Finally, the NMEAR(1) process is detailed in Section 4.3; this generates a first-order Markovian process with mixed exponential marginals. It is useful for simulating situations in which the observed random variables are correlated and overdispersed relative to an exponentially distributed random variable.

2. Exponential Autoregressive Markovian Sequences

We give here three methods of generating first-order autoregressive, Markovian sequences with exponential marginal distributions. The first two are defective in terms of their sample path properties (the first more so than the second) while the third, NEAR(1) and its generalization GNEAR(1), is satisfactory in this respect and is a very rich model. The defect of the first two sequences is also highlighted by the simulation procedures used; they can be generated from one sequence of exponential variables.

The word "autoregression" in the context of a stochastic sequence $\{X_n\}$ is often used rather vaguely. In the first place <u>linear</u>, <u>additive</u> autoregression is usually implied. In the second place first-order autoregression can mean that in the defining equation for X_n the previous value enters explicitly. Thirdly, it can mean that the conditional expectation of X_n, given $X_{n-1} = x_{n-1}$, is an additive linear function of x_{n-1};

$$E(X_n | X_{n-1} = x_{n-1}) = a + bx_{n-1}. \tag{2.1}$$

The processes discussed in this paper are autoregressive in the latter two senses and, except in the case of uniform marginals, are autoregressive in a linear additive way. They are also Markovian; the Markovian property (first-order) means that the probability structure of $X_n, X_{n+1}, \dots$, given $X_{n-1} = x_n$ is independent of $X_{n-2}, X_{n-3}, \dots$.

2.1 The Exponential DAR(1) Process

A very simple exponential autoregressive Markovian sequence is generated by the equation (Jacobs and Lewis [14,15])

$$X_n = V_n X_{n-1} + (1 - V_n)E_n, \tag{2.2}$$

where the V_n's, $n = 1,2,\dots$ are i.i.d. with $P\{V_n=1\} = 1 - P\{V_n=0\} = \rho$ and E_n, $n = 1,2,\dots$ are, as throughout the paper, independent exponential random variables with parameter λ and independent of the V_n's; that is

$$\begin{aligned} P\{E_n \leq x\} &= 1 - e^{-\lambda x}, \quad x \geq 0, \quad \lambda > 0 \\ &= 0 \qquad\qquad , \quad x < 0. \end{aligned} \tag{2.3}$$

If X_0 is exponentially distributed with parameter λ and is independent of the E_n's, the process is stationary. Then the serial correlations $\rho_k = \mathrm{corr}(X_n, X_{n+k})$ are

$$\rho_k = \rho^k \tag{2.4}$$

and

$$E(X_n | X_{n-1} = x_{n-1}) = \rho_1 x_{n-1} + (1 - \rho_1)/\lambda. \tag{2.5}$$

This process, which was introduced to model discrete valued variables, is not well suited to modelling continuous data because runs of X_n's with the same value can occur quite frequently in the sample paths of the process. This happens when X_{n-1} is picked successively in (2.2), rather than the innovation E_n. Moreover the lengths of the runs of similar values are geometrically distributed.

2.2 The Exponential EAR(1) Process

Another model is derived from the usual linear model

$$X_n = \rho X_{n-1} + \varepsilon_n \tag{2.6}$$

in which the i.i.d. innovation process $\{\varepsilon_n\}$ is chosen so that the X_n's are marginally exponential(λ). Gaver and Lewis [1] show that for this to be true, one must have $0 \leq \rho < 1$ and

$$\varepsilon_n = \begin{cases} E_n & \text{w.p. } 1-\rho, \\ 0 & \text{w.p. } \rho, \end{cases} \tag{2.7}$$

where $\{E_n\}$, as previously, are i.i.d. exponential(λ). Again $\rho_k = \rho^k$ and $E(X_n | X_{n-1} = x_{n-1}) = \rho_1 x_{n-1} + (1-\rho_1)\lambda$, as at (2.4) and (2.5) for the exponential DAR(1) model. The difference is in the sample paths; the EAR(1) process simulations show runs of geometrically decreasing X_n's, but no runs of constant value. The geometrically distributed runs occur when only ρX_{n-1} is picked in (2.6).

The Markov property of the two sequences implies that if X_0 is chosen to be E_0, an exponential(λ) random variable independent of $E_1, E_2, \ldots$, then $X_1, X_2, \ldots$ forms a stationary sequence.

Naive inspection of the defining equations (2.2), (2.6) and (2.7) suggests that to generate a <u>stationary</u> sequence of length N, $X_1, \ldots, X_N$, (N+1) i.i.d. exponential deviates and N uniform variates (for the selection process) are needed. However, the sequences can be generated from only one exponential sequence; this is possibly related to the degeneracy in the processes. This method uses the memoryless property of exponential(λ) variables, namely that if E_n is given to be greater than a constant γ, then $E_n - \gamma$ is again exponential(λ).

Thus the algorithm for the EAR(1) process is to initialize by setting $X_0 = E_0$; subsequently set $X_n = \rho X_{n-1}$ if $E_n \leq x\rho = -\ln(1-\rho)/\lambda$; otherwise set $X_n = \rho X_{n-1} + (E_n - x_\rho)$. This uses the fact that, from (2.3), $P\{E_n \leq x_\rho\} = \rho$.

Even greater efficiency can be obtained, though this must be qualified by considerations as to whether (i) the X_n's are to be generated one at a time or in an array; (ii) a subroutine is available to generate exponential random variables faster than can be done by taking logarithms of uniform deviates, and (iii) the speed of division in the computer is short compared to the time needed for generation of uniform deviates.

The more efficient scheme recycles uniform variables, i.e., if U is given to be between constants a and b, where $0 < a < b \leq 1$, then $(U-a)/(b-a)$ is a uniform random variable. (Note that its value is not given, only that it is in (a,b)). The expected number of uniform deviates required to generate an EAR(1) process of length N with this algorithm is $1 + (1-\rho)N$, which is less than the number N required to generate an i.i.d. exponential(λ) sequence. Also the expected number of logarithms is $(1-\rho)N$, while N comparisons are always needed.

2.3 The Exponential NEAR(1) Process

A broader two-parameter exponential sequence which is a first-order autoregressive, Markovian process and an additive linear mixture of random variables is given by Lawrance [7] and developed by Lawrance and Lewis [5]. Called NEAR(1), the sequence is defined as

$$X_n = \varepsilon_n + \begin{cases} \beta X_{n-1} & \text{w.p. } \alpha \\ 0 & \text{w.p. } 1-\alpha \end{cases} \qquad n = 1,2,\ldots, \tag{2.8}$$

where $0 \le \alpha \le 1$ and $0 \le \beta \le 1$ but $\alpha = \beta \neq 1$. Also the selection process is done independently for each n. It can be shown that for the X_n to be marginally exponential(λ) the innovation variable ε_n must be generated from an E_n by the exponential mixture

$$\varepsilon_n = \begin{cases} E_n & \text{w.p. } \delta = \dfrac{1-\beta}{1-(1-\alpha)\beta} \\ (1-\alpha)\beta E_n & \text{w.p. } 1-\delta = \dfrac{\alpha\beta}{1-(1-\alpha)\beta} \end{cases} \qquad n = 1,2,\ldots \tag{2.9}$$

providing α and β are not both equal to one. When $\alpha = 0$ or $\beta = 0$ the $\{X_n\}$ are i.i.d. exponential variables, whereas when $\alpha = 1$ the EAR(1) model given at (2.6) and (2.7) is obtained. In fact choosing α as a function of β in a suitable way, e.g., $\beta = \alpha$, gives an exponential model with a full positive range of serial correlation of order one, since it is easily shown that

$$\rho_k = (\alpha\beta)^k. \tag{2.10}$$

Again

$$\begin{aligned} E(X_n | X_{n-1} = x_{n-1}) &= \alpha\beta x_{n-1} + (1-\alpha\beta)/\lambda \\ &= \rho_1 x_{n-1} + (1-\rho_1)/\lambda \end{aligned} \tag{2.11}$$

and $X_0 = E_0$ gives a stationary sequence. Thus the correlations and regressions are the same as for the first two models. However the NEAR(1) process allows one to model a broader class of exponential sequences, as measured either by sample path behavior or higher-order joint moments; see Lawrance and Lewis [5] for details.

A particularly simple case occurs when $\beta = 1$; this model, called TEAR(1), is very tractable analytically and, as will be shown in Section 4.3, extends easily to the case of mixed exponential distributions for the X_n.

Note that in the NEAR(1) process the innovation ε_n is always present i.e., is not zero with positive probability unless $\alpha = 1$ and it is therefore not possible to simulate the stationary process with less than N+1 uniform variates. A detailed algorithm is given in Section 2.5 for a more general case. Since for a stationary array of N X_n's, exactly N+1 uniform deviates are required because of the ability to recycle the uniforms and transform them into exponentials, it could be advantageous to generate these uniform deviates in an array which would be replaced one at a time by the X_n's. Care must be taken with the recycling of the uniform variates U if $\gamma = 1-\alpha$ is close to one or zero. In that case it is probably better for computational reasons to use 2(N+1) uniform variates. Note that $\alpha = 0$ gives the EAR(1) process. When $\beta = 1$ (the TEAR(1) process) a simpler algorithm can be used since E_n is no longer a mixture of two exponentials.

When $\beta = \frac{1}{2-\alpha}$ another one-parameter sub-class of the NEAR(1) process is obtained with strikingly regular sample path properties. For this process $P\{X_n > X_{n-1}) = 1/2$; this is in striking contrast to the EAR(1) process whose sample paths show runs-down and the TEAR(1) process whose sample paths show runs-up. This is illustrated in Figure 1; all sequences have $\rho = 0.75$ and are transformations of the same E_n sequence. The parameter space of the NEAR(1) process is illustrated in Figure 2.

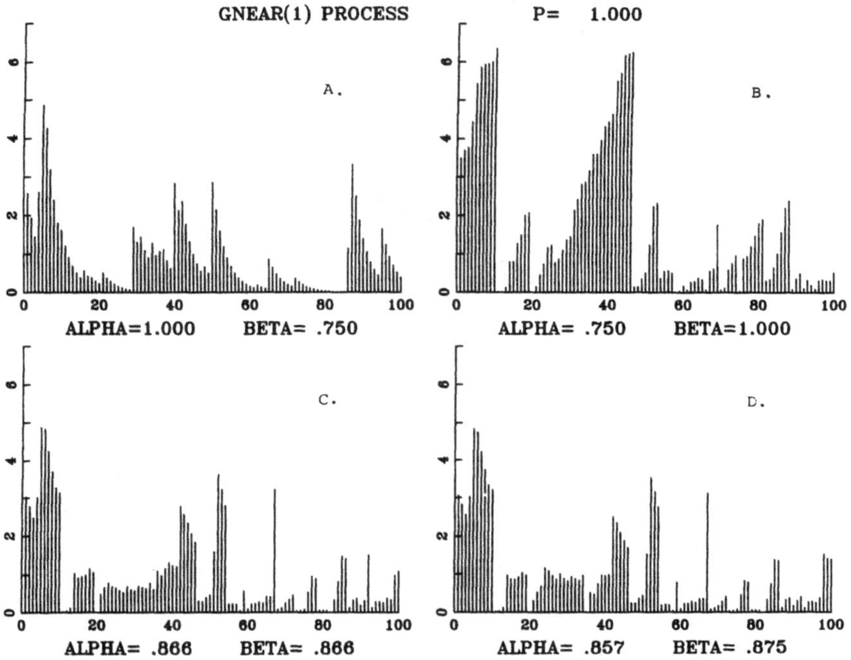

Figure 1. Sample paths of NEAR(1) processes, all with $\alpha\beta = \rho_1 = 0.75$, for different values of α and β. Figure A is the EAR(1) process ($\alpha = 1$, $\beta = 0.75$), Figure B is the TEAR(1) process ($\alpha = 0.75$, $\beta = 1$), Figure C is the PREAR(1) process $\alpha = \frac{1}{2-\beta} = 0.857$, and Figure D is the REAR(1) process $\alpha = \beta = (0.75)^{1/2}$. All the sample paths are transformations of the same i.i.d. exponential sequence E_n, $n = 0,1,\ldots,100$.

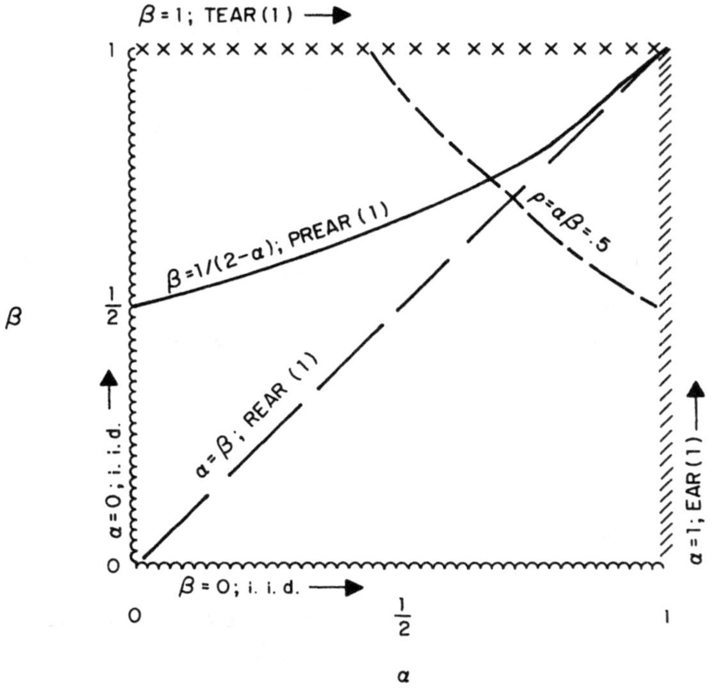

Figure 2. Parameter space for the NEAR(1) exponential autoregressive process. The cases $\alpha = 0$ and/or $\beta = 0$ give i.i.d. exponential sequences while $\beta = 1/(2-\alpha)$ gives the partially reversible PREAR(1) process for which $P(X_n < X_{n-1}) = 1/2$. For $\alpha = 1$ we have the original exponential process EAR(1) which tends to have runs-down. The TEAR(1) case, $\beta = 1$ gives very simple analytics but exhibits runs up. Also shown is a locus of constant ρ_1, in this case $\rho_1 = \alpha\beta = .5$.

2.4 The Generalized Exponential Process GNEAR(1) with Negative Correlation

The exponential processes defined in Sectin 2.1, 2.2, and 2.3 do not exhibit negative correlation or alternation of correlations. Such behaviour is found in, say, a normal linear first-order process for which $\rho_j = \text{cor}(X_n, X_{n+j}) = \rho^j$ and $-1 < \rho < 1$ so that, for instance, ρ_1

can be negative. A scheme for broadening the correlation structure of the EAR(1) process is given in Gaver and Lewis [1]. However in the exponential case a much simpler alternative method is available.

Assume for simplicity that $\lambda = 1$. Now X_{n-1} is a unit exponential variable and $U_n = F(X_{n-1}) = 1 - \exp(-X_{n-1})$ is a uniform (0,1) variable, as is $1 - U_n = \exp(-X_{n-1})$. Then

$$X'_{n-1} = F^{-1}(1 - U_n) = F^{-1}(1 - F(X_{n-1})) = -\ell n(1 - \exp(-X_{n-1}))$$

is a unit exponential variable; in fact it is the antithetic of X_{n-1} which gives the maximum negative correlation attainable in a bivariate exponential distribution:

$$r = \text{corr}(X_{n-1}, X'_{n-1}) = 1 - \pi^2/6 = -.6449. \tag{2.12}$$

Now the process

$$X_n = \begin{cases} \varepsilon_n + \beta X'_{n-1} = \varepsilon_n - \beta\, \ell n(1 - e^{-X_{n-1}}) & \text{w.p. } \alpha \\ \varepsilon_n & \text{w.p. } 1-\alpha \end{cases} \tag{2.13}$$

or

$$X_n = \varepsilon_n + A_n X'_{n-1}, \tag{2.14}$$

where the ε_n's are defined at (2.9) and in which the A_n's are i.i.d. with $P\{A_n=1\} = 1 - P\{A_n=0\} = \alpha$ gives a process with autocorrelations which alternate in sign. In particular $\rho_1 = r(\alpha\beta)$. To combine this with the positive correlation case in a continuous way we introduce a new parameter $p \in [0,1]$ and i.i.d. indicator variables I_n, independent of A_n, from which $P\{I_n=1\} = 1 - P\{I_n=0\} = p$. Then the GNEAR(1) model is defined as

$$X_n = \varepsilon_n + A_n\beta\{I_n X_{n-1} + (1 - I_n)[- \ell n(1 - e^{-X_{n-1}})]\} \tag{2.15}$$

and gives a complete range of first-order serial correlations

$$\rho_1 = \alpha\beta[p + (1-p)r]. \tag{2.16}$$

Higher lag correlations are more complicated and will be given elsewhere. Note, however, that $1-p = 1/(1-r)$ gives a case in which X_n and X_{n-1} are a bivariate exponential pair which are dependent but have zero correlation.

Figure 3 give four sample paths for the case $p = 0$ for values of α and β corresponding to those in Figure 1, which is the case $p = 1$.

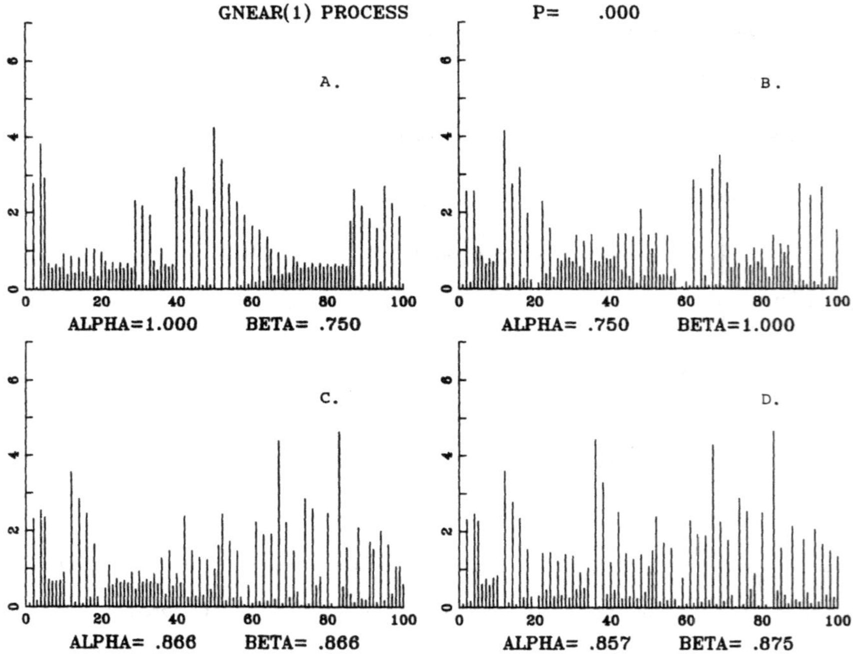

Figure 3. Sample paths for the GNEAR(1) process with $p = 0$, corresponding to those in Figure 1 for the NEAR(1) process ($p = 1$). Here $\rho_1 = -\ 0.6449\alpha\beta$.

2.5 Algorithms for the GNEAR(1) Process

We give here two algorithms for the GNEAR(1) process, one based on generation of uniform deviates one at a time, the other based on the availability of subroutines to generate arrays of uniform and exponential variables. The sequences generated are unit exponentials ($\lambda = 1$). The special case $\alpha = 1$ (EAR(1)) is handled separately as it will cause divides by zero in the main algorithm.

2.5.1 Algorithm GNEAR1A. This algorithm generates a sample of size N from the GNEAR1 process. It is based on the generation of uniform random numbers one at the time, with recycling of these uniform random numbers for further use. It is assumed that a subroutine (UNIFORM) exists that generates faw uniforms. Input values are N, ALPHA, BETA and P, with ALPHA taking values in [0,1] and BETA and P taking values in the closed interval [0,1]. However ALPHA = BETA = 1 is not allowed.

```
INPUT  N, ALPHA, BETA, P
CALL  UNIFORM (U)                          /* Generate uniform U*/
X(0) ← -log_e(U)                           /* convert to exponential*/
C ← (1-ALPHA)*BETA
IF  ALPHA = 0           THEN  D ← 1
                        ELSE  D ← (1-BETA)/(1-C)
END IF
DO    I ← 1 to N
      CALL UNIFORM (U)
      IF U ≤ ALPHA      THEN  CALL UNIFORM (V)
                              IF V ≤ P THEN Y ← BETA* X(I-1)
                                       ELSE Y ← - BETA* log_e(1-exp(-X(I-1)))
                              END IF
                              U ← U/ALPHA                /* Recycle U */
                        ELSE  Y ← 0
                              U ← (U-ALPHA)/(1-ALPHA)    /* Recycle U */
      END IF
```

```
    IF  U ≤ D          THEN  Z ← -log_e(U/D)
                       ELSE  Z ← - C * log_e(U-D)/(1-D))
    END IF
    X(I) ← Z + Y
END DO
```

2.5.2 Algorithm GNEAR1B. This algorithm generates a sample of size N from the GNEAR1 process. It is based on the generation of arrays of uniform and exponentially distributed random numbers. It is assumed that subroutines UNIFORM and EXPON exist that generate arrays of uniform random numbers and exponential random numbers respectively. For each GNEAR1 number generated two raw uniform random numbers are used. The first is recycled in those cases where a third uniform is needed to make the selection with probability P. Input values are N, ALPHA, BETA and P the last three taking values in the closed interval [0,1], except the case ALPHA = BETA = 1.

```
INPUT  N, ALPHA, BETA, P
C ← (1 - ALPHA)*BETA
IF ALPHA = 0           THEN  D ← 1
                       ELSE  D ← (1-BETA)/(1-C)
CALL UNIFORM (U, 2*N)        /*Generate array of 2N uniforms*/
CALL EXPON (E, N+1)          /*Generate array of N+1 exponentials*/
X(0) ← E(N+1)
DO   I ← 1  to  N
    IF  U(I) ≤ ALPHA  THEN  V = U(I)/ALPHA                /* Recycle U */
                            IF V ≤ P THEN Y ← BETA* X(I-1)
                                     ELSE Y ←- BETA* log_e(1-exp(-X(I-1)))
                            END IF
                      ELSE  Y ← 0
    END IF
```

```
    IF  U(I+N) ≤ D    THEN  Z ← E(I)
                      ELSE  Z ← C*E(I)
    END IF
    X(I) ← Z + Y
END DO
```

3. Uniform Markovian Sequences, NUAR(1)

It is convenient to have dependent sequences of random variables with marginal distributions other than exponential. Before discussing other solutions to Equation (2.8) we show that a simple transformation of the NEAR(1) process gives a two-parameter family of Markovian random variables with uniform marginal distributions. It is well-known that if E is a unit exponential random variable, then the transformation exp{-E} gives a uniformly distributed random variable. Thus we have from (2.8) and (2.9) the multiplicative model for a uniform Markovian sequence $\{X_n\}$, n = 1,2,... , called NUAR(1),

$$X_n = \begin{cases} \varepsilon_n X_{n-1}^{\beta} & \text{w.p. } \alpha \\ \varepsilon_n & \text{w.p. } (1-\alpha) \end{cases} \qquad n = 1,2,\ldots , \tag{3.1}$$

where

$$\varepsilon_n = \begin{cases} U_n & \text{w.p. } \gamma = \dfrac{1-\beta}{1-(1-\alpha)\beta} \qquad (3.2) \\ U_n^{(1-\alpha)\beta} & \text{w.p. } 1-\gamma = \dfrac{\alpha\beta}{1-(1-\alpha)\beta} \qquad (3.3) \end{cases} \qquad n = 1,2,\ldots$$

for U_n, n = 1,2,..., i.i.d. uniformly distributed, providing that α and β are not both equal to one. Again if X_0 is uniformly distributed and independent of U_1, U_2, ... the sequence is stationary.

The sequence is clearly quite simply extended to give negative correlation, as in the GNEAR(1) process; in fact X_{n-1} in (3.1) is just replaced by $(1-X_{n-1})$. Algorithms are easily obtained by adaptation of those given in Section 2E for the exponential case. It remains to find the correlation structure and the regression of X_n on X_{n-1}.

To do the former, let X_n^* be a NEAR(1) sequence with $\lambda = 1$, so that the sequence X_n at (3.1) is given by $X_n = \exp\{-X_n^*\}$. Now the joint Laplace-Stieltjes transform of X_n^* and X_{n-k}^*, $\phi_{X_n^*,\, X_{n-k}^*}(s,t) = E\{\exp[-\, sX_n^* - tX_{n-1}^*]\}$, is given by Lawrance and Lewis [5]. Setting $s = t = 1$ in $\phi_{X_n^*,\, X_{n-k}^*}(s,t)$ gives,

$$\phi_{X_n^*,\, X_{n-k}^*}(1,1) = E\{\exp(-X_n^*)\ \exp(-X_{n-k}^*)\} = E(X_n X_{n-k}). \tag{3.4}$$

Then using the fact that for a uniform random variable $E(X) = 1/2$ and $\mathrm{var}(X) = 1/12$ we have, after simplification,

$$\rho_k = \mathrm{corr}(X_n, X_{n-k}) = \frac{3}{2 + \beta^k} \prod_{i=1}^{k} \left(\frac{\alpha\beta}{1 + (1-\alpha)\beta^i} \right)$$

$$k = 1,2,\ldots\ . \tag{3.5}$$

Note that this is not simply a geometrically decaying correlation sequence, as for the NEAR(1) process. However, for the important special case when $\beta = 1$ we get

$$\rho_k = \left(\frac{\alpha}{2-\alpha}\right)^k, \qquad k = 1,2,\ldots\ , \tag{3.6}$$

and thus the serial correlations ρ_k are the kth power of ρ_1, which takes on any value between 0 and 1. Thus we have a particularly simple uniform Markovian sequence, although the sample paths will tend to have runs-up.

A similar analysis given in Lawrance and Lewis [5] shows that

$$E(X_n | X_{n-1} = u) = \frac{1}{2}\, \frac{1 + \beta}{\{(1 + (1-\alpha)\beta\}}\, \{1 - \alpha + \alpha u^{\beta}\} \tag{3.7}$$

so that the regression is not linear for this Markov process with uniform marginals, unless $\beta = 1$.

This uniform sequence could form the basis, via a probability integral transform, of many other sequences with given marginals. The parametrization $\beta = \frac{1}{2-\alpha}$ is a good choice for a one parameter model since this case gives a sample path which is partially time-reversible (see Lawrance and Lewis [5]), with a balance of runs-up and runs-down. However, marginal transformations do not preserve correlation structure, as shown at (3.5), and it is therefore useful to see whether sequences with marginals other than exponential can be generated from (2.8); this requires finding, if possible, a suitable choice of innovation sequence ε_n. The result will be a simple process with autoregressive Markovian structure and the desired marginal distribution.

4. Markovian Sequences with Some Other Marginals

Although an exponential distribution is a common assumption for positive random variables met with in problems in operations research, it is too narrow an assumption to encompass many real situations. Therefore parametric distribution models are invoked which include the exponential as a special case and which allow for the modelling of data which has greater or lesser dispersion than exponentially distributed data. Two commonly used models are

(i) the Gamma(k,λ) distribution whose probability density function is

$$f(x) = \frac{\lambda(\lambda x)^{k-1} e^{-\lambda x}}{\Gamma(k)}, \quad k > 0; \quad \lambda > 0; \quad x \geq 0, \tag{4.1}$$

where $\Gamma(k)$ is the complete gamma function, and

(ii) the (convex) mixture of exponential random variables

$$f(x) = \pi_1\lambda_1 e^{-\lambda_1 x} + (1-\pi_1)e^{-\lambda_2 x}, \quad 0 < \lambda_1 < \lambda_2;$$

$$x \geq 0,\ 0 \leq \pi_1 \leq 1. \tag{4.2}$$

The Gamma distribution has dispersion, measured by the coefficient of variation $C(X) = \sigma(X/E(X))$, which is greater than the exponential value of 1 if $k < 0$ and less than 1 if $k > 1$. The mixed exponential always has $C(X) \geq 1$, the equality occurring when the special case of an exponential random variable with parameters λ_1 or λ_2 holds.

4.1 The Gamma GAR(1) Process

The solution of the standard first-order autoregressive equation (2.6) with stationary gamma marginals defines the GAR(1) process. Using Laplace-Stieltjes transforms with (2.6) shows that for X_n to be Gamma (k,λ), we must have

$$\phi_\varepsilon(s) = E(e^{-s\varepsilon}) = \{\rho + (1-\rho)\,\frac{\lambda}{\lambda+s}\}^k. \tag{4.3}$$

For k integer this has an explicit inverse. For example, for $k = 2$ the innovation ε is zero with probability ρ^2, is exponential(λ) with probability $2\rho(1-\rho)$ and is Gamma$(2,\lambda)$ with probability $(1-\rho)^2$. It is easy to show in general that ε is zero with probability ρ^k, so that the "zero defect" is not serious for large k. A method of simulating a random variable whose Laplace-Stieltjes transform is equation (4.3) was derived by Lawrance [7], using the fact that this sequence arises in a particular type of shot noise process. From this we have the

4.1.1 <u>Gamma Innovation Theorem</u>. Let N be a Poisson random variable with parameter $\theta = -k\,\ell n(\rho)$. Let $U_1, U_2, \ldots, U_N$ be uniformly distributed over (0,1) and independent. Let $Y_1, \ldots, Y_N$ be exponential(λ) and independent. Then ε can be simulated using

$$\varepsilon = \begin{cases} \sum_{m=1}^{N} Y_m \rho^{U_m} & \text{if } N \quad 0, \\ 0 & \text{if } N = 0. \end{cases} \tag{4.4}$$

A proof is not given here. Note that ε is zero with probability $\exp\{-k\ell n(\rho) = \rho^k$. Also the Poisson number N of uniform and exponential random variables which must be generated for each ε has expected value $\theta = -k\ \ell n(\rho)$. This will be prohibitively large, and the simulation will be very inefficient, if k is large and/or ρ is close to zero. Neither of these cases is serious, however. If k is large, say greater than 50, the sequence is almost normal and the usual normally distributed, AR(1) linear process can be used. If ρ is as small as 0.001 then E(N) is only k × (6.9078) which is still reasonable. However, for ρ this small the sequence is approximately i.i.d. Gamma and acceptance-rejection techniques for simulating Gamma variables are known.

It is quite simple to write algorithms for the GAR(1) case analogous to those in Section 2.5. It would pay to have a built-in routine for generating the Poisson variable which will bypass further calculations if $N = 0$. In other words routines for generating Poisson variates which start by searching at the median of a table of cumulative Poisson probabilities will be inefficient.

Unfortunately the NEAR(1) process does not appear to extend to the Gamma case; it can be shown explicitly that there is no innovation ε_n in equation (2.8) which will make X_n have a Gamma distribution with $k = 2$ if $\alpha \neq 1$.

There is, however, another model, the Gamma-Beta model, inspired by an example in Verwaart [16] and discovered independently by Fishman [17], which is quite broad and which can be simulated using the Gamma Innovation Theorem.

4.2 The Gamma-Beta Model, GBAR(1)

This model is a linear autoregressive process with random coefficients which includes the GAR(1) process but is of limited practical use in data analysis because its likelihood function is analytically untractable. Nevertheless it could be useful in simulations, particularly if the zero-defect in the GAR(1) process is unacceptable.

Thus, we define the stationary sequence as

$$X_n = \beta B_n X_{n-1} + \varepsilon_n \qquad 0 \le \beta \le 1, \quad n = 0, \pm 1, \pm 2, \ldots, \tag{4.5}$$

where the B_n's are i.i.d. and independent of X_{n-1}, and B_n, for X_n to have a Gamma (k,λ) distribution, has a Beta$(k-b,b)$ distribution with mean $E(B_n) = (k-b)/k$. The density is

$$f_{B_n}(x) = \frac{\Gamma(k)}{\Gamma(k-b)\Gamma(b)} x^{k-b-1} (1-x)^{b-1}. \qquad 0 \le x \le 1 . \quad 0 < b \le k . \tag{4.6}$$

The distribution of the i.i.d. ε_n's to make X_n Gamma (k,λ) is still to be determined. To do this and to see the rationale behind the model, it is simplest to obtain the distribution of $B_n X_{n-1}$.

Now B_n can be generated as $Z_1/(Z_1+Z_2)$ where Z_1 and Z_2 are independent and Gamma $(k-b,\lambda)$ and Gamma (b,λ) respectively. Moreover, B_n is independent of (Z_1+Z_2), which could be used to generate X_{n-1}. Then, $B_n X_{n-1} = \{Z_1/(Z_1+Z_2)\}\{Z_1+Z_2\} = Z_1$ is Gamma $(k-b,\lambda)$. Using this fact, which can also be shown analytically, and the defining equation (4.3) we have

$$E(e^{-sX_n}) = \phi_{X_n}(s) = \phi_{B_n X_{n-1}}(\beta s)\phi_{\varepsilon_n}(s), \tag{4.7}$$

so that on using the fact that for a Gamma(k,λ) variable the Laplace-Steiltjes transform is $(\lambda/\lambda+s)^k$, we have

$$\phi_{\varepsilon_n}(s) = \phi_{X_n}(s)/\phi_{B_nX_{n-1}}(\beta s)$$

$$= (\frac{\lambda}{\lambda+s})^b (\frac{\lambda+\beta s}{\lambda+s})^{k-b} \quad (4.8)$$

$$= (\frac{\lambda}{\lambda+s})^b [\beta + (1-\beta)\frac{\lambda}{\lambda+s}]^{k-b} . \quad (4.9)$$

Thus ε_n is generated as the sum of a Gamma (b,λ) variate and, from (4.3), a Gamma innovation variable which can be generated from (4.4).

For this model

$$\rho_j = \text{corr}(X_n X_{n+j}) = (\beta \frac{k-b}{k})^j \qquad j = 0,1,2,... \quad (4.10)$$

and

$$E(X_n | X_{n-1} = x_{n-1}) = \rho_1 x_{n-1} + (1-\rho_1)E(X_n).$$

Also when $b = 0$ we have the GAR(1) model. If $\beta = 1$ the model requires only a Gamma (b,λ) variate for ε_n and a Beta$(k-b,b)$ variate for B_n in the simulation. The case $k = 1$ gives an exponential process which is not the same as the NEAR(1) process.

It should be noted that while the Gamma Innovation Theorem makes the Gamma-Beta model tractable from a simulation viewpoint, it is still difficult to write down a likelihood function or to get negative correlation. Thus further developments are needed from the Gamma case. An algorithm is given for the Gamma-Beta Model below.

4.2.1 Algorithm GBAR1 (GAMMA-BETA MODEL). This algorithm generates a sample of size N from the GBAR1 process using array generators POISSON, UNIFORM, EXPONENTIAL and GAMMA to produce Poisson, Real Uniform (0,1), Exponential and Gamma distributed random numbers.

The following restrictions exist on the input parameters:

```
    0 ≤ B ≤ K
    0 < K
    0 < BETA ≤ 1
INPUT  N, BETA, B, K
TH ← -(K-B) * log_e(BETA)              /*Initialization*/
CALL POISSON(PSN,N,TH)    /*Generate N poisson deviates with parameter TH*/
CALL GAMMA(Z1,N,K-B)      /*Generate N Gammas with parameter K-B*/
CALL GAMMA(Z2,N,B)        /*Generate N Gammas with parameter B*/
CALL GAMMA(X(0),1,K)      /*Initialize X(0)*/
DO  I ← 1  to  N
    CALL UNIFORM(U,PSN(I))             /*Generate PSN(I) real (0,1) uniform*/
    CALL EXPON(E,PSN(I))               /*Generate PSN(I) Unit Exponentials*/
    Y ← 0
    DO  J ← 1  TO  PSN(I)
        Y ← Y + E(J) * BETA ** U(J)
    END DO
    BN ← Z1(I)/(Z1(I) + Z2(I))         /*Compute BETA Deviate*/
    X(I) ← BETA * BN * X(I-1) + Y
END DO
```

4.3 Mixed Exponential Markovian Processes MEAN(1) and TMEAR(1)

In addition to Gamma processes, first-order autoregressive Markovian processes with mixed exponential marginal distributions can be obtained from equations (2.8) and (2.9) in two special cases, and these sequences should be widely useful in modelling stochastic systems.

(i) The case $\alpha = 1$; MEAR(1).

In Gaver and Lewis [1] it is shown that the solution to the Laplace transform of ε_n for the linear model (2.6) is a constant ρ plus a (generally) non-convex mixture of three exponential functions. This can be shown to be a proper density function if $\rho \leq \lambda_1/\lambda_2$, but it can also be shown that it is not a density function for all ρ less than one and greater than or equal to zero. More particularly, Lawrance [6] showed that unless λ_1 is much smaller than λ_2 (and thus the X_n are very over-dispersed relative to an exponential random variable) a solution exists for ε_n for all ρ. Thus we have a useful process, although again the zero-defect of order ρ is a problem. However, one case which cannot be simulated this way occurs when $1/\lambda_2 = 0$, i.e., X is zero with probability $1-\pi_1$, and exponential (λ_1) otherwise. This kind of situation occurs in practice as, e.g., the waiting time for an item in an inventory system. Fortunately it can be handled in the next case.

(ii) The case $\beta = 1$; TMEAR(1).

When $\beta = 1$ in equation (2.8), a mixed exponential process TMEAR(1) is obtained which is extremely simple to simulate since the innovation ε_n is just the mixture of two exponentials for all $0 \leq \rho < 1$. Moreover, the process has no zero-defect. As discussed above, the sample paths will tend to "run up," but this is no great problem unless ρ is fairly large. Thus we have the following Theorem (Lawrance and Lewis [18]) which we state without proof:

TMEAR(1) Theorem. Let the first-order autoregressive, Markovian sequence $\{X_n\}$ be defined by

$$X_n = \varepsilon_n + V_n X_{n-1}, \qquad n = 1,2,3,\ldots$$

where for the i.i.d. sequence V_n, $n = 1,2,3,\ldots$, $P\{V_n=1\} = 1 - P\{V_n=0\} = \alpha$ for $0 \leq \alpha < 1$. Then the sequence $\{X_n\}$ is stationary and has a (convex) mixed exponential marginal distribution with probability density function

$$f_X(x) = \pi_1\lambda_1 e^{-\lambda_1 x} + \pi_2^{-\lambda_2 x}, \qquad x \geq 0 \tag{4.11}$$

where $0 < \lambda_1 < \lambda_2$; $0. < \pi_1 < 1$; $\pi_1 + \pi_2 = 1$; $\mu_2 = \frac{1}{\lambda_1}$ and $\mu_2 = \frac{1}{\lambda_2}$, if ε_n is i.i.d. and has a (convex) mixed exponential distribution given by

$$f_{\varepsilon}(x) = \eta_1\gamma_1 e^{-\gamma_1 x} + \eta_2\gamma_2 e^{-\gamma_2 x} \qquad x \geq 0$$

with

$$\gamma_1 > \gamma_2 > 0;\ \eta_1,\ \eta_2 > 0,\ \eta_1 + \eta_2 = 1 \tag{4.12}$$

where

$$\mu = E(X) = \pi_2\,\mu_2 + \pi_1\mu_1; \tag{4.13}$$

$$b = \mu_1 + \mu_2 - \alpha\mu; \tag{4.14}$$

$$\beta = \mu_1 + \mu_2 - \mu; \tag{4.15}$$

$$a = (1-\alpha)\mu_1\mu_2; \tag{4.16}$$

$$\gamma_1,\gamma_2 = \frac{1}{2}\,\{b \pm \surd\,[b^2-4a]\}; \tag{4.17}$$

$$\gamma_0 = \pi_2\mu_1 + \pi_1\mu_2; \tag{4.18}$$

$$\eta_1 = (\gamma_1-\gamma_0)/(\gamma_1-\gamma_2);\ \eta_2 = (\gamma_2-\gamma_0)/(\gamma_2-\gamma_1) \tag{4.19}$$

and X_0 is independent of ε_1, ε_2, ... and has probability density function (4.11).

Note that the special cases where $\pi_1 = 0$ or $\pi_2 = 1$ give NEAR(1) exponential processes with parameters λ_2 and λ_1 respectively. Thus they should be handled by Algorithm 2 since they will cause computational problems. The case $\lambda_1 = \lambda_2$ also gives a NEAR(1) process and is excluded for similar reasons. This guarantees that $\gamma_1 > \gamma_2$. The algorithm below implements this theorem.

4.3.1 Algorithm TMEAR1. This algorithm generates a sample of size N from the TMEAR1 process. It is based on the generation of arrays of uniform and unit exponential numbers. Subroutines UNIFORM and EXPON are assumed to exist to generate such arrays. For each TMEAR1 number generated two raw uniform random numbers and one exponential number are needed.

```
INPUT  N, PI1, PI2, MU1, MU2, ALPHA
MU ← PI1 * MU2 + PI2 * MU2                                /* Initialization */
 B ← MU1 + MU2 - ALPHA * MU
 A ← (1 - ALPHA) * MU1 * MU2
 T ← SQRT(B * B - 4 * A)
G2 ← .5 * (B - T)
G1 ← .5 * (B + T)
G0 ← PI2 * MU1 + PI1 * MU2
E1 ← (G1 - G0)/G1 - G2)
R1 = 1/G1
R2 = 1/G2                                                 /* Initialization */
CALL UNIFORM(U,2*N+1)              /* Generate 2N+1 uniforms */
CALL EXPON(E,N+1)                  /* Generate N+1 unit exponentials */
IF U(2N+1) ≤ PI1          THEN  X(0) ← MU1 * E(N+1)
                          ELSE  X(0) ← MU2 * E(N+1)
END IF
DO   I ← 1  to  N
     If  U(I) ≤ E1        THEN  Z ← R1 *E(I)
                          ELSE  Z ← R2 *E(I)
     END IF
     IF  U(I+I) ≤ ALPHA   THEN  Y ← (X(I-1)
                          ELSE  Y ← 0
     END IF
     X(I) ← Y + Z
END DO
```

5. Generalizations

In all of the processes discussed here except the exponential GNEAR(1) the correlations are non-negative and geometrically decreasing.

A particular scheme for obtaining negative correlation is given for the exponential case. Another scheme for obtaining alternating correlations which are possibly negative and which is broadly applicable is given in Gaver and Lewis [1] and in Lawrance and Lewis [5]. Another problem is that different types of dependence and higher-order Markovian dependence might be encountered in data. Schemes for obtaining mixed autoregressive moving average exponential sequences where the autoregression has order p and the moving average has order q are given in Lawrance and Lewis [4]. The mixed exponential process TMEAR(1) is easily extended to give a process with this type of extended correlation structure. This will be discussed elsewhere.

6. References

[1] Gaver, D. P. and Lewis, P. A. W., "First Order Autoregressive Gamma Sequences and Point Processes," Adv. in Appl. Probab., 12, 1980, 727-745.

[2] Lawrance, A. J. and Lewis, P. A. W., "A Moving Average Exponential Point Process (EMA1)," J. Appl. Probab., 14, 1977, 98-113.

[3] Jacobs, P. A. and Lewis, P. A. W., "A Mixed Autoregressive-Moving Average Exponential Sequence and Point Process (EARMA 1,1)," Adv. in Appl. Probab., 9, 1977, 87-104.

[4] Lawrance, A. J. and Lewis, P. A. W., "The Exponential Autoregressive Moving Average EARMA(p,q) Process," J. Roy. Statist. Soc. Ser. B, 42, 1980, 150-161.

[5] Lawrance, A. J. and Lewis, P. A. W., "A New Autoregressive Time-Series Model in Exponential Variables (NEAR(1))," Adv. in Appl. Probab., 13, 1981, 826-845.

[6] Lawrance, A. J., "The Mixed Exponential Solution to the First-Order Autoregressive Model," J. Appl. Probab., 17, 1980, 546-552.

[7] Lawrance, A. J., "Some Autoregressive Models for Point Processes," Point Process and Queuing Problems, Colloquia Mathematica Societates Janos Bolyai, 24, P. Bartfai and J. Tomko eds., North Holland, 1981, Amsterdam, 257-275.

[8] Lewis, P. A. W., "Simple Models for Positive-Valued and Discrete-Valued Time Series with ARMA Correlation Structure," Proc. Fifth International Symposium Mult. Anal., Multivariate Analysis-V, P. R. Krishnaiah, ed. North-Holland, Amsterdam, 1980, 151-166.

[9] Lawrance, A. J. and Lewis, P. A. W., "Simulation of some Autoregressive sequences of positive random variables." Proc. 1979 Winter Simulation Conference, Highland, H.J., Spiegel, M. G. and Shannon, R. eds., 1979. IEEE Press, N.Y., 301-308.

[10] Cox, D. R. and Lewis, P. A. W., Statistical Analysis of Series of Events, Methuen, London, 1966.

[11] Lewis, P. A. W. and Shedler, G. S., "Analysis and Modelling of Point Processes in Computer Systems," Bull. ISI, XLVII (2), 1978, 193-219.

[12] Jacobs, P. A., "A Closed Cyclic Queueing Network with Dependent Exponential Service Times," J. Appl. Probab., 15, 1978, 573-589.

[13] Jacobs, P. A., "Heavy traffic results for single-server queues with dependent (EARMA) service and interarrival times," Adv. in Appl. Probab., 12, 1980, 517-529.

[14] Jacobs, P. A. and Lewis, P. A. W., "Discrete Time Series Generated by Mixtures, I: Correlational and Runs Properties," J. Roy. Statist. Soc. Ser. B., 40, 1978, 94-105.

[15] Jacobs, P. A. and Lewis, P. A. W., "Discrete Time Series Generated by Mixtures II: Asymptotic Properties," J. Roy. Statist. Soc. Ser. B, 40, 1978, 222-228.

[16] Vervaat, W., "On a stochastic difference equation and a representation of non-negative infinitely divisible random variables," Adv. in Appl. Probab., 11, 1979, 750-783.

[17] G. S. Fishman, personal communication.

[18] Lawrance, A. J. and Lewis, P. A. W., "A mixed exponential time series model, NMEAR(p,q)," Naval Postgraduate School Technical Report NPS55-80-012, 1980. To appear in Management Sci.

The work of the authors was supported by the Office of Naval Research under Grant NR-42-284.

† Dept. Statistics, University of Birmingham, Birmingham, England.

†† Dept. Operations Research, Naval Postgraduate School, Monterey, CA.

TESTING FOR INITIALIZATION BIAS IN THE MEAN OF A SIMULATION OUTPUT SERIES

Lee W. Schruben

Abstract

An initial state for a computer simulation model must be specified each time the program is run. Often convenient starting conditions (such as an empty system) are unlikely to occur naturally and introduce initialization bias in the output. One of the most important manifestations of initialization bias is a change in the output process mean near the beginning of a run while the program "warms up". Sometimes a change in the output mean is obscured by random variations in the process being simulated. An approach to measuring the significance of initialization bias in a output series is presented in [1]. The method is general in that the supporting assumptions are not very restrictive.

Let $Y_1, Y_2, \ldots, Y_n$ represent the output process and let μ_i denote the mean of Y_i. The sequence $\{S_n(0) = 0,\ S_n(k) = \bar{Y}_n - \bar{Y}_k;$ $k = 1,2,\ldots,n\}$ is used to test for a change in μ_i. If μ_i is constant for the entire process then the scaled test sequence

$$T_n(t) = \frac{[nt]S_n([nt])}{\sqrt{n}\sigma} \quad ; \ t \in [0,1]$$

will behave as a standard Brown bridge process as the run duration, n, becomes large. The scaling constant σ^2 is equal to $\lim_{n\to\infty} n\ \mathrm{VAR}(\bar{Y}_n)$ and may be estimated in a variety of ways.

If μ_i is not constant then the limiting Brownian bridge process is shifted away from a zero mean. A one-sided test for initialization bias of a particular sign is developed by comparing the observed maximum of $T_n(t)$ and its location with a standard Brownian bridge maximum and the location of the maximum. In [1] the performance of this test is examined using several simulation programs. The test appears to be valid and to have good power in detecting the presence of initialization bias.

If the change in the process mean can be assumed to follow a particular functional form, i.e., $\mu_i = \mu(1 - f_i)$ where $f_i \to 0$, then the most powerful test (based on $S_n(k)$) of the hypothesis $f_i = 0$ against the alternative, f_i, can be developed. This test is presented in [2]. The most powerful test against a quadratic transient mean function is studied with several simulation models. This test was also apparently valid and usually about as powerful as the test in [1] in detecting bias. In situations where the transient mean is approximately quadratic the test in [2] was considerably more powerful than the test in [1]. Very little computation is involved in either test.

These tests should not be used by rote (as with any statistical test) but should be helpful in assessing whether or not initialization bias is a serious problem in a particular model. The tests also might be useful in supporting arguments that a particular initialization strategy has been effective in controlling bias; again, not as a substitute for common sense or intuition.

1. References

[1] Schruben, L., (1981) "Detecting Initialization Bias in Simulation Output", Technical Report 444, School of Operations Research and Industrial Engineering, Cornell University, Ithaca, New York 14853.

[2] Schruben, L., Singh, H., and Tierney, L. (1980) "A Test of Initialization Bias Hypotheses in Simulation Output", Technical Report 471, School of Operations Research and Industrial Engineering, Cornell University, Ithaca, New York 14853.

Cornell University, Ithaca, New York 14853

QUEUEING MODELS IN PERFORMANCE ANALYSIS, I

Daniel Heyman, Chairman

A. E. Eckberg

J. M. Holtzman

RESPONSE TIME ANALYSIS FOR PIPELINING JOBS
IN A TREE NETWORK OF PROCESSORS

A. E. Eckberg, Jr.

Abstract

A commonly recurring system architecture for many real-time applications consists of a main processor with several front end processors attached. A rationale for such an architecture is that simpler operations, not requiring a centralized information base, can often be more economically implemented if they are off-loaded to simpler, albeit not necessarily slower, processors. Consequently there is more processing power available in the main processor for those operations requiring centralized information.

In this paper we consider response times in the generalization of such an architecture: a tree network of processors. Jobs enter the network at low level front end processors, and exit after processing at the main processor. Exogenous arrivals are modeled as occurring according to Poisson processes, and processing requirements at each processor level are assumed to be deterministic. Exact response time distributions, as well as simple mean response time expressions, are obtained for the situation, commonly resulting from "standard engineering practice," where the occupancies of all processors are equal. Comparisons are made with the approximation that would result were the individual nodes to be modeled as M/D/1 queues. In particular it is shown that this approximation results in significant performance analysis errors if the processor occupancies are high and/or if the performance criterion involves the tails of the response time distribution.

The basic model can also be used to compare the performance resulting from different design alternatives in a number of situations. For example, it is shown that a First-Come-First-Served discipline at each processor in the above described tree network results in a performance far superior to that resulting from a Processor-Sharing discipline. Also, the effectiveness of a standard flow control scheme for a packet switching node is evaluated using the basic model.

1. Introduction and Summary

The pervasive use of modular design methodologies in computer-based systems has made network architectural issues extremely important. From fairly simple architectures consisting of a main processor and several peripheral (or front end) processors, to local computer networks, to global, and often geographically dispersed, computer networks, it is often the case that architectural structure plays a fundamental role in determining a system's performance.

The real-time performance analysis of such systems often centers on the response times of jobs, either at a single node or along a particular path segment in the network. Several models have been proposed and analyzed for estimating the characteristics of such response times. In particular, the fairly generalized queueing networks of Baskett, Chandy, Muntz, and Palacios [1] have served as the vehicle for analyzing numerous network structures. However, there are situations where the assumptions inherent in such models are likely to result in significant errors.

This paper is motivated by an architecture, arising in many real-time applications, in which jobs are "pipelined" through a "tree network" of processors; see Figure 1.1. Jobs enter the network at low

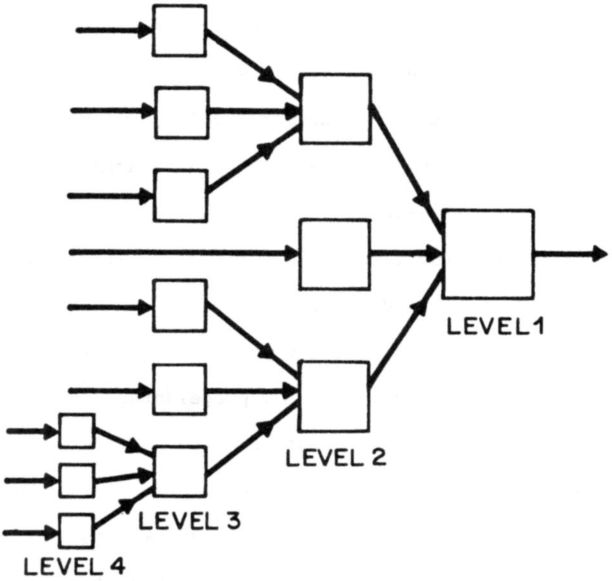

Figure 1.1. A Tree Network of Processors

level front end processors, and exit after processing at the single main processor. A rationale for such an architecture is that simpler operations, not requiring a centralized information base, can often be more economically implemented if they are off-loaded to simpler, albeit not necessarily slower, processors. For ease of exposition, we shall refer to "levels" of processors, where the main processor is at level 1 and the inputs of any level i processor are either the outputs of a subset of the level $i+1$ processors or exogenously arriving jobs. We do not assume that all of the front end processors reside at the same level.

In many real-time applications utilizing such an architecture, the **total** processing times (i.e., real-time processing requirements) of all jobs are nearly deeterministic and equal; or, there may be K classes of jobs such that within a given class all jobs have a nearly deterministic and identical total processing time, although these requirements may differ from class to class. In such a situation it is usually possible to segment the jobs into tasks and arrange that each task of a given job be implemented at a different processor level in an architecture such as that of Figure 1.1. Moreover, it usually can be arranged that for $i \geq 1$ and for each level i processor:

(i) the processing times of all tasks at this processor are deterministic and equal;

(ii) the input to this level i processor is either the output from one or more level $i+1$ processors, or an exogenous input, but not a combination of both of these;

(iii) if there are level $i+1$ processors converging on this level i processor, the number of such processors is no larger than the ratio of the smallest of the processing times at these level $i+1$ processors to the processing time at the level i processor.

For example, Figure 1.2 illustrates the possible implementation of

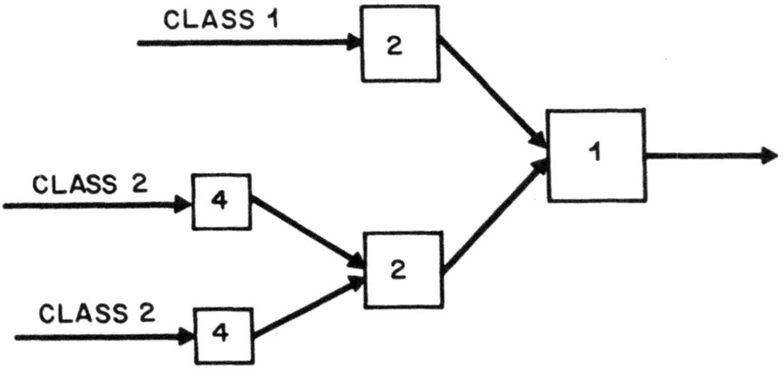

Figure 1.2. Implementation of Two Job Classes, Requiring 3 and 7 Processing Time Units

two classes of jobs, where class 1 and class 2 jobs require, respectively, 3 and 7 total processing time units. Similarly, Figure 1.3 illustrates the case for three classes of jobs with total processing times of 11, 9, and 4 units.

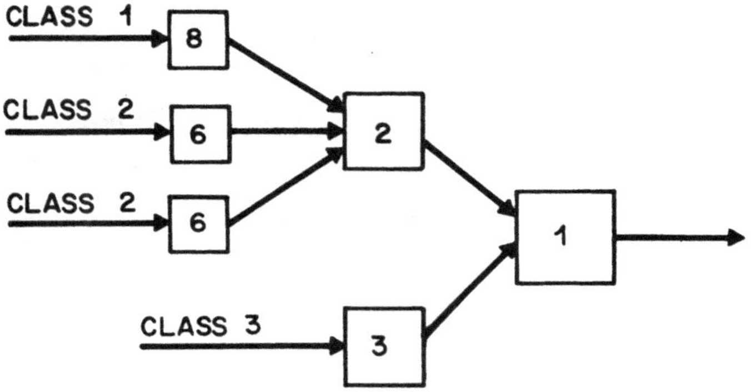

Figure 1.3. Implementation of Three Job Classes, Requiring 11, 9, and 4, Processing Time Units

Condition (i) reflects the determinism of real-time processing requirements common in many applications, and also the likely event that certain basic tasks are key elements for several different job classes, thereby allowing the type of convergent job flow illustrated in Figures 1.2 and 1.3. Note that if we replace condition (iii) with a more restrictive one (implying (iii)):

(iii') if there are level $i+1$ processors converging on this level i processor, then the processing times at all of these level $i+1$ processors are equal, and the number of such processors equals the ratio of their (common) processing time to the processing time at the level i processor,

and if the arrival rates of jobs are adjusted so that all front end processors have the same occupancy ρ, then the occupancies of <u>all</u> processors in the tree will equal ρ. Architectural design is often based on equalizing load in this manner. An example of such a "balanced" network is depicted in Figure 1.2.

The objectives of this paper are to develop the methodology for analyzing the response times of jobs in such tree networks of processors satisfying the above conditions (i)-(iii), and where the processing discipline at each processor is First-Come-First-Served (FCFS). For such arrangements, condition (iii) is seen to imply that processing delays at any level i processor receiving input from level $i+1$ processors must be bounded. This fact suggests that approximations based on traditional queueing network models are likely to result in considerable errors.

In the remaining sections we develop the required methodology for determining processing response times, both at a single processor, and along any path segment in the network. Explicit expressions are obtained both for means and for complementary distributions, with the former being expressed in terms of the well known Erlang loss function. It is shown, for example, that the processing delays, as seen at individual processors by the tasks of a given job as it traverses the network, are independent; thus, to obtain the response time distribution along a path segment of the network, one has only to convolve the individual response time distributions at the nodes along this path. Finally, the results are illustrated with several numerical examples.

The results reported here could form the basis of a useful architectural design methodology; however, certain extensions would increase their usefulness. In particular, condition (i) can be slightly relaxed; the remark following Theorem 2.3.1 gives one simple result. Next, it would be desirable to relax condition (ii), thereby allowing a combination of exogenous inputs and outputs from other processors to be the input to a given processor; reference [2] presents some results which might possibly be extended. Finally, it is desirable to attempt easing condition (iii). Reference [3] intersects somewhat with this paper, and contains some interesting results for a slight relaxation of (iii) when the "fan-in" at a processor equals two.

2. Analysis

The analysis of this section is focused on a single interior (i.e., other than front end) node in a network of the type described in the Introduction. Our objective is to determine the characteristics of response times of tasks processed at this node, i.e., the sum of the actual processing time and the time spent waiting for processing. Since we are assuming that tasks at this node have identical processing times and are processed in a FCFS manner, we model the node as a FCFS queue with unit deterministic service times,[1] and concentrate on the waiting time. We obtain explicit and easily computable expressions for both the mean and the complementary distribution of the waiting time, and also show that the waiting time of a job at this interior node is independent of waiting times of that job at previously encountered nodes in the network. This property allows computation of the distribution of the total waiting time of a job in the network via convolution. These results depend on fairly loose assumptions about the stream of tasks arriving at the node, which are always justified when the node is an interior node of a tree network satisfying the assumptions in the Introduction, but which are valid in other settings as well, such as that illustrated by the flow control example in Section 3.

2.1 Problem Statement

Let $S_1, S_2, \ldots, S_N$ be N independent stationary arrival streams, and denote the rate of stream S_i by λ_i. Also, let S denote the superposition of the streams S_i. We assume that there exists a constant T such that:

(i) $T > N$;

(ii) for each i, the interarrival times of S_i are no smaller than T, with probability 1.

Aside from the above, we make no further assumptions regarding the S_i; however, one can picture, as in the case for an interior node in a tree network, that S_i is the output of a FCFS single server queue with deterministic service time T_i, so that a logical choice for T is: $T = \min_i T_i$. Note that there is no unique choice for T, but as a consequence of (ii) we must have $\lambda_i \leq T^{-1}$ for all i.

Now consider a FCFS single server queue with unit deterministic service times, and for which the input process is S, the superposition of the streams S_i, $1 \leq i \leq N$ (Figure 2.1). Our objective is to analyze

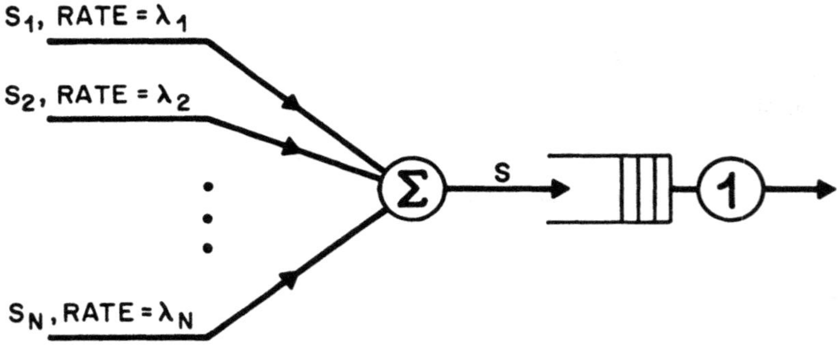

Figure 2.1. The Basic Queueing Model

the waiting times in this system. To this end we introduce the following notation:

W_i = waiting time of an arbitrary arrival selected from S_i

W = waiting time of an arbitrary arrival selected from S

$F_i(x) = P[W_i > x]$

$F(x) = P[W > x]$

$w_i = E[W_i]$

$$w = E[W].$$

It will be shown that closed form expressions for the above distributions and means can be obtained, expressed only in terms of N, $\lambda_1, \lambda_2, \ldots, \lambda_N$, provided conditions (i) and (ii) above are satisfied.

2.2 Some Fundamental Results

The distinguishing feature of the queueing system depicted in Figure 2.1 is that no more than N arrivals can occur in any interval of length T. Consequently, because $T > N$, no arrival at the queue can experience an unbounded waiting time; in fact it is easily seen that no individual waiting time can exceed $N-1$, and this bound can be achieved only if N arrivals occur simultaneously. The following result strengthens this observation.

Proposition 2.2.1. Tag an arbitrary arrival from the stream S, and position the time origin at this epoch. Let $X_0 = 0$, and for $n \geq 1$, let X_n denote the length of time since the n'th previous arrival, so that the arrival times during $(-\infty,0]$ are given by $..,-X_{n+1},-X_n,-X_{n-1},\ldots,-X_1$, 0.[2] Define n^* as

$$n^* = \min\{n \geq 0 | X_{n+1} \geq T\}.$$

Then, $n^* \leq N-1$, and W, the waiting time of the tagged arrival, is given by

$$W = \max(0,1-X_1,2-X_2,\ldots,n^*-X_{n^*}). \tag{2.2.1}$$

Proof. That $n^* \leq N-1$ follows trivially, since at most N arrivals can occur in $(-T,0]$, i.e., $X_N \geq T$. Now, let $W(n)$ denote the waiting time of the n'th previous arrival. Because all service times equal 1, the Lindley recurrence ([4]) for a FCFS single server queue becomes

$$W = \max(0,W(1)+1-X_1)$$

$$W(n) = \max(0,W(n+1)+1-(X_{n+1}-X_n)),\ n \geq 1,$$

or simply,

$$W = \max(0,1-X_1,2-X_2,\ldots,n-X_n+W(n)),\ n \geq 1.$$

In particular,

$$W = \max(0,1-X_1,\ldots,n^*-X_{n^*},n^*+1-X_{n^*+1}+W(n^*+1)). \quad (2.2.2)$$

Now observe that, because there can be at most N arrivals in any interval of length T, there is at least one arrival in (-T,0] for which the waiting time is zero. Let this be the k'th previous arrival, i.e., $W(k) = 0$. Note that k may equal 0, but in any case we have $k \leq n^*$. Thus, for some k, $0 \leq k \leq n^*$, $W = \max(0,1-X_1,2-X_2,\ldots,k-X_k)$ which, together with (2.2.2), proves (2.2.1).

The essence of Proposition 2.2.1 is that the waiting time of any arrival is <u>functionally</u> expressible in terms of the relative arrival epochs during the previous T-interval. There is always at least one server idle period which buffers the waiting time of the tagged arrival from any causal influence of arrivals prior to this T-interval. Also, it should be observed that none of the arrivals during the previous T-interval can have originated from the same stream as the tagged arrival. We thus have the following results.

<u>Corollary 2.2.2</u>. Conditioned on the arrival epochs during (-T,0], W is statistically independent of the arrivals prior to this interval.

<u>Corollary 2.2.3</u>. W is statistically independent of any other arrivals originating from the same stream as the tagged arrival.

<u>Corollary 2.2.4</u>. If stream S_i is the output process of a queueing system (this system may itself be a network of queues), and if the tagged arrival in S originates from S_i, then W is statistically independent of the waiting time and service time experienced by the tagged arrival in this previous queueing system.

Proof. Corollary 2.2.2 is a direct consequence of (2.2.1). Now, while W is not statistically independent of all arrival epochs prior to (-T,0], any such statistical dependence that exists is purely a result of the dependence of $X_1, X_2, \ldots, X_{n*}$ on these prior arrivals. However, it is clear that the X_i, $1 \leq i \leq n*$, are independent of all arrivals from the tagged arrival's stream, and this proves Corollary 2.2.3. Finally, Corollary 2.2.4 follows directly from Corollary 2.2.3.

2.3 Mean Waiting Times

After having established the above results, it is a fairly easy matter to derive expressions for the quantities w and w_i.

<u>Theorem 2.3.1</u>. For $1 \leq i \leq n$, define the functions $s_{n,i}: R^n \to R$ as

$$s_{n,i}(\alpha_1, \ldots, \alpha_n) = \text{sum of products of all distinct } i\text{-tuples taken from } \{\alpha_1, \ldots, \alpha_n\} \tag{2.3.1}$$

$$= \sum_{j_1=1}^{n-i+1} \sum_{j_2=j_1+1}^{n-i+2} \cdots \sum_{j_i=j_{i-1}+1}^{n} \alpha_{j_1} \alpha_{j_2} \cdots \alpha_{j_i} .$$

Then,

$$w = \frac{1}{2} [s_{N,1}(\lambda_1, \ldots, \lambda_N)]^{-1} \sum_{j=2}^{N} j! \; s_{N,j}(\lambda_1, \ldots, \lambda_N) \tag{2.3.2}$$

and

$$w_i = \frac{1}{2} \sum_{j=1}^{N-1} j! \; s_{N-1,j}(\lambda_1, \ldots, \lambda_{i-1}, \lambda_{i+1}, \ldots, \lambda_N). \tag{2.3.3}$$

Proof. Consider the set $\underline{N} = \{1, \ldots, N\}$, and all subsets of $\underline{N}$. For a non-null $\eta \subset \underline{N}$, consider the system of Figure 2.2, i.e., where only the streams S_i, $i \in \eta$, are superposed and used as input to the queue. For this situation, let

L_η = mean number in queue

v_η = mean virtual waiting time of the system

w_η = mean waiting time in queue,

where L_η and v_η are "time averages" and w_η is a "customer average." Also, let

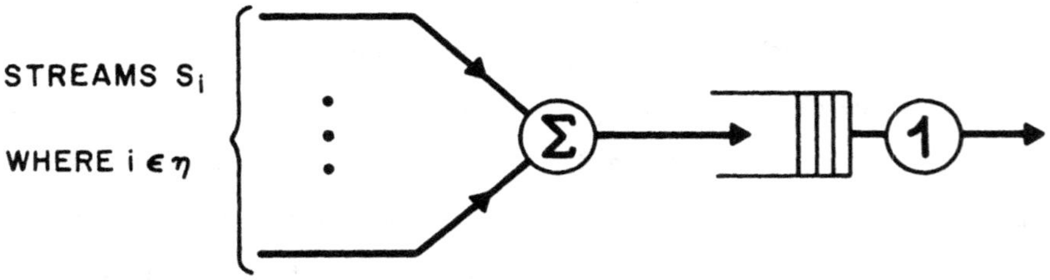

Figure 2.2. Superposing a Subset of the Streams S_i (Proof for Thm. 2.3.1)

$$\lambda_\eta = \sum_{i\varepsilon\eta} \lambda_i .$$

Note that because of the unit deterministic service times, λ_η is the probability that the server is busy at an arbitrary point in time; and if the server is busy, then the mean remaining service time of the task is service is just 1/2. Therefore, $v_\eta = L_\eta + \frac{1}{2}\lambda_\eta$. Moreover, from Little's Law, $L_\eta = \lambda_\eta w_\eta$, and so

$$v_\eta = \lambda_\eta[w_\eta + \frac{1}{2}] . \qquad (2.3.4)$$

Now, for any non-null $\eta \neq \underline{N}$, and for $i \notin \eta$ define

$w_{\eta,i}$ = mean customer waiting time for arrivals from S_i when the queue input is the superposition of S_i and $\{S_j,\ j \varepsilon \eta\}$.

From Proposition 2.2.1 it is clear that the waiting times experienced by the arrivals from S_i are caused solely by the streams S_j, $j \varepsilon \eta$. Indeed, the actual customer waiting times (random variables) for S_i can be realized by sampling, at the S_i arrival epochs, the virtual waiting time sample function resulting from the streams S_j, $j \varepsilon \eta$. Because the streams are stationary, and S_i is independent of S_j, $j \varepsilon \eta$, it can be easily shown that the mean customer waiting time for S_i is just the mean (time average) virtual waiting time produced by S_j, $j \varepsilon \eta$, i.e.,

$$w_{\eta,i} = v_\eta . \qquad (2.3.5)$$

Finally, w_η is just the weighted sum

$$w_\eta = \sum_{i\varepsilon\eta} \frac{\lambda_i}{\lambda_\eta} w_{\eta-\{i\},i} \; . \tag{2.3.6}$$

Combining (2.3.4), (2.3.5), and (2.3.6) we arrive at the following recursion

$$w_\eta = \sum_{i\varepsilon\eta} \frac{\lambda_i}{\lambda_\eta} \lambda_{\eta-\{i\}}[w_{\eta-\{i\}} + \frac{1}{2}] ,$$

with the initial conditions: $w_{\{i\}} = 0$, for all $i \varepsilon \underline{N}$. A simple inductive argument will show that the solution to this recursion is just

$$w_\eta = \frac{1}{2} \lambda_\eta^{-1} \sum_{j=2}^{n} j! \; s_{n,j}(\{\lambda_i, \; i \; \varepsilon \; \eta\})$$

where n is the cardinality of η; moreover,

$$w_{\eta,i} = v_\eta = \frac{1}{2} [\lambda_\eta + \sum_{j=2}^{n} j! \; s_{n,j}(\{\lambda_i, \; i \; \varepsilon \; \eta\})]$$

Finally, the proof is completed by noting that

$$\lambda_\eta = s_{n,1}(\{\lambda_i, \; i \; \varepsilon \; \eta\}), \; w = w_{\underline{N}}, \text{ and } w_i = w_{\underline{N}-\{i\},i} \; .$$

Remark. It is possible to generalize the above somewhat to incorporate nondeterministic service times. Thus, if the mean service time equals 1, if the second moment of the service time equals m_2, and if there exists a constant τ satisfying $P[\text{service time} \leq \tau] = 1$ and $N \tau \leq T$, then (2.3.2) and (2.3.3) need only be multiplied by m_2 to obtain valid expressions for w and w_i.

Remark. The values of $s_{n,i}(\alpha_1,\ldots,\alpha_n)$, $1 \leq i \leq n$, can be easily computed via the following recursion. Let $\beta_{1,1} = \alpha_1$, and for $1 \leq k \leq n-1$ let

$$\beta_{k+1,1} = \beta_{k,1} + \alpha_{k+1}$$

$$\beta_{k+1,i} = \beta_{k,i} + \alpha_{k+1} \beta_{k,i-1}, \text{ for } 2 \leq i \leq k$$

$$\beta_{k+1,k+1} = \alpha_{k+1} \beta_{k,k} \; .$$

Then $s_{n,i}(\alpha_1,\ldots,\alpha_n) = \beta_{n,i}$.

Finally, considerable simplification results if $\lambda_i = \lambda$ for all i. Then it is easily seen that $s_{n,i}(\lambda_1,\ldots,\lambda_n) = \binom{n}{i}\lambda^i$, and we have the following.

Corollary 2.3.2. When $\lambda_i = \lambda$ for all i,

$$w = w_i = \frac{1}{2N\lambda}\sum_{j=2}^{N} j!\binom{N}{j}\lambda^j = \frac{1}{2}\sum_{j=1}^{N-1} j!\binom{N-1}{j}\lambda^j$$

$$= \frac{(N-1)\lambda}{2}\,[B(N-2,\ \lambda^{-1})]^{-1}$$

where $B(\cdot,\cdot)$ is the Erlang loss function:

$$B(M,a) = \frac{a^M/M!}{\sum_{i=0}^{M} a^i/i!}$$

2.4 Waiting Time Distributions

Proposition 2.2.1 provides the necessary machinery for determining the complementary distributions $F(x)$ and $F_i(x)$. We begin with the following result.

Lemma 2.4.1. For an arbitrarily selected arrival from the stream S_i, let n_i^* denote the number of arrivals in the previous T-interval, just as n^* was defined in Proposition 2.2.1. Then, the conditional complementary distribution $P[W_i > x \mid n_i^* = k]$ is given by

$$P[W_i > x \mid n_i^* = k] = T^{-k}P_k(T,x)$$

where, for a given $x \geq 0$, $\{P_i(t,x);\ i \geq 0,\ t \in R\}$ is a family of functions satisfying

$$P_0(t,x) = 0$$

$$\frac{\partial}{\partial t} P_i(t,x) = iP_{i-1}(t,x),\ i \geq 1$$

$$P_i(i-x,x) = [\max(0,i-x)]^i,\ i \geq 1.$$

For fixed $x \geq 0$ and $i \geq 1$, $P_i(t,x)$ is a polynomial of degree i-1 in t, with the representation

$$P_i(t,x) = \sum_{j=0}^{i-1} q_{i,j}(x)[t - i + x]^j$$

where the coefficients are computed via

$$q_{i,0}(x) = [\max(0,i-x)]^i, \quad i \geq 1$$

$$q_{i,j}(x) = \frac{i}{j} \sum_{k=j-1}^{i-2} \binom{k}{j-1} q_{i-1,k}(x), \quad i \geq 2, \text{ and } 1 \leq j \leq i-1.$$

Proof. From Proposition 2.2.1 we have that

$$W_i = \max(0, 1-X_1, \ldots, n_i^*-X_{n_i^*}).$$

Now, because none of the streams can produce more than one arrival in any T-interval, each of the n_i^* arrivals in the previous interval must have originated from a separate stream, and none of these can have originated from S_i. Since S_i is independent of (and thus totally out of synchronization with) the remaining streams, each of these n_i^* arrivals occurs at random (i.e., uniformly distributed) within this T-interval. Consequently, conditioned on $\{n_i^* = k\}$, the random variables $X_1,\ldots,X_k$ are the order statistics of k independent random variables, each uniformly distributed over $[0,T)$. Explicit calculations ([5]) then complete the proof.

Remark. The above recursion for the coefficients $q_{i,j}(x)$ can be explicitly solved to yield

$$q_{i,j}(x) = j\binom{i}{j} \sum_{k=1}^{i-j} \binom{i-j}{k} (i-k)^{i-j-k-1}[\max(0,k-x)]^k,$$

$$\text{for } i \geq 1 \text{ and } 1 \leq j \leq i-1.$$

However, as will be seen shortly, to calculate $F(x)$ and $F_i(x)$ one usually needs to evaluate $P_k(T,x)$, for $0 \leq k \leq N-1$. Thus usually the recursive computation of the $q_{i,j}(x)$ is most convenient.

We next consider the distribution of n_i^*.

<u>Lemma 2.4.2</u>. Define parameters ν_j as

$$\nu_j = \lambda_j T, \quad 1 \leq j \leq N, \tag{2.4.1}$$

and assume that each $\nu_j < 1$. Then the distribution of n_i^* is given by

$$P[n_i^* = k] = [\prod_{\substack{j=1 \\ j\neq 1}}^{N} (1-\nu_j)]\, s_{N-1,k}(\frac{\nu_1}{1-\nu_1},\ldots,\frac{\nu_{i-1}}{1-\nu_{i-1}},\frac{\nu_{i+1}}{1-\nu_{i+1}},\ldots,\frac{\nu_N}{1-\nu_N}),$$

for $0 \leq k \leq N-1$.

In the above, $s_{N-1,0}(\ldots) = 1$, and $s_{N-1,k}(\ldots)$ is given by (2.3.1) for $k \geq 1$.

Proof. Each of the streams S_j, $j \neq i$, independently contributes arrivals to the T-interval preceding the tagged arrival epoch. Moreover, S_j contributes either one arrival or zero arrivals; and, as the mean number of arrivals in any T-interval equals $\nu_j = \lambda_j T$, the probability of the former equals ν_j, and of the latter, $1-\nu_j$. Thus

$$P[n_i^* = k] = \sum_{\substack{i_1=1 \\ i_1\neq i}}^{N-k+1} \sum_{\substack{i_2=i_1+1 \\ i_2\neq i}}^{N-k+2} \cdots \sum_{\substack{i_k=i_{k-1}+1 \\ i_k\neq i}}^{N} \nu_{i_1}\nu_{i_2}\cdots\nu_{i_k} \prod_{\substack{j=1 \\ j\neq i,i_1,\ldots,i_k}}^{N} (1-\nu_j)$$

which can be manipulated into the above expression.

We are now in a position to determine $F(x)$ and $F_i(x)$.

Theorem 2.4.3. The complementary distributions $F_i(x)$ and $F(x)$ are given by

$$F_i(x) = [\prod_{\substack{j=1 \\ j\neq i}}^{N} (1-\nu_j)] \sum_{k=0}^{N-1} T^{-k}\, s_{N-1,k}(\frac{\nu_1}{1-\nu_1},\ldots,\frac{\nu_{i-1}}{1-\nu_{i-1}}, \frac{\nu_{i+1}}{1-\nu_{i+1}},\ldots,\frac{\nu_N}{1-\nu_N})\, P_k(T,x) \tag{2.4.2}$$

$$F(x) = \frac{\prod_{j=1}^{N} (1-\nu_j)}{\sum_{j=1}^{N} \nu_j} \sum_{k=1}^{N} kT^{1-k}\, s_{N,k}(\frac{\nu_1}{1-\nu_1},\ldots,\frac{\nu_N}{1-\nu_N})\, P_{k-1}(T,x) \tag{2.4.3}$$

where the ν_j's are given by (2.4.1).

Proof. Equation (2.4.2) follows directly from Lemmas 2.4.1 and 2.4.2. Moreover, the probability that the tagged arrival comes from S_i

is just $\nu_i / \sum_{j=1}^{N} \nu_j$, and so

$$F(x) = \frac{1}{\sum_{j=1}^{N} \nu_j} \sum_{i=1}^{N} \nu_i F_i(x),$$

which reduces after some manipulation to (2.4.3).

Remark. The restriction that $\nu_j < 1$ in Lemma 2.4.2 and Theorem 2.4.3 can be relaxed, but the resulting formulae are not as notationally clean. However, the following result removes the need for such specialized formulae, as it indicates that T can be reduced so that each $\nu_j < 1$.

Theorem 2.4.4. The expressions (2.4.2) and (2.4.3) for F(x) and $F_i(x)$ are invariant to the value of T, as long as $N < T \leq \min_j \lambda_i^{-1}$.

Proof. This can be proved directly (and laboriously) by differentiation of (2.4.2) and (2.4.3) with respect to T, and using the properties of $\{P_k(T,x)\}$ given in Lemma 2.4.1. But a simpler proof follows by noting first that there exist streams S_i with rates $\lambda_1, \ldots, \lambda_N$ such that the interarrival times of each S_i are no smaller than $\min_i \lambda_i^{-1}$, and second that for such a situation any value of T in the range $N < T \leq \min_i \lambda_i^{-1}$ satisfies the assumptions set forth in Section 2.1. F(x) and $F_i(x)$ can be computed using any of these values of T, and the invariance follows.

As a result of Theorem 2.4.4 we have the following Corollaries, the first of which considerably simplifies the computations for $F_i(x)$ and F(x) for the case where each $\lambda_i = \lambda$, and the second of which has significance when the results of this Section are applied to tree networks of processors.

Corollary 2.4.5. If each $\lambda_i = \lambda$, then

$$F(x) = F_i(x) = \lambda^{N-1} P_{N-1}(\lambda^{-1}, x)$$

where $P_{N-1}(T,x)$ is as defined in Lemma 2.4.1.

Proof. From Theorem 2.4.4 we can take $T = \lambda^{-1}$. But then n_i^* equals N-1 with probability 1, and the result follows from Lemma 2.4.1 and the observation that $F(x) = F_i(x)$.

Corollary 2.4.6. If each stream S_i is the output process of a FCFS single server queue with deterministic service time, then the magnitudes of these deterministic service times have no effect on the complementary delay distributions F(x) and $F_i(x)$ so long as the i'th of these, t_i, falls in the range $N < t_i \leq \lambda_i^{-1}$.

3. Examples

3.1 Behaviors of F(x) and w, and the M/D/1 Approximation

The purpose of this example is simply to quantify the delay characteristics at a single interior processor node in a tree network. For simplicity we shall limit our attention to the symmetric case, i.e., a processor node to which there are N arrival processes, each with the same rate λ, and for each of which the minimum spacing between arrivals is no smaller than N. As in the preceding section, we assume unit deterministic processing times at the processor. It will be convenient to express F(x) and w in terms of the processor occupancy, ρ, where

$$\rho = N \lambda .$$

We wish to illustrate the behaviors of F(x) and w for various values of N and ρ. Also, because it is natural to suspect that the behavior of the delay at such a node can be approximately modeled by means of an M/D/1 queue, at least for large values of N for which the superposition of the arrival streams can be argued to be approximately Poisson (see [6]), another objective of this example is to numerically evaluate the quality of this M/D/1 approximation.

From Corollaries 2.3.2 and 2.4.5, w and F(x) are given in terms of N and ρ as

$$w = \frac{\rho}{2}\left(1 - \frac{1}{N}\right)\left[B(N-2, N/\rho)\right]^{-1}$$

and

$$F(x) = (\rho/N)^{N-1}\, P_{N-1}(N/\rho, x).$$

Figure 3.1 illustrates the behavior of w as a function of ρ and N, as do Figures 3.2 and 3.3 for F(x). Also given in these figures are the approximations that result from an M/D/1 model. A convergent trend towards the M/D/1 approximants is clear in these figures; but, expecially for larger values of ρ, the rate of convergence is seen to be extremely slow. This slow convergence can be quantified by deriving the asymptotic behavior of w, as $N \to \infty$, and as ρ is held constant. Known asymptotic behaviors of the Erlang loss function (see [7], Theorems 13 and 14) result in the following:

$$w \sim \frac{\rho}{2(1-\rho)}\left[1 - \frac{1}{N(1-\rho)^2} + \frac{\rho(2+\rho)}{N^2(1-\rho)^4}\right],$$

as $N \to \infty$, for $\rho < 1$, and

$$w \sim \left(\frac{\pi N}{8}\right)^{1/2}\left[1 - \left(\frac{32}{9\pi N}\right)^{1/2} + \frac{1}{24N}\right],$$

as $N \to \infty$, for $\rho = 1$.

3.2 Mean Response Times in Tree Networks

The results developed in Section 2 allow the analysis of fairly general tree network structures. However, due to space limitations, here we restrict our attention to the consideration of tree networks that are symmetrically configured and loaded. Thus, consider a network as in Figure 3.4, where there are L levels of processors, and where the fan-in at each level i processor, $1 \leq i \leq L-1$, equals N, i.e., there are N level $i+1$ processors whose outputs form its input. For simplicity, we assume that the speeds (e.g., in basic operations per second) of all processors are identical. We also assume: (i) at each of the level L, i.e., front end, processors there is an independent Poisson stream of

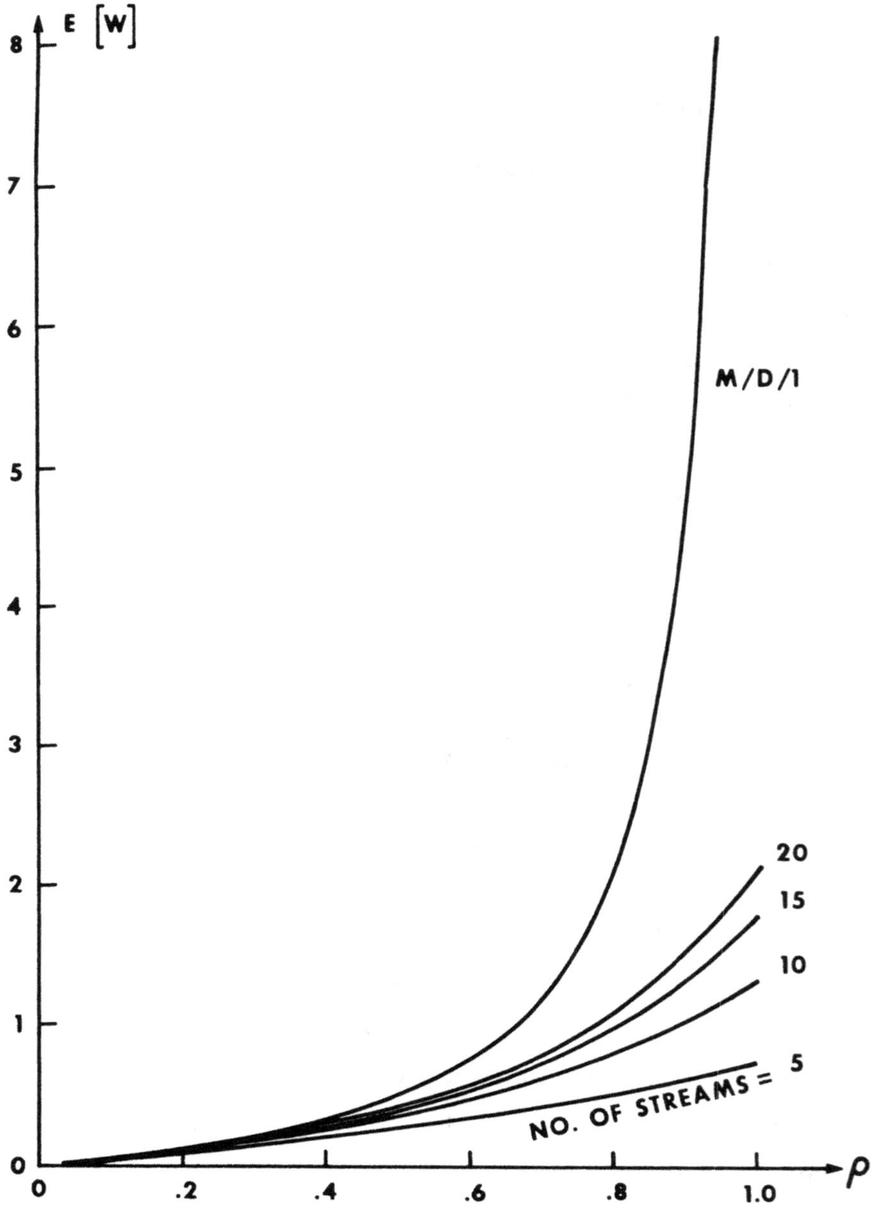

Figure 3.1. Mean Waiting Time Vs N and Processor Occupancy

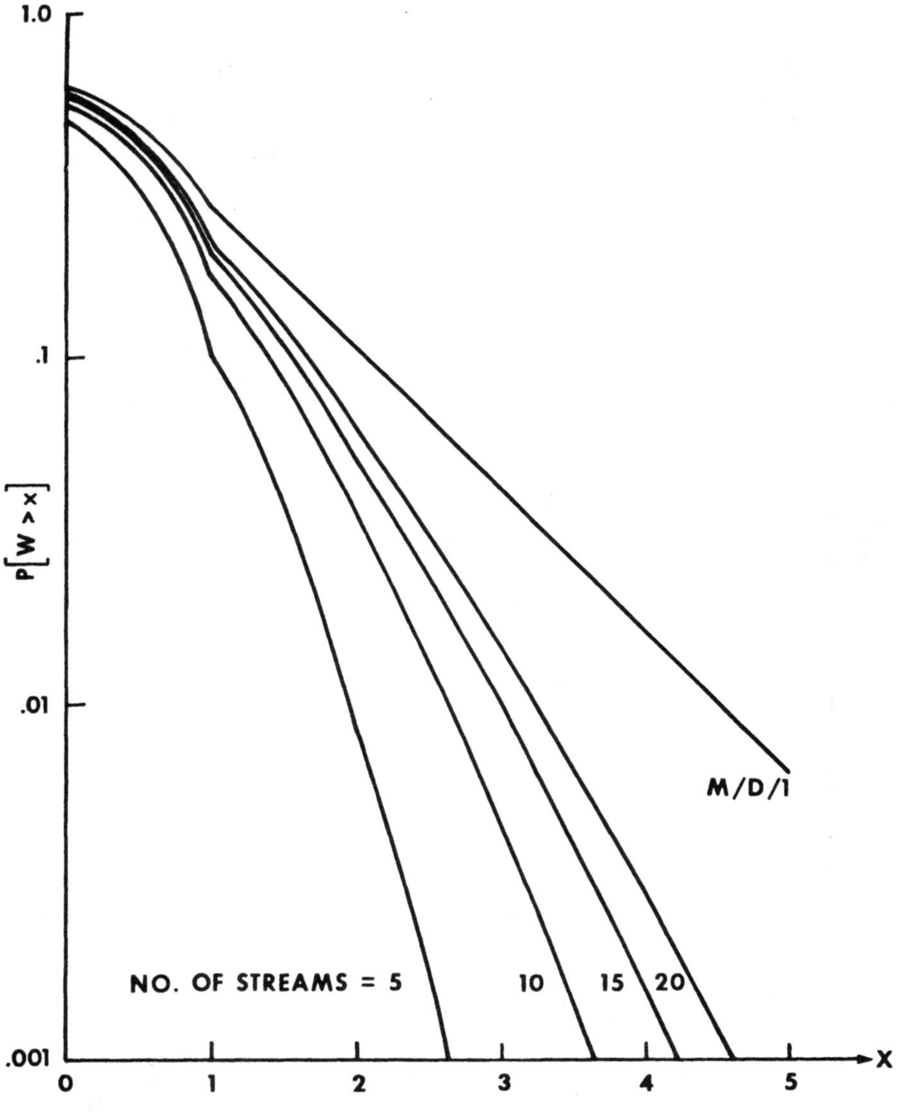

Figure 3.2. Waiting Time Complementary Distributions for 60% Processor Occupancy (ρ=.6)

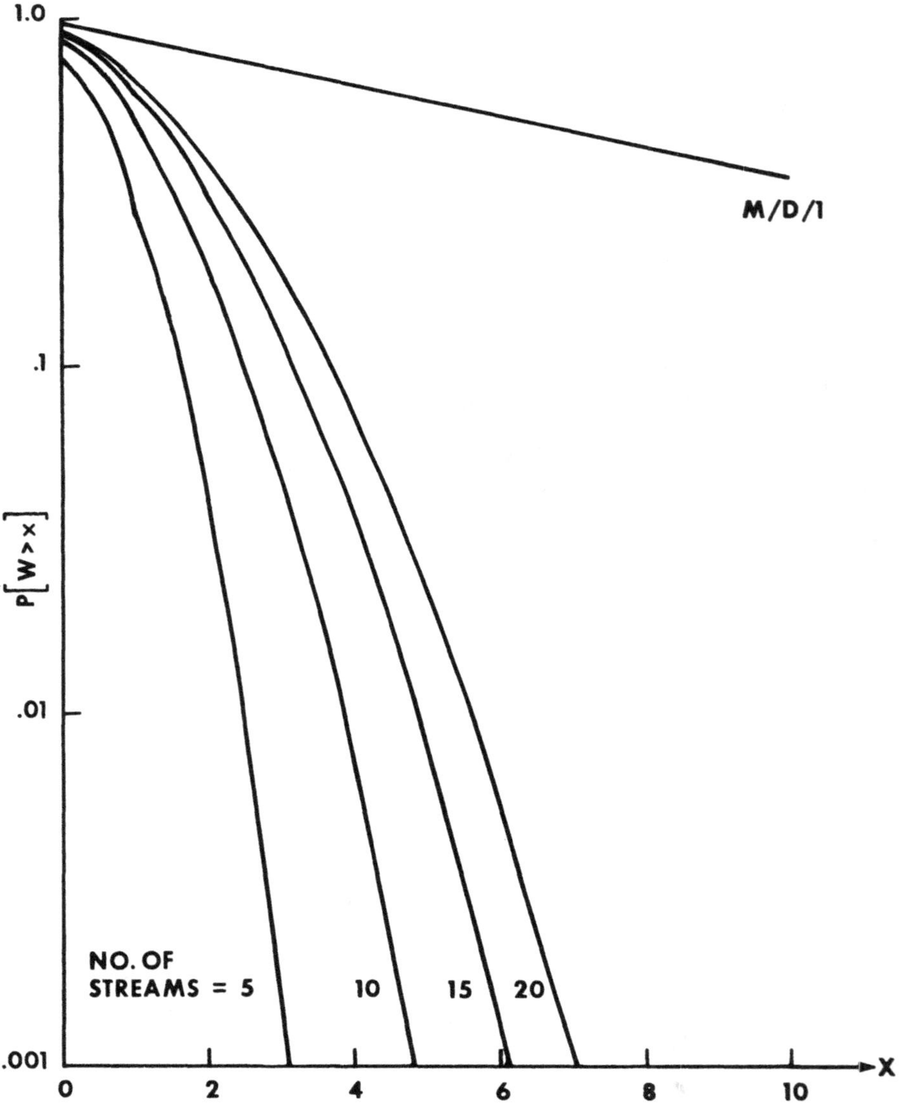

Figure 3.3. Waiting Time Complementary Distributions for 95% Processor Occupancy (ρ=.95)

arriving jobs, where the arrival rate is λ and each job brings with it a total processing requirement of exactly T time units; (ii) each such job is segmented into L tasks, $\tau_1,\ldots,\tau_L$; and (iii) task τ_i is assigned to the appropriate level i processor. It is further assumed that this segmentation of jobs is such that for the case of equal arrival rates thus described, the occupancies of all processors in the network are equal. This implies that the processing requirement, T_i, of task τ_i is

$$T_i = \frac{N-1}{N^L-1} N^{i-1} T.$$

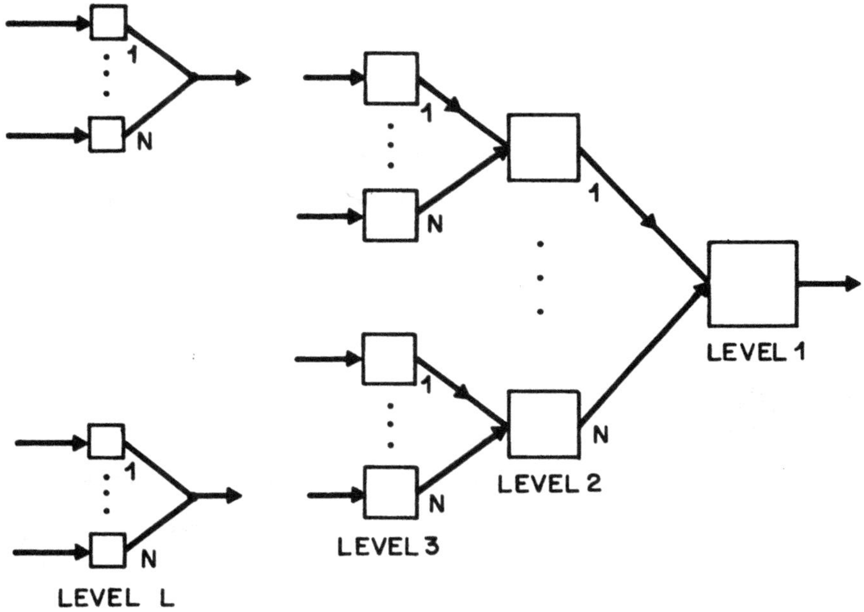

Figure 3.4. Symmetric L-level Network with Fan-in of N

A design question that developers of such an architecture might face is the question of service discipline at individual processing nodes. We present here a quantitative comparison between FCFS and Processor-Sharing (PS) disciplines. Thus, if the service at each processor is FCFS, it is easily seen that the total mean response time, i.e., processing plus delay, of a job traversing the network is

$$R_{FCFS} = T + \frac{\rho}{2(1-\rho)} T_L + \sum_{i=1}^{L-1} \frac{\rho}{2} (1 - \frac{1}{N}) B^{-1}(N-2, N/\rho) \; T_i$$

$$= T\{1 + \frac{N-1}{N^L - 1} \frac{\rho}{2(1-\rho)} [N^{L-1} + (1-\rho)\frac{N^{L-1}-1}{N} B^{-1}(N-2, N/\rho)]\}$$

On the other hand, if the discipline is PS, it is known (see, e.g., [1]) that the mean delay at each node is the same as if it were an M/M/1 queue. Thus

$$R_{PS} = T + \frac{\rho}{1-\rho} \sum_{i=1}^{L} T_i$$

$$= \frac{T}{1-\rho} \quad .$$

A comparison R_{FCFS} and R_{PS} is given in Figure 3.5. The superiority of FCFS over PS <u>in this example</u> has two explanations: first, FCFS service of <u>deterministic</u> tasks is superior to PS, for virtually any arrival process; and secondly, FCFS service at level i processors results in a smoother arrival process at the level i-1 processors (these are all Poisson if the service disciplines are PS), thus further reducing the overall mean response time.

3.3 A Flow Control Application

In many computer networks, both local and global, interprocessor communications is effected via packet switching. In packet switching, individual packet switches are charged with receiving packets from source processors and forwarding them to destination processors. The ability to provide acceptable and equitable service to all sources depends on all the sources implementing a flow control scheme to limit the rate at which packets arrive at the packet switch to a level the packet switch can handle. One popular flow control scheme monitors the number of sources actively engaged in sending packets, and requires each to provide a minimum spacing between the packets it sends to the packet switch.

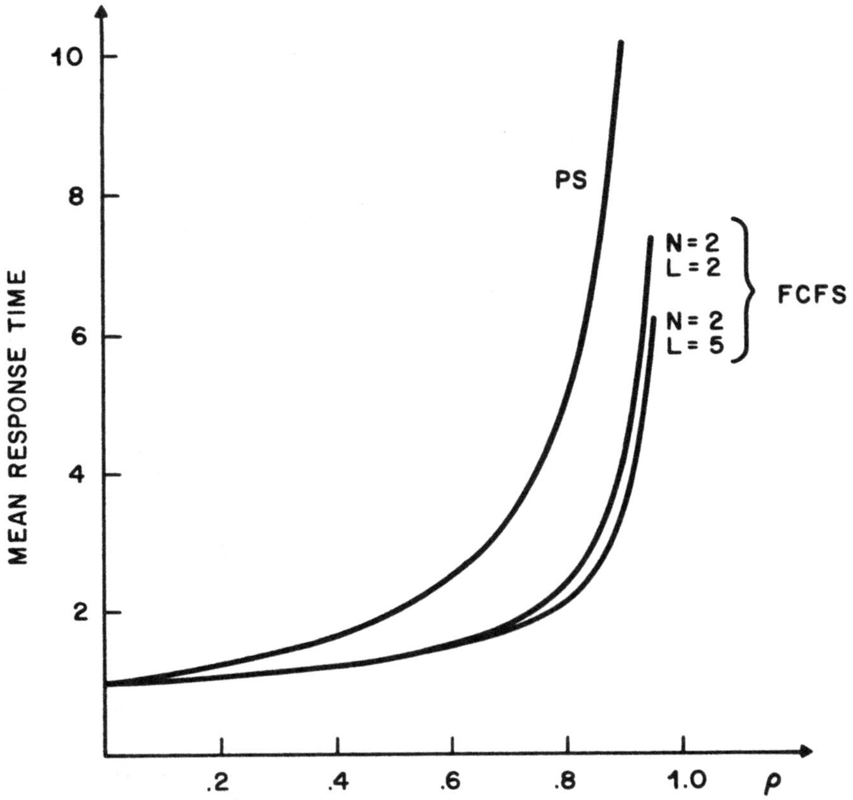

Figure 3.5. Mean Network Response Time vs Load for PS and FCFS Disciplines When T = 1

The results of Section 2 can now be immediately applied to evaluate the effectiveness of such a scheme. If the packet switch can process one packet per time unit, and if each of N sources is limited to sending only one packet in any interval of length N, then the resulting packet switching delay is as given in Figure 3.1. The M/D/1 delay in this figure can be interpreted as the delay with no flow control. It can be seen that such a flow control scheme effectively improves the packet switching delays, even when the number of processors being flow controlled is large.

4. Acknowledgements

The author wishes to thank A. A. Fredericks, B. T. Doshi, and D. L. Jagerman for useful discussions.

5. References

[1] Baskett, F., Chandy, K. M., Muntz, R. R., and Palacios, F. G. (1975), "Open, Closed, and Mixed Networks of Queues with Different Classes of Customers," J. Assoc. Comput. Mach., 22, pp. 248-260.

[2] Kaplan, M. (1980), "A Two-fold Tandem Net with Deterministic Links and Source Interference," Oper. Res., 28, pp. 512-526.

[3] Ziegler, C. and Schilling, D. L. (1980), "Waiting Times at Fast Merger Nodes," Proceedings ICC, 23.2.1-6.

[4] D. V. Lindley (1952), "The Theory of Queues with a Single Server," Proceedings Cambridge Phil. Soc., 48, pp. 277-289.

[5] A. E. Eckberg, Jr. (1979), "The Single Server Queue with Periodic Arrival Process and Deterministic Service Times," IEEE Trans. Comm., COM-27, pp. 556-562.

[6] E. Çinlar (1972), "Superposition of Point Processes," Stochastic Point Processes (P. A. W. Lewis, ed.), Wiley, pp. 549-606.

[7] D. L. Jagerman (1974), "Some Properties of the Erlang Loss Function," Bell System Tech. J., 53, pp. 525-551.

6. Endnotes

[1]The assumption of unit service times simplifies the notation; nonunit service times are easily accommodated via time-scaling in the final results.

[2]If the n'th and (n+1)'th previous arrivals occur simultaneously, then $X_n = X_{n+1}$. If the tagged arrival occurs in a batch of K arrivals, then we shall consider the individual arrivals within this batch to be ordered arbitrarily and the position of the tagged arrival relative to

this ordering to be a random variable (conditionally) uniformly distributed over $\{1,2,...,K\}$. Consequently, some of the X_i's for $i \geq 1$ may equal 0.

Bell Laboratories, Holmdel, New Jersey 07733.

Discussant's Report on
"Response Time Analysis for Pipelining Jobs
In a Tree Network of Processors,"
by A. E. Eckberg

The queueing situation examined in this paper is clearly of interest in computer applications and perhaps other hierarchical queueing structures. The result on the independence of the delays experienced by a given job at individual processors is very significant in the analysis of such systems. However, the validity of the major results of the mean waiting times and the waiting time distributions is questionable. This is due to two unproven assertions make by the author which are essential to the proofs in question:

1) "... the customer waiting time for S_i is just the mean (time average) virtual waiting time produced by S_j, $j \in \eta$ ..." stated in the proof of Theorem 2.3.1, and

2) "Since S_i is independent of ... the remaining streams, each of these n_i* arrivals occurs at random (i.e., uniformly distributed) within this T-interval." stated in the proof of Lemma 2.4.1. Both of the above would clearly be true if the arrivals formed a Poisson process. Since this is not the case and proofs have not been provided, these assertions, and therefore the subsequent results, should be seriously questioned.

Author's Response to Discussant's Report

First, I would like to respond to the Discussant's Report by acknowledging that some of the proofs in my paper are not overly detailed, but depend to a great extent on a good feeling on the part of the reader for what the notions of independence and stationarity of random processes really "mean" (i.e., how they can be exploited). Although I hope to provide additional substantiation for the validity of these results with my comments that follow, I also recognize that these comments do not supply the rigor that my Discussant would like to see. Indeed, I believe that the application of "heavy machinery" to the proofs of results which should be "obvious" after some careful thought sometimes serves to obfuscate simple and potentially useful concepts - something to be avoided if modeler-analysts hope to maintain contact and communication with the real world.

The results in question are (1) the sentence preceding Eq. (2.3.5), or equivalently, Eq. (2.3.5) itself; and (2) the assertion in the proof to Lemma 2.4.1 that, conditioned on $\{n_i^* = k\}$, these k arrivals in the T-interval immediately preceding the arbitrarily selected arrival from stream S_i occur at independent and uniformly distributed points within this T-interval. My Discussant then implies that if the arrival processes were Poisson then the validity of these results would be evident. This is clearly true for the latter result, but (2.3.5) would but be true in this case. The correct version of (2.3.5) when the arrivals are Poisson is $W_{\eta,i} = v_{\eta+\{i\}}$, i.e., the mean customer waiting time for S_i would equal the mean virtual waiting time produced by <u>all</u> the streams: S_i and S_j, $j \in \eta$. However, it is worth noting that (2.3.5) would be valid for the <u>finite source</u> single server queueing system where service times are exponentially distributed and where each of the arrival streams consists of arrivals from one of N identical sources, each of which has an exponentially distributed idle period (see, e.g., Chapter 3 of R. B. Cooper, <u>Introduction to Queueing Theory</u>, 1972).

Now addressing the first questioned result, it has been demonstrated in the Proof that (as a result of Proposition 2.2.1) the actual waiting times experienced by arrivals from stream S_i are precisely the values of the sample function $V_\eta(\cdot)$ at these arrival epochs, where $V_\eta(\cdot)$ is the virtual waiting time sample function that would result were the queue input to consist solely of the streams S_j, $j \in \eta$. So the basic question is whether the mean value of the (stationary) random process $V_\eta(\cdot)$, i.e., $v_\eta = E[V_\eta(t)]$, equals the mean value of the random variable $V_\eta(\tilde{t}_j)$, where $\tilde{t}_j$ is an "arbitrary" arrival epoch of S_i. It appears clear from the fact that S_i and $V_\eta(\cdot)$ are independent and that $V_\eta(\cdot)$ is stationary, that the answer to this question is "obviously yes." It is not even necessary that S_i be stationary to conclude that $E[V_\eta(t)] = E[V_\eta(\tilde{t}_j)]$.

As for the second questioned result, first note that the time origin can be positioned at the arrival epoch of the arbitrarily selected arrival from stream S_i without affecting the stationarity of the remaining streams; this follows from the independence of the streams. Now note that for each $j \neq i$, and for any $0 < \nu \leq T$, S_j will have one arrival in the interval $[-\nu, 0)$ with probability $\lambda_j\nu$, and no arrival in this interval with probability $1 - \lambda_j\nu$; this is because the number of arrivals in an interval of length T or less is zero or one, while (from stationarity) the mean number of arrivals in any interval of length ν is $\lambda_j\nu$. It follows that the distribution function of the random variable X_j, defined as $X_j = -\sup\{t < 0 | S_j$ has an arrival at epoch $t\}$, satisfies $P[X_j \leq x] = \lambda_j x$, for $x \leq T$. Thus, given that S_j has an arrival in $[-T, 0)$, the location of this arrival is uniformly distributed over this interval. Finally, given that a total of k arrivals from the streams S_j, $j \neq i$, occur in this interval, the independence of these k arrival epochs follows from the independence of the streams S_j, $j \neq i$.

Discussant: Dr. Linda Green, Graduate School of Business, Columbia University, Uris Hall, New York, New York 10027.

MEAN DELAYS OF INDIVIDUAL STREAMS INTO A QUEUE: THE $\Sigma GI_i/M/1$ QUEUE

J. M. Holtzman

Abstract

Arrivals in traffic streams with different characteristics can experience drastically different delays when offered to a common server. Such is particularly the case in computer communication networks because of the differences between data traffic types, such as batch and interactive. To analyze mean delays for this problem, an approximate approach is presented for the superposition of independent renewal processes offered to an exponential server.

The solution for the $\Sigma GI_i/M/1$ queue is reduced to the solution of individual GI/M/1 queues. The approach is based on two approximations, one of which is suggested by heavy traffic limit theory. For 2 input streams, numerical results indicate that the approximation is accurate in heavy traffic and that it underestimates at lower loads. It is surprisingly accurate in some cases of lower loads. The extension to more than 2 streams is also discussed.

1. Introduction

A problem of some practical importance is the analysis of queues and queuing networks with more than one traffic stream offered to a server. A natural approach is to analyze, either exactly or approximately, the delay experienced by an arbitrary arrival in the superposition of the streams. However, the arrivals in the different streams can experience drastically different delays than the delay averaged over all arrivals. Such is the case particularly in computer communication networks because of the differences between data traffic types, such as batch and interactive.

Two papers evaluating delays seen by individual traffic streams are [1] and [2], which consider single server queues with Poisson plus renewal input. In view of the present limited applicability of available results and the need for such results, this paper offers an approximation to mean delays for this problem. In particular, the $\Sigma GI_i/M/1$ queue with first-come first-served service is considered (stationarity is assumed throughout). The basic approach is presented for the superposition of two renewal processes offered to an exponential server. Section 2 presents the basic results.

The approximation is based on two basic approximations:

(A1) Each stream's queue is first treated separately with the other stream contributing to an inflated service time.

(A2) Each stream experiences delays due to its own customer-averaged[1] queue and due to the time-averaged queue of the other stream. This is motivated by the independence of the two streams: each is considered to arrive randomly with respect to the other.

The form of the approximation represented by (A1) is suggested by heavy traffic results in Section 3. Using the heavy traffic results for this system, it is shown that the mean delay can be expressed as a sum of two terms, each of which represents the solution to a GI/M/1 system, with one of the renewal processes as input. The individual GI/M/1 solutions are coupled only through service rate reductions caused by the server serving the other renewal process. Inflating service times have been used by other approximators, e.g., in [3] (actually, offered stream arrival rates are increased there). Inflating service times (or related approximations) also show up, perhaps more naturally, in priority queueing systems; see [9], p. 319. Also see [10], p. 567.

In Section 4, we introduce (A2) and derive the approximation for $n = 2$ streams. In Section 5, we present some numerical comparisons with known exact results. Section 6 discusses the extension to $n > 2$ input streams. Concluding remarks are given in Section 7.

2. Terminology and Basic Result

The basic approach will be developed for two independent renewal processes offered to one exponential server. The interarrival time random variables are positive and nonlattice. First-come first-served discipline is assumed. Define for $i = 1,2$

λ_i = mean arrival rate of renewal process i

ϕ_i = Laplace-Stieltjes transform of the interarrival time distribution of renewal process i

μ^{-1} = mean service time

c.v.(s) = coefficient of variation of service time

c.v.(a_i) = coefficient of variation of interarrival time of renewal process i

$\rho_i = \lambda_i/\mu$

$\rho = \rho_1 + \rho_2$

D_i = approximation for mean delay experienced by i-arrivals.

It is assumed that $\rho < 1$.

The basic result to be presented is:

$$D_1 = \mu^{-1}\left[\frac{\omega_1}{1-\omega_1} + \frac{\bar{\rho}_2}{1-\omega_2}\right] \tag{2.1}$$

$$D_2 = \mu^{-1}\left[\frac{\bar{\rho}_1}{1-\omega_1} + \frac{\omega_2}{1-\omega_2}\right] \tag{2.2}$$

where

$$\bar{\rho}_1 = \frac{\rho_1}{1-\rho_2} \quad , \tag{2.3}$$

$$\bar{\rho}_2 = \frac{\rho_2}{1-\rho_1} \quad , \tag{2.4}$$

$$\omega_1 = \phi_1[\mu(1-\rho_2)(1-\omega_1)] \ , \tag{2.5}$$

$$\omega_2 = \phi_2[\mu(1-\rho_1)(1-\omega_2)] \ . \tag{2.6}$$

3. Heavy Traffic Limits

Although heavy traffic results for superpositions of renewal inputs are available (see, e.g., [4], which has further references), we shall give a simple presentation here and manipulate the results into a form suitable for our purpose.

Since we are dealing with superpositions of renewal processes, which are not generally renewal, we need a heavy traffic result for more general processes. Such a result is given in [5], p. 40, which is valid under quite weak conditions. The needed result, restated in our terminology, is that

$$\lim_{\rho\to 1-} \text{Prob.}\left(W(\rho) \geq \frac{y}{1-\rho}\right) = e^{-2y/\sigma^2} \tag{3.1}$$

where W is the virtual waiting time and σ^2 is obtained from

$$\text{Var}[X(t) - X(0)] = \sigma^2 t + o(t) \ , \tag{3.2}$$

X(t) is given in terms of S(t), the summation of all the service times of requests arriving up to time t:

$$X(t) = S(t) - t \ . \tag{3.3}$$

For the superposition of two independent renewal processes arriving with identically distributed service time (s) requests,

$$\mathrm{Var}[X(t)-X(0)] = t\left[\frac{\mathrm{Var}(s)}{E(a_1)}+\frac{E^2(s)\mathrm{Var}(a_1)}{E^3(a_1)}+\frac{\mathrm{Var}(s)}{E(a_2)}+\frac{E^2(s)\mathrm{Var}(a_2)}{E^3(a_2)}\right] + o(t) \tag{3.4}$$

where the a_i are the interarrival times of the renewal processes (see [5], pp. 41, 47). Thus,

$$\sigma^2 = \mathrm{Var}(s)(\lambda_1+\lambda_2) + E^2(s)\left[\lambda_1^3\mathrm{Var}(a_1)+\lambda_2^3\mathrm{Var}(a_2)\right] \tag{3.5}$$

From (3.1) and (3.5), we have

$$E(W) = \frac{E(s)\frac{\rho_1}{1-\rho_2}}{2\left(1-\frac{\rho_1}{1-\rho_2}\right)}\{[c.v(s)]^2 + [c.v.(a_1)]^2\}$$

$$+\frac{E(s)\frac{\rho_2}{1-\rho_1}}{2\left(1-\frac{\rho_2}{1-\rho_1}\right)}\{[c.v.(s)]^2 + [c.v.(a_2)]^2\} \tag{3.6}$$

where c.v. denotes coefficient of variation. (3.6) is the mean service time, E(s), multiplied by the sum of two terms, each of which can be heuristically given the "interpretation" as the (heavy traffic limit) mean queue (including the one in service) of one of the renewal processe processes offered to a server with mean service rate $\mu(1-\rho_i)$ where i corresponds to the other input renewal process[2]. That is, the service rate is reduced by the fraction of time the server is occupied with the other input. This interpretation is the basis for our first approximation (A1) mentioned in Section 1.

4. Use of Approximation (A2) and Derivation of Results

To use approximation (A2) we note that in a GI/M/1 queueing system, the time-averaged queue size (including the one in service) is $\rho/(1-\omega)$ and the customer-averaged queue size is $\omega/(1-\omega)$; ([6], Chap. 2). ρ is the traffic intensity and $\omega (0 < \omega < 1)$ satisfies

$$\omega = \phi(\mu(1-\omega)) ,$$

where ϕ is the Laplace-Stieltjes transform of the renewal interarrival time distribution.

Using (A2), we obtain

D_1 = mean service time · [customer-averaged queue size, stream 1 + time-averaged queue size, stream 2].

Then using the relations in this section's first paragraph and (A1) yields (the service time inflation is displayed below in (4.3) to (4.6))

$$D_1 = \mu^{-1}\left[\frac{\omega_1}{1-\omega_1} + \frac{\bar{\rho}_2}{1-\omega_2}\right] , \tag{4.1}$$

$$D_2 = \mu^{-1}\left[\frac{\bar{\rho}_1}{1-\omega_1} + \frac{\omega_2}{1-\omega_2}\right] , \tag{4.2}$$

where

$$\bar{\rho}_1 = \rho_1/(1-\rho_2), \tag{4.3}$$

$$\bar{\rho}_2 = \rho_2/(1-\rho_1), \tag{4.4}$$

$$\omega_1 = \phi_1[\mu(1-\rho_2)(1-\omega_1)] , \tag{4.5}$$

$$\omega_2 = \phi_2[\mu(1-\rho_1)(1-\omega_2)] , \tag{4.6}$$

and ϕ_i is the Laplace-Stieltjes transform of the interarrival time distribution of renewal process i. (4.1)-(4.6) are the results given in Section 2.

Remark. When either λ_1 or λ_2 is zero, or both processes are Poisson, the approximations are exact.

5. Numerical Example

The example given in this section is to show the credibility of the approach and also to show the extreme discrepancies in delays that can be experienced by renewal processes with different characteristics.

The example is a GI+M/M/1 queue for which analytical results are available for comparison ([1], pp. 1318-1321). The GI stream is an interrupted Poisson process (IPP) - a Poisson process modulated by the turning on and off a switch (on- and off-times of the switch are independent exponential random variables). Our approximation requires solution for ω for an IPP/M/1 System (see [7], p. 536).

We assume that $\mu = 1$ (or that time is normalized so that the mean service time is the time unit) and we use the following terminology:

λ_1 = mean rate of arrivals in IPP (i.e., rate out of switch; see previous paragraph)

λ_s = mean rate of Poisson process entering switch of interrupted Poisson process ($\lambda_s > \lambda_1$)

N = mean number of arrivals per on-time of IPP switch

λ_2 = mean rate of Poisson process

$f = \frac{\lambda_1}{\lambda_1+\lambda_2}$ = fraction of total traffic that is IPP

$\rho = \lambda_1+\lambda_2$

D_1, D_2 = approximations for mean delays

W_1, W_2 = exact mean delays

Figures 1 and 2 show some results for N = 200 and λ_s = 1.0 and 0.2.[3,4] These results are displayed for the following reasons. Figure

1 shows the extreme discrepancy in the delays seen by the two different streams. It also shows the excellence of the approximation at all loads in this case. On the other hand, Figure 2 shows the type of errors observed. The approximation underestimated in other examples, also.

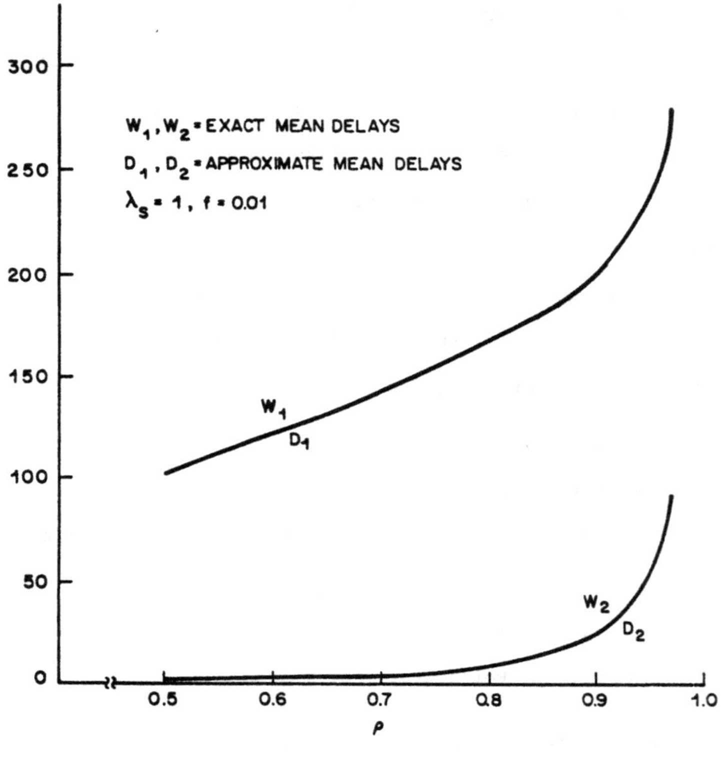

FIGURE 1 EXACT & APPROXIMATE MEAN DELAYS VS. UTILIZATION

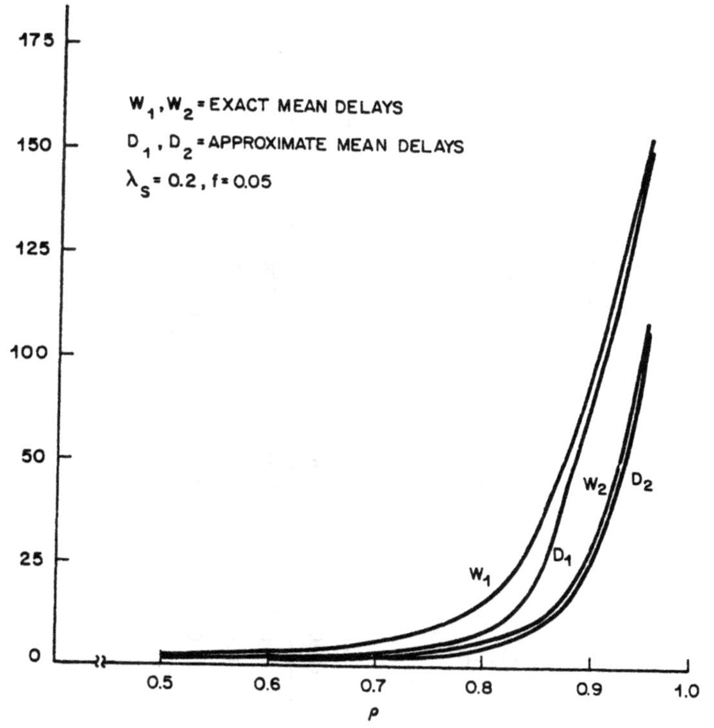

FIGURE 2 EXACT & APPROXIMATE MEAN DELAYS VS. UTILIZATION

6. More Than Two Arrival Streams

The approach can be generalized in an obvious way to the $\sum_{i=1}^{n} GI_i/M/1$ queue with $n > 2$. It was tried out in determining the mean queue length[5], when each interarrival density function is of the form

$$f(x) = p_1 e^{-\lambda_1 x} + p_2 e^{-\lambda_2 x} \quad .$$

The superposed renewal processes are identical and, thus, the expected queue lengths are equal. Tables 1 and 2 show results for interarrival squared coefficient of variation $c^2 = 2$, and 6, respectively.

Table 1. Comparison of Approximation with Simulation, $c^2=2$

n		$\rho=0.3$	$\rho=0.5$	$\rho=0.7$	$\rho=0.8$	$\rho=0.9$
2	Approximation	0.458	1.137	2.907	5.274	12.637
	Simulation	0.460	1.139	2.913	5.244	13.273
	(Standard Deviation)	(0.003)	(0.011)	(0.037)	(0.055)	(0.320)
4	Approximation	0.443	1.076	2.694	4.898	12.000
	Simulation	0.442	1.128	2.834	5.137	12.915
	(Standard Deviation)	(0.003)	(0.018)	(0.032)	(0.074)	(0.326)
8	Approximation	0.436	1.040	2.536	4.549	11.169
	Simulation	0.436	1.055	2.769	5.037	12.217
	(Standard Deviation)	(0.003)	(0.009)	(0.037)	(0.082)	(0.321)
16	Approximation	0.432	1.020	2.491	4.304	10.356
	Simulation	0.434	1.036	2.588	4.762	11.668
	(Standard Deviation)	(0.002)	(0.017)	(0.040)	(0.093)	(0.305)

Table 2. Comparison of Approximation with Simulation, $c^2=6$

n		$\rho=0.3$	$\rho=0.5$	$\rho=0.7$	$\rho=0.8$	$\rho=0.9$
2	Approximation	0.500	1.422	4.667	9.844	26.847
	Simulation	0.505	1.521	5.268	11.053	27.202
	(Standard Deviation)	(0.007)	(0.038)	(0.143)	(0.380)	(1.179)
4	Approximation	0.463	1.197	3.500	7.483	22.982
	Simulation	0.466	1.353	4.375	9.159	25.914
	(Standard Deviation)	(0.006)	(0.029)	(0.097)	(0.231)	(1.369)
8	Approximation	0.445	1.094	2.876	5.688	17.721
	Simulation	0.454	1.180	3.858	8.544	24.962
	(Standard Deviation)	(0.005)	(0.008)	(0.105)	(0.274)	(0.960)
16	Approximation	0.437	1.046	2.591	4.788	13.325
	Simulation	0.438	1.106	3.272	6.948	21.725
	(Standard Deviation)	(0.004)	(0.022)	(0.070)	(0.154)	(0.938)

Simulated results and standard deviations are taken from [8]. For $n = 2$, the approximation is good but as n increases the approximation loses accuracy - underestimates more. The effect becomes more pronounced as c^2 increases.

An implication is that if one wants to use the approach for $n > 2$ streams, one might analyze each stream separately and lump the other (n-1) streams together; i.e., use some approximate approach to represent the (n-1) superposition by a renewal process. This is left to future work.

7. Concluding Remarks

The results are encouraging enough to warrant further investigation - further study of accuracy (preferably error bounds), improvements, and possible extensions. One possible improvement is to use the approach presented here as a supplement to an exact or good approximate solution to the delays experienced by the arbitrary arrival in the superposition (see [11], [12]). That is, the approximation of this paper would be used only for the ratio of the individual delays.

8. Acknowledgments

I gratefully acknowledge the comments of Al Fredericks, Dan Heyman, and Ward Whitt and programming help from Dave D'Angelo and Tony Anastasio.

9. References

[1] A. Kuczura, "Queues with Mixed Renewal and Poisson Inputs," Bell System Tech. J., Vol. 51, pp. 1305-1326, July-August 1972.

[2] I. Sahin, "Equilibrium Behavior of a Stochastic System with Secondary Input," J. Appl. Probab., 8, pp. 252-260, 1971.

[3] I. Rubin, "An Approximate Time-Delay Analysis for Packet-Switching Communication Networks," IEEE Trans. Comm., Vol. COM-24, pp. 210-222, February 1976.

[4] D. L. Iglehart and Whitt, W., "Multiple Channel Queues in Heavy Traffic," Adv. in Appl. Probab., Vol. 7, pp. 150-177, 1970.

[5] A. A. Borovkov, Stochastic Processes in Queuing Theory, Springer-Verlag, 1976 (translation of 1972 Russian book).

[6] L. Takács, Introduction of the Theory of Queues, Oxford Univ. Press, 1962.

[7] H. Heffes, "On the Output of a GI/M/N Queuing System with Interrupted Poisson Input," Oper. Res., Vol. 24, pp. 530-542, May-June 1976.

[8] S. Albin, "Approximating Superposition Arrival Processes of Queues," unpublished work.

[9] M. Reiser, "Interactive Modeling of Computer Systems," IBM Systems J., Vol. 15, pp. 309-327, 1976.

[10] K. C. Sevcik, "Priority Scheduling Disciplines in Queueing Network Models of Computer Systems," Information Processing 1977, IFIP, pp. 565-570.

[11] P. J. Kuehn, "Approximate Analysis of General Queueing Networks by Decomposition," IEEE Trans. Comm., COM-27, pp. 113-126. January 1979.

[12] W. Whitt, "Approximating the Superposition of Renewal Processes by a Renewal Process," ORSA/TIMS Joint National Meeting, New Orleans, Louisiana, April 30-May 2, 1979.

10. Endnotes

[1] Customer averages are averages experienced by arriving customers (as opposed to time averages).

[2] This "interpretation" is valid for Poisson processes.

[3] In a computer communication system this IPP could correspond roughly to a batch type call originating on a 56 or 9.6 Kbs line, respectively, and offered to a 56 Kbs trunk.

[4] The parameters λ,ω,γ of the IPP are $\lambda = \lambda_s$, $\gamma = \lambda_s/N$, $\omega = \lambda_1\gamma/(\lambda_s-\lambda_1)$, with $\lambda_1 = f\rho$. Also,

$$[\text{c.v.}(a_1)]^2 = 1 + \frac{2(\lambda_s-\lambda_1)^2}{\gamma\lambda_s} \quad ,$$

so that the coefficient of variations of these interarrival times are of the order 15-20.

[5]Mean waiting times can be obtained via Little's law.

Bell Laboratories, Holmdel, New Jersey 07733

PROBABILISTIC MODELS IN PERFORMANCE ANALYSIS OF COMPUTER SYSTEMS AND COMMUNICATION NETWORKS

Donald Gaver, Chairman

A. A. Fredericks
D. Hunter
P. J. Schweitzer

ANALYSIS AND DESIGN OF PROCESSOR SCHEDULES FOR REAL TIME APPLICATIONS

A. A. Fredericks

Abstract

This paper is intended to provide some background on the concept of scheduling for systems with real time applications and to introduce the reader to some of the key issues in the design and analysis of these schedules. It illustrates the potential impact that scheduling in a stochastic environment can have on system performance, the need for constructing adequate probabilistic models and the difficulty in doing so. Several quidelines for processor scheduling are also given. These guidelines emerged from the analysis and design of a variety of actual systems and their use has resulted in substantial performance improvements and capacity increases for these systems.

1. Introduction

In this paper we briefly discuss some of the key issues in the analysis and design of processor schedules for systems with real time applications. The format is to consider a simplified example system (but one which has the main ingredients and potential problems of actual systems), synthesize some processor scheduling alternatives and evaluate their performance. This will illustrate the potential impact that scheduling in a stochastic environment can have on system performance, the need for constructing adequate probabilistic models and the difficulty in doing so. In the process of evaluating three scheduling alternatives, several general guidelines for scheduling real time systems in a stochastic environment will emerge. These guidelines resulted from the analysis and design of many actual systems and their use has resulted in substantial performance improvements and capacity increases for these systems.

2. System Description and Problem Formulation

The example system which we will consider is shown on Figure 1.

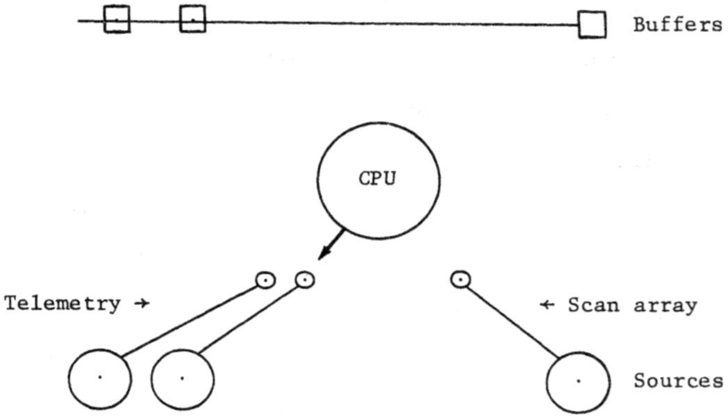

Figure 1. Example System

The system consists of N_s sources which at various times will originate requests to send data to the computer. The central processing unit (CPU) is required to scan (poll) these sources to detect requests, assign buffers, inform the source that it is ready to receive data, and then collect the data. We will assume that no futher actions are required after each data stream is successfully received. Although extremely idealized, this example system will illustrate most of the ingredients and potential problems of scheduling for actual real time systems.

Thus we need to schedule (at least) three distinguishable tasks that the processor must execute:

Task 1) detecting requests for service,

Task 2) preparing to collect data (including buffer assignment and informing the source of readiness),

Task 3) collecting data.

Our objective is to develop a schedule for executing these tasks, in particular, one that will maximize capacity. Before proceeding, however, we need at least two additional items; first, some criteria by which we can gauge adequate performance and second, workload characterization to allow us to predict performance. To be concrete we assume the following criteria:

Criteria 1: The mean delay in responding to a request (i.e., completing Tasks 1 and 2 above) shall not exceed 5 sec.

Criteria 2: The probability that the time to respond to a request exceeds 10 sec. shall be less than 10^{-2}.

Criteria 3: The probability that an error occurs during data collection shall be less than 10^{-4}. (What constitutes an error will be made clearer shortly when we further define the "data stream".)

While these criteria are only for illustrative purposes, they are indicative of typical criteria for many real time systems. Note that there is both a mean delay requirement for recognizing requests and a "tail of the distribution" requirement. Also, there is a much more stringent requirement for correctly handling data than for delays to respond to new requests.

Workload characterization for this system consists of two parts, first, specifying the nature of the underlying demands and second specifying the system resources needed to satisfy these demands. In our case the latter will only be CPU time, for simplicity we will assume that buffers can be engineered so that they are never a problem (e.g., one per source). For the underlying demand we define

$\frac{1}{\alpha}$ = mean time (sec) from completion of one request to send data until a new request is initiated.

We further assume that data consist of a stream of "on-off" pulses and that to properly "collect" the data means that we must successfully count the number of pulses, i.e., recognize each state transition. (For concreteness, our example system may be interpreted as a microprocessor responsible for dial pulse reception on incoming trunks in a switching machine.)

For the CPU work times we define

a = time (ms) to scan each source (i.e., for a new request or to detect pulse state changes.)

b = time (ms) to assign a buffer and inform a source of readiness to receive data.

c - time (ms) to process each pulse state change.

Finally, we are now in a position to address our basic question of specifying a schedule, i.e., when the processor should execute each of the tasks so that the system can handle the greatest number of sources while meeting all of the stated criteria.

3. Preliminary Sizing

Before addressing the scheduling question, we begin by obtaining a simple estimate of the maximum capacity we might expect. During an arbitrary (long) time interval, T(sec), we can decompose the total work time of the processor, W(T) into

$$W(T) = W_s(T) + W_b(T) + W_p(T) \tag{1}$$

where $W_s(T)$ is the work done scanning, $W_b(T)$ is the "set up" work, i.e., assigning a buffer and sending a start signal, and $W_p(T)$ is the work done collecting pulses. We can approximate these work times on average by

$$\bar{W}_s = \left(\frac{T}{\bar{t}_c}\right)(aN_s) \times 10^{-3}$$
$$\bar{W}_b = (T\alpha N_s)(b) \times 10^{-3} \qquad (2)$$
$$\bar{W}_p = (2\bar{n}_p T\alpha N_s)(a+c) \times 10^{-3}$$

where $\bar{t}_c$ is the mean interscan time for each source, $\bar{n}_p$ denotes the mean number of pulses per request and the factor 10^{-3} converts the work times to sec. Also, note, the approximation for $\bar{W}_s$ ignores the "finite source" effect assumed earlier for new request generation - we will look at this more carefully later.

The best we might hope for is that the interscan times are all equal to a constant, $\bar{t}_c$. In this case the mean scan delay as seen by a source would be $\frac{\bar{t}_c}{2}$ and hence a value of $\bar{t}_c = 10$ sec <u>might</u> be acceptable. Using this value and the obvious requirement that $W(T) \leq T$, equations (1) and (2) yield the following approximate bound on N_s

$$N_s \leq \frac{1}{a \times 10^{-4} + \alpha b \times 10^{-3} + 2\alpha(a+c)\bar{n}_p \times 10^{-3}} \qquad (3)$$

For concreteness, if we assume that $\frac{1}{\alpha} = 100$ sec, $\bar{n}_p = 50$ pulses, $a = .1$ ms, $b = 5$ ms, $c = .05$ ms, equation (3) yields

$$N_s \leq \frac{1}{10^{-5} + 5 \times 10^{-5} + 1.5 \times 10^{-4}} \approx 4760 \qquad (4)$$

While equation (4) gives us some idea of the capability of our system, we do not know if we can construct a <u>realizable</u> schedule that would even approach this value of 4760 sources and meet Criteria 1, let alone the other criteria. (E.g., (4) totally ignores delays due to contention.)

4.0 Schedule Design Alternatives

We begin by looking at three, increasingly complex, schedules for our system. These will be compared in this section only via the same type of approximate bounding done above, and again only considering the mean scan time. Later we will address the question of schedule evaluation.

4.1 Schedule 1 - Single Thread Schedule

Perhaps the simplest "schedule" that could be used is the following. Scan each source sequentially looking for a request to send data. When one is found, stop scanning and completely service this request - i.e., assign a buffer, send a ready signal, and then wait until all pulses are collected. At that point, reinitiate scanning for another request.

Arguments similar to those above lead to the following approximate bound analogous to (3)

$$N_s \leq \frac{1}{a\times 10^{-4}+\alpha b\times 10^{-3}+2\alpha\bar{n}_p\times 10^{-3}\times\max((a+c),15\text{ms})} \tag{5}$$

where the $\max(\cdot,\cdot)$ term in the denominator incorporates the fact that with this schedule, the processor is unavailable for work during the elapsed time of pulse collection.

Using the same parameter values as above we obtain

$$N_s \leq \frac{1}{10^{-5}+5\times 10^{-5}+1.5\times 10^{-2}} \approx 67 \tag{6}$$

We see from (6) that the main item limiting capacity is the time spent waiting for pulses to occur.

Knowing the cause of the limitation we can readily design a new schedule which will improve things.

4.2 Schedule 2 - Multiprogrammed Schedule

In an attempt to make effective use of the "idle" time inherent in the single thread schedule, we will consider the following schedule. Sources are scanned, looking for new requests, but now when one is found only the set up work is done and then the source is left. While "waiting" for the state change, the processor will continue to look for new sources and to detect state changes on previously recognized requests to send data. Of course, in order not to miss a state change on an active source (which would result in a miscount of the number of pulses) the processor must return to each source that is active at least once every 15 ms. This results in a completely analogous situation to that obtained for the single thread schedule, but now the minimum mean cycle (interscan time), $\bar{t}_c$, must be 15 ms as opposed to 10 sec. The equation analogous to (3) and (5) is

$$N_s \leq \frac{1}{67a\times10^{-3}+\alpha b\times10^{-3}+2\alpha n_p c\times10^{-3}} \tag{7}$$

and to (4) and (6) (with the assumed parameter values)

$$N_s \leq \frac{1}{6.7\times10^{-3}+5\times10^{-5}+5\times10^{-5}} \approx 150 \tag{8}$$

Note that while we now have a mean scan time of 15 ms as opposed to 10 sec we have doubled the capacity of our system! This illustrates a rather simple use of the following important guideline.

Guideline 1. Design schedules that are efficient at (and above) capacity.

For example, while the multiprogrammed schedule clearly has a great deal of additional (scanning) overhead at light loads, it is operating efficiently at heavy loads - this will become even clearer shortly. However, 150 is still not close to 4760. Can we do much better?

4.3 Schedule 3 - Clocked Schedule

A glance at Equation (8) quickly reveals that the item limiting capacity is the first term in the denominator which corresponds to the scanning task. The problem is that we are scanning all sources at a

15 ms rate when clearly we need only do this for those sources that are active. Indeed, we have violated the following important guideline.

Guideline 2. Be careful of combining two functions with disparate criteria into one work task.

There are a variety of ways of alleviating the apparent problem with the multiprogrammed schedule. We give one alternative - a clocked schedule. We divide time into "slots" of length T and schedule a certain sequence of tasks to be executed in each such slot. Specifically, we begin each slot by executing a "high priority" task (H) which consists of scanning all active sources for pulses, i.e., those sources for which a request to send data has been previously acknowledged but for which data collection is not yet complete. We next schedule a "low priority" task (L) which consists of doing set up work for new requests that were previously recognized. Finally we schedule a "fill" task (S) which consists of scanning as many requests as possible until the T ms timer indicates a new slot. Figure 2 shows such a clocked schedule. The area of each slot is meant to correspond to T ms; the amount of time spent processing each task, for a hypothetical realization, is similarly indicated by area.)

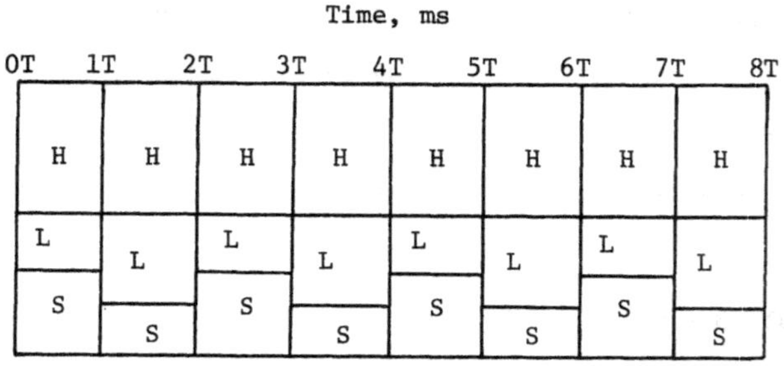

Figure 2. Clocked Schedule

Thus, a new request is first recognized while the CPU is executing task S, at this point an entry is made to indicate work for task L. In the next slot, task L will execute, readying this source for data collection which will be done in subsequent executions of task H. This type of schedule is common in many real time systems from small microprocessor applications to large switching systems.

To obtain a capacity estimate for this schedule we proceed as follows. First define $t_a^{(T)}$ as the average time available in each slot for task S, $\bar{W}(X)$ as the average work done each slot executing task X, and N_a as the mean number of active sources. Then clearly

$$t_{(a)}^{T} = T - \bar{W}(H) - \bar{W}(L)$$

where

$$\bar{W}(H) = N_a(a + \frac{T}{15}c)$$

$$\bar{W}(L) = \alpha N_s Tb$$

Note that $\frac{T}{15}$ is the fraction of active sources that will have a state change to process - assuming $T \le 15$ ms which is clearly necessary. Now from Little's Law we have that $N_a = \alpha N_s(30\bar{n}_p \times 10^{-3})$ - again ignoring the finite source effect. Using the requirement that the time to scan all sources must be less than 10 sec. yields the (approximate) bound

$$N_s \le \frac{1}{a\times 10^{-4} + \frac{3\alpha a\times 10^{-2}}{T} + 2\alpha\bar{n}_p c\times 10^{-3} + (\alpha b)\times 10^{-3}} \qquad (9)$$

And, for the parameter values assumed for the other schedule, $T \le 15$ ms implies that

$$N_s \le \frac{1}{10^{-5}+10^{-4}+5\times 10^{-5}+5\times 10^{-5}} \approx 4760 \qquad (10)$$

Thus this schedule appears to be capable of achieving the schedule independent bound achieved earlier! However, given the simplified estimates made above, it is not clear whether any of these schedules would actually meet the mean delay criteria, let alone the other criteria for load levels even near the approximate capacity bounds obtained. We now look more closely at the performance of these schedules.

5. Evaluation of Schedule Alternatives

We will not attempt to obtain accurate estimates of performance, but rather only refine our analysis to the extent that is necessary to make a point. To simplify our discussion and allow more concreteness, we will assume that the time for a source to go active has an exponential distribution (mean $\frac{1}{\alpha}$ as above) and that all data streams consist of exactly $n_p = \bar{n}_p$ pulses.

5.1 Evaluation of Schedule 1

Schedule 1 can be modeled as a cyclic queue with a ms overhead associated with changing work queues (scanning the next source) and d ms for servicing a new request. (See Figure 3)

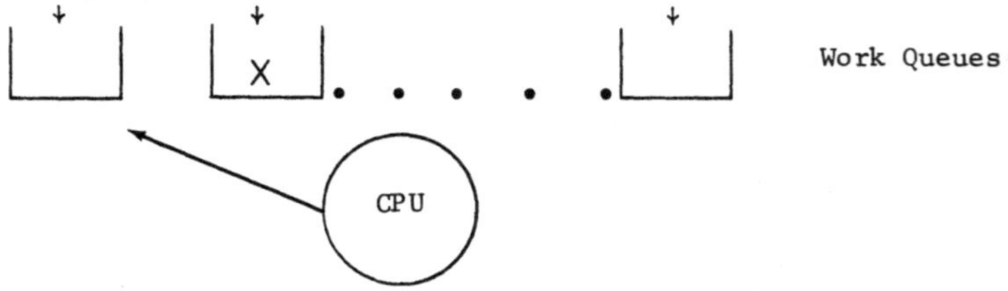

Figure 3. Cyclic Schedule

Queueing models of this type arise in the analysis of a variety of systems with real time applications. Many exact results have been obtained for the case of Poisson input with exhaustive or gated service, however, applications to more complex models for real time systems usually necessitate approximations. (See, for example, [1] as well as the references contained there). If we assume Poisson input for our system, i.e., the "infinite" source approximation we have been using we find that the mean cycle time (i.e., time to visit all sources once), $\bar{t}_c$, is given by

$$\bar{t}_c = \frac{aN_s}{1-d\alpha N_s} \tag{11}$$

provided the equilibrium condition $N_s < \frac{1}{d\alpha}$ is met. Using our standard parameter values (recall that for Schedule 1 d $\approx$ 1500 ms, i.e., the <u>elapsed</u> time for pulse collection), the requirement that $\bar{t}_c$ = 10 sec. results in a value of 66.62 for N_s, however, since $\bar{t}_c \underset{N_s \to 66\ 2/3}{\to} \infty$ we might very well be reluctant to load our system to 66 sources. This sensitivity may be due to our infinite source assumption since for the actual finite source case we must have that $\bar{t}_c \leq (a+d)\ N_s$ which is obviously finite for any N_s. We will now briefly look at the finite source case and at least obtain an approximation for $\bar{t}_c$. While the method of generating functions is usually more productive for attacking these problems, we will stay in the time domain since our main objective is to provide some insight into the nature of the problem. Keeping this objective in mind we will avoid the details of a more rigorous treatment.

Defining $t_{c,j}$ as the jth cycle time we can write

$$t_{c,j} = \sum_{i=1}^{N_s} t_{i,j} + a\ N_s \tag{12}$$

$$\text{where } t_{i,j} = \begin{cases} d - \text{if source i has a request on the } j^{th} \text{ scan} \\ 0 - \text{otherwise.} \end{cases}$$

Now the expectation of $t_{1,j}$ conditioned on $\{t_{2,j-1},\ldots,t_{N_s,j-1}\}$ is

$$E_{j-1}(t_{1,j}) = d\left(1-e^{-\alpha\left(\sum_{i=2}^{N_s} t_{i,j-1}+N_s a\right)}\right) \tag{13}$$

And taking the expected value of (13) we obtain

$$\bar{t}_{1,j} = d\left(1-e^{-\alpha N_s a} E\left\{e^{-\alpha \sum_{i=2}^{N_s} t_{i,j-1}}\right\}\right) \tag{14}$$

If we use Jensen's inequality, take the limit as $j \to \infty$, and recognize that all sources are identical, (14) yields

$$\bar{t}_1 \leq d - d\, e^{-\alpha N_s a} e^{-\alpha(N_s-1)\,\bar{t}_1} \tag{15}$$

Taking the expected value of (12), limit as $j \to \infty$, and using (15) we obtain

$$\bar{t}_s \leq N_s\left(a+d\left(1-e^{-\alpha((1-\frac{1}{N_s})\bar{t}_c+a)}\right)\right) \tag{16}$$

Using (16) as an approximation for $\bar{t}_c$ (i.e., replacing the inequality by an equality) we now find that for $N_s = 66.62$, $\bar{t}_c$ is only .38 sec. However, we will have the same sensitivity for this finite source case since to get $\bar{t}_c = 10$ sec we need only increase N_s to 71. For the parameter values used, this schedule violates the following scheduling guideline.

<u>Guideline 3</u>. Be sure performance is not overly sensitive to system parameters.

Note that when no sources are active, it takes (for $N_s = 67$) .0067 sec. to scan all sources. The mean time from this state until one source goes active is $\frac{1}{N_s \alpha} = 1.5$ sec; while this is short, all sources will have been scanned <u>223</u> times in this period. On the other hand, once we get one request, we are risking an avalanche effect. That is, the mean cycle time of .38 seconds is the result of "averaging" many cycles of length .0067 seconds, with others considerably longer than the average of .38 seconds. The result of this correlation is a tremendous variability in the cycle time t_c, and hence performance. Criteria 2 would clearly be violated at this load level, thus the following guideline.

Guideline 4. Consider tails of distributions of performance measures - means are not enough.

There is another guideline that should be illustrated here. Clearly the mean scan time is a poor measure of performance that the customer sees. On the other hand, it may have been a candidate for a performance measurement, particularly since it would be easy to implement. One simply counts the number of scan completions in a known period of time and divides this into the total time period. Thus,

Guideline 5. Measurements should include items that can be directly related to performance "as viewed by the customer".

Beware of easy to implement but poor performance measurements.

5.2 Evaluation of Schedule 2

The following simple example shows that the 10^{-4} requirement on accurate data collection (Criteria 3) is violated with 150 sources. If there is one active source and we scan a pulse in the last third of its on-time, then with probability $(1-e^{-\alpha(N_s-1)(15\times10^{-3})}) \approx .022$ the set up time for a new request will cause us to miss the next pulse. (See Figure 4).

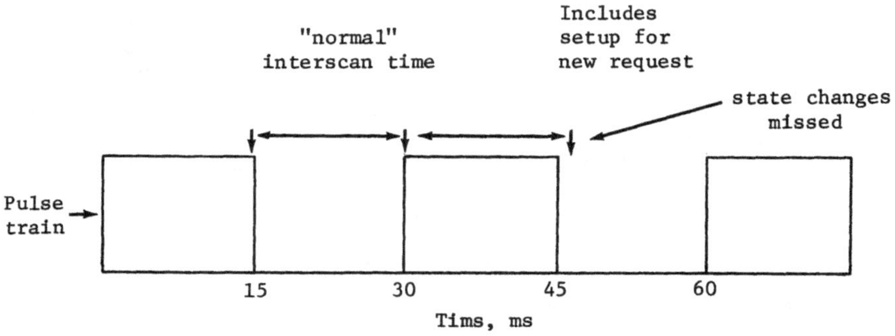

Figure 4. Detection of Pulses

Hence, clearly Criteria 3 is violated. Indeed, a little thought leads to the conclusion that we cannot support more than 99 sources, a number which at least requires 2 new requests during a scan cycle (9.9 ms) to miss a pulse. Actually, having a schedule limit on a "hard" criteria such as Criteria 3 is usually less desirable than having it limit on the "softer" delay criteria, that is, although it cannot always be adhered to, one should consider the following guideline.

Guideline 6. The first criteria to be violated under "overload" should be "soft".

Indeed this schedule violates the Golden Rule for processor scheduling for real time systems - Ensure Graceful Degradation.

We note before proceeding that Schedule 2 is another example of a (more complex) cyclic schedule.

5.3 Evaluation of Schedule 3

Clearly the clock tick value of T = 15 ms used above to obtain an approximate bound would lead us into the same difficulty as with Schedule 2 with respect to pulse collection. To determine an appropriate clock tick and a load level that would meet all of our criteria one needs to construct a model for predicting delays. Generalizations of this type of schedule are very important for real time applications although they are extremely difficult to analyze exactly. We briefly outline, in the context of Schedule 3, one approximation method that has been found useful. (A discussion of the application of this method to more general clocked schedules is given in [2] and results of a more comprehensive study will be available shortly). We can view this schedule in the following manner.

At deterministic time points T,2T,... two tasks arrive for processing, one of which has preempt resume priority over the other. We approximate the service times for the high (low) priority tasks by independent, identically distributed random variables with distribution function $F_h(s_h)(F_\ell(s_\ell))$ and assume that the high priority service times. (S_h) are independent of the low priority service times (S_ℓ). Clearly delays associated with execution of the high priority work can be obtained by analyzing a $D/G_h/1$ queue. The problem we address is finding the distribution function $F_D(d)$ of the delay in starting an arbitrary low priority task. (Delays until task completion can be handled in a similar manner). We begin by noting that the backlog of work (high and low priority) found by an arbitrary arrival is given by the waiting time in a D/G/1 queue with interarrival time T and service time $S_h + S_\ell$. Denote the distribution function for this work backlog (waiting time) Y by $F_w(y)$.

Consider now the random walk characterized by

$$S_o = y,\ S_1 = y - T_{a,1}, \ldots, S_n = y - \left(\sum_{i=1}^{n} T_{a,i}\right)$$

where the $T_{a,i}$ have the same distribution function as $T_a = T - W_h$. Now conditioned on the value of the backlog Y = y, denote by $\tilde{P}_n(y)$ the probability that the first time that the random walk encounters a barrier at S = o occurs after the nth step. Then the probability, $P_D(nT)$, that the start of an arbitrary low priority task is delayed more than nT is given by

$$\tilde{P}_D(nT) = \int_o^\infty \tilde{P}_n\ (y)\ dF_w\ (y) \qquad (17)$$

Equation (17) is made useful from a practical viewpoint by the approximations of $F_w(y)$ of the form $1 - C\ e^{-ay}$ which are discussed briefly in [2] and elaborated on in [3].

Delays for scanning can be analyzed in the same manner by considering a random walk starting at S_o = time needed to scan all sources once, e.g., $S_o = a\ N_s$ in our example.

Using methods of this type we find that a reasonable clocked schedule for our system is that shown on Figure 5 and that it can support approximately 3000 sources. (Note that the task L has been broken up into subtasks to minimize the probability of overrunning the clock, a possible source of increased overhead.)

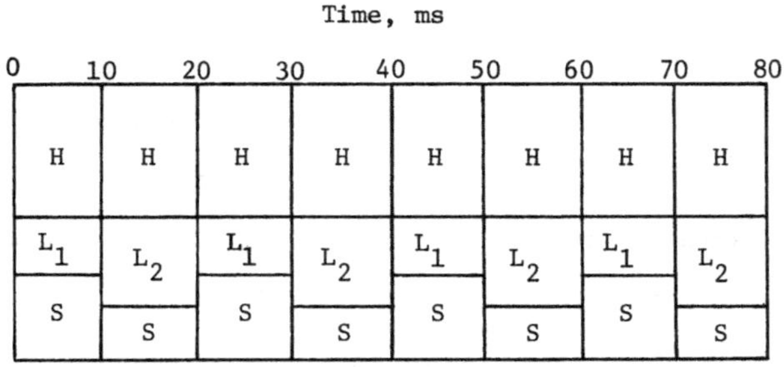

Figure 5

Note that unlike the other two schedules, a temporary work overload will effectively lock out scanning for new requests - a desirable property which adheres to the following guideline.

Guideline 7. The schedule should naturally limit new work under overload.

However, note that while this can usually help for brief "statistical" overloads or a slowly building overload, it is no guarantee of adequate overload performance. Indeed, we need to follow

Guideline 8. Explicit overload controls should be included.

6.0 Concluding Remarks

While the example system considered here is somewhat artificial, studies of actual systems have shown that relatively simple scheduling changes can significantly increase the capacity of existing systems. Moreover, even for our example system, the best schedule depends strongly on the parameter values, rather than just the functions. For example, if there were a much higher degree of activity, i.e., requests from sources, Schedule 2 would be a preferable alternative. On the other hand, if the scanning overhead were high and the elapsed time needed to collect data small, Schedule 1 would be preferable.

Thus, the determination of a good schedule is not always obvious. It depends not only on the functions to be performed and the criteria placed on these, but also on the traffic demand levels and other system parameters. Hence the following final guideline.

Guideline 9. It is desirable to have a schedule that is flexible, particularly if it must be specified at the early stage of a system's development when many system parameters are not yet known precisely.

We reemphasize that there are few results available from the analysis of suitable stochastic models for processor scheduling in a real time applications environment. Clearly a comprehensive theory for the analysis of such processor schedules is highly desirable. Moreover, there are virtually no tools for the synthesis of feasible (much less "optimal") schedules for these systems.

7. References

[1] Kuehn, P. J., "Multiqueue Systems with Nonexhaustive Cyclic Service," Bell System Tech. J., Vol. 58, March, 1979.

[2] Fredericks, A. A., "Analysis of a Class of Schedules for Computer Systems with Real Time Applications," in Performance of Computer Systems, M. Arato, A. Butoimenho, E. Gelenbe, eds.

[3] Fredericks, A. A., "A Class of Approximations for Queueing Systems with Real Time Applications," Bell System Tech. J., March, 1982.

Bell Telephone Laboratories, Holmdel, New Jersey 07733

MODELING REAL DASD CONFIGURATIONS

David Hunter

Abstract

In this paper we survey modeling of Direct Access Storage Devices (DASD) from the point of view of a practitioner attempting to model a real, running system. Our example is DASD subsystems for IBM MVS systems. We examine the features of DASD configurations and the components of DASD service and response times, indicating the important features for performance management. We then treat the incorporation of real performance data in such models and the resulting problems which arise.

1. Introduction

The intent of this paper is to provide a survey of the problems that arise in constructing probabilistic performance models of real computing systems. The particular case considered is that of Direct Access Storage Devices (DASD) connected to IBM System/370 computers running the Multiple Virtual Storage (MVS) operating system. This evaluation occurred as a part of the development of a DASD subsystem tuning aid for such systems.

Since there has been a good deal of work on DASD performance modeling in the literature, our first objective is to give an empirical description of the System/370 DASD environment and comment on what components of DASD request servicing are truly important for performance in a real system. These observations are based on the author's experience with a number of measurement programs in varying detail of real MVS systems. Next, we will discuss the use of existing probabilistic modeling technology in conjunction with available MVS performance data sources to construct a model of an actual running system. The intent here is not describe the modeling effort in detail, but instead to illustrate the inevitable problems and constraints which arise in parameterizing a model from real data. In this discussion, we will point out areas of the transition from data to model which could use further analysis.

2. Environment

2.1 Hardware/Software

MVS is an operating system that contains support for batch jobs, interactive users, and various subsystems for database and interactive applications. It supports multiprogramming and the use of virtual storage. It is usually run on large CPU's and will often have from 50 direct interactive users to several hundred interactive users connected to a major subsystem.

The DASD configuration for an MVS system may range from 40 devices to over 250. These devices usually all have the Rotational Position Sensing (RPS) feature which allows the rotational positioning of the disk device without incurring channel connection time. Moreover, the System 370 I/O architecture provides a number of connection options for these devices which are supported by MVS. Figure 1 shows a portion of a conventional DASD configuration for a System 370 CPU. The major pieces of such a configuration: CPU, channel, control unit, and string are illustrated. However, more typical configurations for MVS system are shown in Figures 2 and 3. Note the presence of an optional channel which may be connected from the CPU to the control unit or head of string, and the possible connections to other systems, usually running MVS as well. This multiplicity of connections is the rule, not the exception in the MVS environment.

```
-------------
| CPU       |
-------------
-------------    -----------------    ---------------
| CHANNEL 1 +----+  Control Unit +---+   Head of    |
-------------    -----------------   |    String    |
                                     ---------------
                                     |   Devices    |
                                     ---------------
```

Figure 1. Simple MVS DASD Configuration

```
-------------
| CPU       |           +----> Other system(s)
-------------           |
-------------    -------+---------    ---------------
| CHANNEL 1 +----+  Control Unit +---+   Head of    |
-------------    ----------------- +-+    String    |
-------------    ----------------- |  ---------------
| CHANNEL 2 +----+  Control Unit +-+ |   Devices    |
-------------    -------+---------    ---------------
                        |
                        +----> Other system(s)
```

Figure 2. Common MVS DASD Configuration

```
-------------           +----> Other system(s)
| CPU       |           |                +----> Other system
-------------           |                |
-------------    -------+---------    ---+-----------
| CHANNEL 1 +----+  Control Unit +---+   Head of    +
-------------    -+---------------   |    String    |
-------------     |                   ---------------
| CHANNEL 2 +-----+                   |   Devices    |
-------------                         ---------------
```

Figure 3. Common MVS DASD Configuration

2.2 Anatomy of a System/370 DASD I/O Request

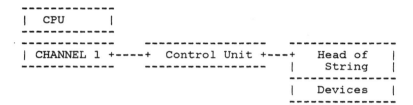

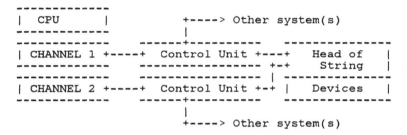

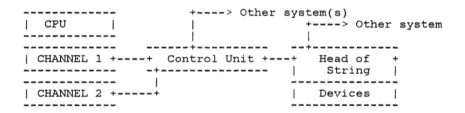

To set the basis for the following discussion, Table 1 gives an operational diagram of the servicing of a disk I/O request. The operations are listed in order and the operator is the location of the primary activity. The channel column indicates whether the channel is required for the operation to take place. The classification column separates the operations into three groups, Queueing, Path Delay, and Service which are discussed in Section 3.

Table 1

Operation	Operator	Channel	Classification
1 Queue for device	CPU	Not Held	Queueing
2 Wait for path*	CPU	Not Held	Path delay
3 Seek transmission	Channel	Held	Service
4 Seek**	Device	Not Held	Service
5 Reconnect delay* **	Device	Not Held	Path delay
6 Sector transmission	Channel	Held	Service
7 Latency	Device	Not Held	Service
8 Reconnect delay (RPS miss)*	Device	Not Held	Path delay
9 Channel program (incl. transfer)	Channel	Held	Service

* will not occur if channel free of other traffic
** will not occur if arm already at correct cylinder

3. What Features Are Important?

3.1 DASD Service Time

Basic DASD service time consists of the fundamental device operations necessary to service an I/O request. These are separated into five operations (3,4,6,7,9) in Table 1, but basically consist of arm positioning, rotational positioning, and channel operations. In discussing the service parameters of a disk subsystem, there is a great difference between the nominal device parameters and the parameters seen in a real system. A nominal description of an IBM 3350 disk drive is the following (IBM (1978)):

Nominal seek time:	25.0 ms
Average latency:	8.3 ms
Transfer rate:	1.2 MB/sec

A conventional view of DASD I/O would indicate that in the absence of interference from other requests, the time to process a disk I/O request for a 4K block would be

Nominal seek time:	25.0 ms	Channel free
Average latency:	8.3 ms	Channel free
Transfer time:	3.3 ms	Channel busy
Total:	36.6 ms	

In reality, the picture is much different.

First, as is well known, the nominal seek time is based on a uniform model of data reference (i.e., the arm moves with equal probability to any other position) while in actuality, sequential data accesses predominate. Moreover, the large number of DASD on a MVS system usually ensures that there is only modest interleaving of requests to different data sets although a number may be open on the device. This implies that disk arms very often do not move - measurements indicate that in real MVS systems the percentage of disk I/O requests that cause arm movement varies from 30 to 50%. Similar results have been reported by Lynch (1972) and Wilhelm (1976) has proposed a disk reference model to incorporate this feature. This casts doubt on the benefits of seek scheduling which is not used in MVS, but has been extensively treated in the literature: a further discussion of this is provided in Section 3.3. Practically, an average seek time for a 3350 drive might well be 10 ms or less.

Second, while data transfer is certainly a part of channel busy time, other time consuming activities are carried out on the channel. To be perfectly accurate, the channel is much more than a set of wires connecting the CPU with the control unit. It is a limited CPU with its own instruction set which contains a large number of search instructions. In fact in terms of busy time, an MVS channel is often more a search engine than it is a transfer engine. A average channel busy time breakdown for I/O requests to 3350's might be as follows:

Search + Overhead	Transfer
6.2 ms	3.3 ms

The search time is the time the channel spends connected to the devices searching for the desired record. The overhead is the time that it takes for the channel to actually transmit the instructions in the channel programs that perform the I/O operations. The transfer is the average time required for actual data transfer.

Finally, the latency time is approximately correct. The uniform model for latency leading to a latency delay of one half rotation, proves to be fairly good in practice, although there are yet some programs which do not make use of the rotational position sensing feature of the devices.

Therefore, a realistic appraisal of the time for servicing an I/O request to a 3350 might look something like this:

Average seek time:	10.0 ms	Channel free
Average latency:	8.3 ms	Channel free
Channel time:	9.5 ms	Channel busy
Total:	27.8 ms	

Notice that this is about 9 ms less than the casual view suggests, but that it provides about three times the channel busy time, through non-device dependent activities. This shift is important for it emphasizes the importance of connection parameters in real systems.

3.2 DASD Path Delays

Path delays are additions to basic DASD service time which occur due to contention with other I/O requests for shared paths. They are labeled as such in Table 1. Here (and in the modeling discussion in

Section 4) they will be treated as a second component of service time which is added to basic service time to form the extended service time for a request. Conceptually, this is because they occur after a request has reached the head of a system's device queue and there is no other request from that system in service. This is also the way they have been modeled in the literature.

In the process of servicing I/0 requests two distinct kinds of path related delays may be encountered. The first is in starting of requests and the second is in responding to the signaling of completion of disconnected services by a device. In the first case, a request at the head of a queue for a given device may be delayed because it encounters a channel, control unit, head of string, or device busy. (The device busy can only occur with a request from another system, since the request will not be started on its own system until there is no prior request in service.) The request will have to wait for the busy condition to be cleared before it can start, or if an alternate path exists (and is effectual for the particular conflict encountered) it can be attempted on that path where it again may be delayed.

In the second case, when a device completes a disconnected seek or position sensing operation, it may be kept from immediately signaling this to the CPU by a busy condition back along the path to the CPU. (Note that it will only try the path on which the initial connection was made.) After a seek the device will signal as soon as the path becomes free, but after a position sensing operation the device will have to complete one full rotation before it can signal again. This latter action is called a RPS miss.

In terms of magnitude, the RPS miss delay is usually thought to be the greatest, since even one RPS miss will cause a delay of one rotation which is 16.7 ms. (for most DASD on MVS systems) - a sizeable fraction of the basic service time. However, in practice the wait for path may also be very large in configurations connected to multiple systems. In particular, part of the multiple system support is a locking mechanism called Reserve/Release which will prevent any other system from using a device until released by the reserving system. This and outright usage of a device by another system may well overshadow RPS miss time. Generally, usage of all parts of the path by a given system and other sharing systems has to be monitored to keep these delays small, and these effects must be modeled to show the true effect of the connection parameters on performance.

3.3 DASD Queueing Delays

Queueing of requests for DASD I/O in MVS is managed in software on a FCFS basis. (Some of the more current releases have a non-preemptive priority queue based on the requesting job's priority.) In real MVS systems, long queue lengths are very rare with an average of .3 being about the limit for heavily used devices. This is due largely to a distinct effort on the part of the systems support staff to reduce queueing delays. When large queues are found to be occurring on a device an examination is made to find whether it is long service times or overutilization that is causing the problem.

In the case of long service times the culprits are usually path delays such as high channel utilizations causing a large RPS miss delay, or excessive contention for a device with another system. Seek times are occasionally a problem, usually when multiple frequently used system datasets are on the same device. The solution here is usually a device reorganization to move these datasets together or the movement of some of the datasets to other devices.

There is usually also constant monitoring to detect overutilized devices. In real MVS systems there is generally a very great skew of access rates to devices. It is not uncommon for 90% of the nonpaging DASD I/O requests to go to 20% of the connected devices, or for 33% to go to just three of the connected devices. This skew is generally not a dynamic random phenomenon - the overused devices generally remain so throughout an operational shift, although the precise ordering of devices by access rate may change. The solution here is to take heavily used data off the busy devices and move it to less used devices. Generally no attempt is made to "perfectly" balance devices, but instead to reduce the impact to a tolerable level. The statistics quoted above are for reasonably well tuned systems.

Faced with this kind of situation, it does not appear that there is much application for seek scheduling techniques as described in Denning (1967); Teorey and Pinkerton (1972); Coffman, Klimko, and Ryan (1972); and Oney (1975). Since all of these methods require a queue to operate on, they are not appropriate for an environment where queues are maintained at a low level and seek time is usually not the major service delay. The basic management philosophy is that the queue itself is the problem and it should be decreased directly.

4. Using Real Data for DASD Modeling

4.1 Existing Modeling Technology

There exist two major approaches to analytic DASD subsystem modeling; standalone and as part of a central server system model. The standalone model uses fixed Poisson inputs to the DASD devices and treats each as a M/G/1 queue with the (extended) service time built up of basic service and path delays calculated from the specified flow rates over various portions of the connection path. The various components of extended service are assumed to be independently distributed and only be specified to the level of a mean and coefficient of variation. Recent treatments of the standalone model are Wilhelm (1977), Zahorjan et al. (1978), and Bard (1980). Of special interest is Bard (1980) which contains modeling techniques applicable to the multiple connection configurations. In general, the basic I/O structure we have described above differs little from that in these papers except that we have made a more complete accounting of path delays, one or more of which is not considered by these authors.

A recent paper by Hofri (1980) points out that seek times are by their nature not independent, and thus the M/G/1 formulation is not correct. Based on the seek model of Wilhelm (1976), Hofri is able to use the methods of Neuts (1977) and Coffman and Hofri (1978) for queues with general semi-Markov service in this application. In this author's experience, for the high zero seek fractions encountered in practice this refinement makes little difference. That is, the dependence is not that strong when the percentage of zero seeks is high. This observation is also made by Hofri.

In the central server model, the basic standalone approach is maintained except that the flow rates are now determined by the rest of the network and the devices are usually treated as having exponential service to simplify analytical solutions. Moreover, since the device service times are dependent on flow rates through the path delays, an iterative calculation of analytic solutions is required. A recent proof of the convergence of the iterative procedure is given in Garner et al. (1980).

Throughout this discussion the standalone M/G/1 model will be used although most comments are applicable to both.

4.2 The Resource Management Facility (RMF)

The most difficult problem encountered in any real system performance modeling activity is usually obtaining data to construct and parameterize the model, not calculating solutions. To a great extent the features and the applicability of model are determined by the data available. Therefore, the modeler is bound by the constraints of most system monitoring tools which are basically: a low overhead requirement which leads to curtailed function and a design suited to other purposes than those of system modelers. A large amount of effort is therefore devoted to reworking the data so that it can be incorporated into a model.

The most common source of systems performance measurement data for MVS systems is the Resource Measurement Facility (RMF) which is described in detail in IBM (1980). This is a widely installed software monitor which records a variety of system performance data. It accumulates data by either direct counting of system events or periodic sampling of system states, and at intervals writes it out. RMF also includes a postprocessor to format the data into reports on the various system components. For the DASD subsystem, it provides I/O rates, queue lengths, and utilizations for channels and devices as well as "utilizations" for various path delays. These latter utilizations are the fraction of time the I/O requests for various devices were delayed for path busies of different kinds. Note carefully that no times are actually collected directly, instead they all have to be inferred by the use of Little's formula in this context; that is dividing the time average population (or utilization) in a particular component by the measured I/O rate.

The problems of using RMF data for parameterizing DASD models run the gamut from mundane to significant. An example of the mundane is that little used devices may show a positive I/O rate (since accesses are counted directly), but have no reported utilization since that is sampled and sampled at a rate significantly slower than the I/O service rate. This makes service time calculation impossible or suspect for low rate devices, but these are unimportant from a performance perspective so there is little loss.

Of more significance are the following. First, RMF often provides a statistic for a group of devices, such as all those connected to a channel, instead of for individual devices. An example of this aggregation problem is channel service time which is based on all devices connected to a channel rather than by device. A second problem is device invisibility, whereby many device operations are invisible to the CPU where RMF is running and thus cannot be timed separately. For instance, RMF gives no direct information on seek times.

The solution to the first problem is inevitably, to use the average of the aggregated statistics for each of the aggregated devices and hope for sufficient device homogeneity to prevent distortions. Solutions for the second problem are to infer wherever possible the parameters of interest, and use information from other sources for events that are largely invariant across environments. We do not intend in this paper to describe how to infer from RMF data distributional parameters for each of the operations in satisfying a DASD I/O request which are given in Table 1. Instead, we will illustrate the concepts above with only a subset of the overall problem which is discussed in the next section.

4.3 Inference for Components of DASD Service Time

To determine mean values for the distributions of the components of basic service time some kind of inferential procedure is required. RMF provides an individual device utilization that corresponds to a time average occupancy of I/O requests for the device in operations 3-9 of Table 1. Using the RMF device I/O rate, a average total passage time may be calculated. To break this time up into the separate components, inference based on a model structure is necessary. As a brief aside, this is much the same procedure followed in statistics where a structure is hypothesized and parameters estimated. What remains to be derived in the performance modeling area are the statistical properties of such estimates and the wealth of tests for goodness of fit and assumption checking which are available in most statistical problems.

To numerically illustrate the process we will assume that the device is a 3350 and start with a calculated service time of 35 ms (see Table 2). The first step in breaking out service components is to subtract latency. This requires reliance on the uniform model for latency and hopefully complete use of RPS by the channel programs. For a 3350 DASD the rotation time is 16.67 ms. so the mean latency time according to the uniform model is 8.3 ms. Next, average total channel time may be subtracted. This channel time is only available from RMF for the entire collection of devices on the channel or channels to which the chosen device is connected. It is derived from the channel busy statistic and the channel I/O rates. Therefore, using the standard aggregation fixup an average or weighted average of the service times on all of the connecting channels is assumed to be the channel time for the given device. For the numerical example assume that is 9.5 ms. Thus after subtracting latency and channel time this leaves (Table 2) 17.2 ms to be accounted for among seek time and reconnect delays (operations 4, 5, and 8 in Table 1).

Since reconnect delay after the seek and the seek itself are directly related by whether the arm has to move or not, these will be temporarily considered together and the reconnect delay after the sector transmission will be removed first. The reconnect delay after the sector transmission is the usual RPS miss problem and conventionally the average RPS miss penalty is treated as some function of channel utilization multiplied by the device rotation time (e.g., Wilhelm (1977)). A conventional model of this phenomenon has been to assume that the device is independent of other devices on the channel and to assume that each attempt at reconnection is an independent Bernoulli trial with probability of failure given by the channel utilization conditional on the given device not using the channel. (Some less appropriate measures such as unmodified channel utilization are sometimes seen in the literature.) If this probability is denoted by p then the total number of misses per I/O request is a geometric variable with mean $p/(1-p)$.

Table 2

Total RMF service time	35.0 ms	Ops 3-9
- Avg. Latency	8.3 ms	Op 7
- Avg. Channel Time	9.5 ms	Ops 3,6,9
Seek and Reconnect delays	17.2 ms	Ops 4,5,8
- RPS miss	3.2 ms	Op 8
Seek and post-seek delay	14.0 ms	Ops 4,5

There are two problems with this view of the RPS miss delay. First, while this view may be adequate for single path connections, for multiple system and multiple path connections some careful accounting must be done to determine the probability of encountering path busies. An enlightening solution of this problem is given in Bard (1980), which reduces to the conventional estimate in the case of a single path. Second, empirical observations and data from simulations reported in Zahorjan et al. (1978) and Bard (1980) indicate that the Bernoulli model underestimates the mean number of RPS misses. Bard suggests that the calculated probability, p, may be correct for the probability of the first miss, but subsequent misses are then conditionally more likely to occur. He suggests that the expression for mean number of RPS misses given above is more accurate if p is increased to $p + p^2/2$.

Following Bard's methodology and assuming that p is calculated for the particular configuration to be .15, then the mean number of misses is .19 and the mean RPS miss delay is therefore 3.2 ms. Subtracting this yields 14 ms as the total for seek and post-seek reconnect delay time.

At this point, all components have been separated with the exception of the channel busy operations (3,6,9 in Table 1) and the seek/post-seek operations (4,5 in Table 1) which are separately grouped. The channel busy operations may be separated based on hardware measurements since the seek and set sector transmissions are effectively deterministic events. Subtracting these from the channel time yields the average time for the bulk of the channel program including data transfer.

Considering the seek and post-seek reconnect delay, first recall that they are definitely not independent events - if the arm is already in position, no seek will occur and thereby there will be no reconnect delay. When the reconnect delay does occur, it may be treated as in Wilhelm (1977) and a residual life calculation made to determine mean wait after seek based on the channel busy process. Our more detailed treatment here separates the channel busy services into three distinct kinds and requires some modification of Wilhelm's argument for this reason and to support the multipathing flow rates. Practically, the seek and set sector transmissions are negligible compared to the channel program execution and may well be ignored. We will assume that the distribution of this wait time, when it occurs, is independent of the seek distribution, and denote the mean wait time as w.

Then to fully separate the seek components requires a further model to parameterize the seek distribution sufficiently to permit estimation.

One possibility is to adopt the model of Wilhelm (1976), and assume that with probability $1-q$ the I/O request does not require an arm movement, and with probability q will move with equal probability to the other disk cylinders (i.e., a nominal seek). Then the average combined seek and seek delay time from Table 2, S, is given by:

$$S = q(w + N)$$

where N is the device's nominal seek time and w is the reconnect delay described above. Solving for q, yields

$$q = S/(w + N).$$

as the zero seek probability. Thus, the average seek time is qN and the average post seek delay time is qw. This is effectively a proportional allocation of the observed combined seek time between seek and post-seek delay. At this point, the extended service time has been fully decomposed.

4.4 Further Considerations

We have given an example of the estimation of mean times for various constituents of extended service time from real data based on imposed modeling structures. Similar considerations apply to determining the mean wait for path and queueing delay. One topic not treated is determination of coefficients of variation for the components of service in the M/G/1 model. RMF provides some higher moment information for DASD, but again there has to be some resort to a model or to hardware measurement data to determine appropriate values along the above lines. Also recall that some of the operations in Table 1 are definitely not independent and some modifications must be made to the usual determination of an overall coefficient of variation for service.

Clearly, much remains to be done in this area, particularly in determining the realm of validity of the aggregation and other assumptions, and in deriving the statistical properties of the estimates. One fundamental assumption which was not discussed is the character of the I/O request process and the accuracy of the Poisson process model. We are tentatively using this empirical methodology with success in some modeling efforts from real data, and an assessment of general utility with a larger body of data is under way. Practically, the basic worth of any probabilistic modeling procedure may be measured in its ability to answer real questions about real problems.

5. References

Bard, Y., (1980). A model of shared DASD and multipathing, Comm. ACM, 23, pp. 564-572.

Coffman, E.G., Klimko, L.A., and Ryan, B., (1972). Analysis of scanning policies for reducing disk seek times, SIAM J. Comput., 1, pp. 269-279.

Coffman, E.G., and Hofri, M., (1978). A class of FIFO queues arising in computer systems. Oper. Res., 26, pp. 864-880.

Denning, P., (1967). Effects of scheduling on file memory operations, Proc. Spring Joint Computer Conference, 30, pp. 9-21.

Garner, R.L., Jacobson, P.A., and Lazowska, E.D., (1980). The method of surrogate delays: simultaneous resource possession in analytic models of computer systems, Tech. Report 80-04-03, Dept. of Computer Science, Univ. of Washington.

Hofri, M., (1980). Disk scheduling: FCFS vs SSTF revisited. Comm. ACM, 23, pp. 645-653.

IBM, (1978). Introduction to IBM Direct-Access Storage Devices and Organization Methods, GC20-1649-10, IBM Corp.

IBM, (1980). OS/VS2 MVS Resource Measurement Facility (RMF) Reference and User's Guide, SC28-0922-3, IBM Corp.

Lynch, W.C., (1972). Do disk arms move?, Performance Evaluation Review, 1, December, pp. 3-16.

Neuts, M.F., (1977). Some explicit formulas for the steady-state behavior of a queue with semi-Markovian service times, Adv. in Appl. Probab., 9, pp. 141-157.

Oney, W.C., (1975). Queuing analysis of the scan policy for moving head disks, J. Assoc. Comput. Mach., 22, pp. 397-412.

Teorey, T. and Pinkerton, T.B., (1972). A comparative analysis of disk scheduling policies, Comm. ACM, 15, pp. 117-184.

Wilhelm, N.C., (1976). An anomaly in disk scheduling: a comparison of FCFS and SSTF seek scheduling using an empirical model for disk accesses, Comm. ACM, 19, pp. 13-17.

Wilhelm, N.C., (1977). A general model for the performance of disk systems, J. Assoc. Comput. Mach., 24, pp. 14-31.

Zahorjan, J., Hume, J.N.P., and Sevcik, K.C., (1978). A queueing model of a rotational position sensing disk system, INFOR, 16, pp. 199-216.

Written while the author was the the IBM Thomas J. Watson Research Center, Yorktown Heights, New York 10598. The author is now located at the IBM Data Systems Division Laboratory, Poughkeepsie, New York 12602.

Discussant's Report on
"Modeling Real DASD Configurations,"
by D. Hunter

Dr. Hunter's paper gives an interesting description of a real computer system, the nature of the data available to fit analytic models, and the analytic models used. There are two striking aspects of the paper. One is the complexity of resource sharing in the real system versus the simplicity of the analytic model. The paper by Bard referenced by Hunter indicates that the analytic model appears to capture the performance of the DASD system Bard considered quite well under normal operating loads, but not under heavy loading. This model breakdown may result from the fact that, under heavy loading, dependence induced by the resource sharing has greater influence on the performance of the system. A challenge for models of the next generation is to comprehend the reason for the discrepancy, and to model heavy traffic behavior more accurately.

The other striking aspect of the paper is the apparent coarseness of the data avilable to fit the analytic model. Dr. Hunter points out that software monitors used to collect the data are usually designed for purposes other than modelling. As a result some of the statistics for model parameters are inferred rather than observed directly. For example the software monitor does not sample service times; mean service times are inferred by the use of Little's formula. Thus, the distributions of service times cannot be directly estimated. These distributions may make a difference in system performance at high traffic levels, accounting for the discrepancy mentioned earlier. Also the software monitor will often provide a statistic for a group of devices rather than for individuals; this is Dr. Hunter's aggregation problem; the size of this effect is unknown, but a sensitivity check with hypothetically different individuals could be informative.

There are at least three sources of error possible in fitting probabilistic models. One is that the model is inadequate. Another is the sampling error and bias in the parameters that can be sampled directly; this error may be the result of an incorrect understanding of the sampling procedure used by the software monitor. The third source of error is in the estimation of derived parameters. For example, an estimate of mean service time derived from Little's formula could be seriously biased by nonstationarity.

The building of a probability model is one of the early steps in the scientific method. Other steps include the estimation of parameters in the model and model validation. Since these last two steps involve data collection and analysis, they are messy and difficult. However, Dr. Hunter's paper points out the need for research in these difficult areas, and should stimulate further progress. It is one of only a few papers that attempt to explicitly combine data analysis and statistics with probability modeling.

Discussant: Dr. Patricia A. Jacobs, Department of Operations Research, Naval Postgraduate School, Monterey, California 93940

BOTTLENECK DETERMINATION IN NETWORKS OF QUEUES

Paul J. Schweitzer

Abstract

Procedures are given for identifying saturated servers in open networks of queues, or closed networks of queues with customer populations approaching infinity. The former leads to simultaneous functional equations for the throughputs, while the latter leads to a non-linear complementarity problem which in important special cases may be interpreted as a Kuhn-Tucker set. Efficient computational procedures are supplied for both the open and closed network cases.

1. Introduction

This paper provides procedures for bottleneck determination in open and closed networks of queues (NOQs). Bottlenecks refer to servers operating at 100% utilization. For open NOQs, this can occur if the external arrival rates are sufficiently high. For closed NOQs, this can occur as the population sizes approach infinity.

In both the open and closed NOQ cases, a first moment analysis usually suffices - ergodic arguments based upon mean service times. In the open NOQ case, bottleneck determination reduces to solving a set of <u>simultaneous functional equations</u>. In the closed NOQ case, bottleneck determination reduces to a <u>non-linear complementarity problem</u>. Efficient computational procedures are given for both cases.

This paper summarizes results from two technical reports [11], [12] for the open and closed NOQ cases. The interested reader is referred to these reports for proofs and further information.

The results presented for open NOQs are believed new. The reader will note the solution methods described here for solving the functional equations (linear programming, successive approximation in policy space, successive substitution) are familiar from transient/discounted Markov decision programming [2], [3], [4], which deals with a similar set of functional equations.

The results presented for closed NOQs with large populations, in terms of a non-linear complementarity problem which can be reduced to an optimization problem in some cases, are also believed new. The model employed here differs from Lavenberg's analysis [5] of product-form networks, which employs one IS service center with infinitely many parallel servers, whose service times are scaled upwards as the population increases. The model also differs from that of Pittel [6], which assumes a finite waiting room at each server. Closely related techniques were used by Bard [1].

This paper deals with networks of queues with static routing, one constant-rate server at each service facility node, FCFS service discipline, and unlimited waiting room. Weakening any of these assumptions would be of great interest, but would significantly alter the system behavior and associated bottleneck analysis.

The results presented here assume neither Poisson external arrivals nor "local-balance" service distributions. The results are therefore more general than those for product-form networks, but at the expense of using assumptions which are harder to validate.

2. Bottleneck Determination in Closed Networks of Queues

We first present the descriptive equations for <u>finite</u> populations of each customer class. In this case, no server can be expected to be a bottleneck (100% utilization), since there is generally a positive probability it is idle, with all customers elsewhere. Bottlenecks arise only when population sizes approach infinity, and the results obtained in this manner are an approximation to the behavior for large, but finite, populations.

2.1 Notation and Assumptions

- R customer classes, numbered $r = 1,2,\ldots,R$
- S servers, numbered $s = 1,2,\ldots,S$.
- All servers are FCFS with unlimited waiting room.
- $P(r)_{st}$ = routing matrix for class-r customers

  $1 \leq r \leq R$; $1 \leq s, t \leq S$.

$$P(r)_{st} \geq 0, \qquad \sum_{t=1}^{S} P(r)_{st} = 1$$

- $\pi(r) = [\pi(r)_s]$ = unique equilibrium distribution for $P(r)$:

$$\pi(r)\, P(r) = \pi(r), \qquad \sum_{s=1}^{S} \pi(r)_s = 1, \qquad \pi(r)_s \geq 0$$

(Uniqueness of $\pi(r)$ holds if $P(r)$ has exactly one closed, communicating set of states, and possibly transient states as well. If $P(r)$ had two or more closed communicating set of states, our analysis holds provided class r customers are split into separate classes for each set of states, and R increased accordingly.)

- $\frac{1}{\mu_{rs}}$ = mean service time of a class-r customer at server s

  (Distribution of service time is unspecified.)

- K_r = number of customers of class r. $K_r \geq 1$ for each r.

The <u>3RS unknowns</u> of interest are

N_{rs} = mean number of class-r customers at server s (time average)

λ_{rs} = throughput (departures per unit time) of class-r customers at server s. (Let $\lambda(r)$ denote the S-vector with s-th component λ_{rs}.)

W_{rs} = mean sojourn time (queueing and service) of class-r customers at server s,

$1 \leq r \leq R, \quad 1 \leq s \leq S$

The 3RS equations for these quantities are

$$N_{rs} = \lambda_{rs} W_{rs} \qquad \text{(Little's formula)} \tag{2.1}$$

$$\lambda_{rs} = \frac{\pi(r)_s K_r}{\sum_{t=1}^{S} \pi(r)_t W_{rt}} \tag{2.2}$$

(Since $\lambda(r) = \lambda(r) P(r)$, $\lambda(r)$ is proportional to $\pi(r)$, with proportionality constant fixed by $\sum_s N_{rs} = K_r$.)

$$W_{rs} = \frac{1}{\mu_{rs}} + \sum_{t=1}^{R} N_{ts}^{(r)} \frac{1}{\mu_{ts}} - \text{(correction for residual service time)} \tag{2.3}$$

where

$N_{ts}^{(r)}$ = mean number of class-t customers at server s when a class-r customer arrives.

This becomes a closed system of equations when $N_{ts}^{(r)}$ is specified. Mean value analysis [8] uses the result $N_{ts}^{(r)}(K_1,\ldots,K_R) = N_{ts}(K_1,K_2,\ldots,K_r-1,\ldots,K_R)$ for product-form networks and solves (2.1, 2.2, 2.3) by a finite recursion. This approach is impractical if $K_1 x K_2 x \ldots x K_R$ is large, and unjustified if product-form is lacking, so approximation methods [7] are sought instead. This author has proposed the approximation [9] [10]

$$N_{ts}^{(r)} \sim N_{ts} \frac{K_r - \delta_{rt}}{K_r} \tag{2.4}$$

which has strong computational benefits, since (2.1, 2.2, 2.3) may be reduced from 3RS equations for 3RS variables down to R + S equations for R + S variables: $\sum_{s=1}^{S} \pi(r)_s W_{rs}$, $1 \leq r \leq R$, and $\sum_{r=1}^{R} N_{rs}/\mu_{rs}$, $1 \leq s \leq S$.

The approximation (2.4) has the following heuristic justification. If $t \neq r$, (2.4) says $N_{ts}^{(r)} \sim N_{ts}$, i.e., an arrival sees time-average mean queue length for classes other than his own. When $t = r$, $N_{ts}^{(r)} \sim N_{rs} \frac{K_r - 1}{K_r}$, i.e., an arrival sees time average queue-lengths for his own class, scaled down by $(K_r-1)/K_r$ since he does not see himself on queue. Alternatively, pretend that all K_r customers of class r have independent, identical, probabilities $q_{rs} = N_{rs}/K_r$ of being at server s. The mean number of class r servers at server s will be $K_r \, q_{rs} = N_{rs}$, as it must, and the mean number of class r customers seen by an arrival to server s is $(K_r-1)q_{rs} = N_{rs}(K_r-1)/K_r$.

2.2 Bottleneck Analysis

Set $K = \sum_{r=1}^{R} K_r$ = total population of customers

$$f_r = \frac{K_r}{K} = \text{fraction of population belonging to class } r.$$

Let $K \to \infty$ with the proportions $\{f_r\}$ remaining fixed. Assume that

$$|N_{ts}^{(r)} - N_{ts}| = o(K) \quad 1 \leq r,t \leq R, \quad 1 \leq s \leq S \qquad \text{(a1)}$$

the correction term in (2.3) is O(1) in K. (a2)

Then (2.3) implies W_{rs} is independent of r, to terms of O(1). Assume

$$\frac{W_{rs}}{K} \to u_s^* \quad 1 \leq r \leq R, \quad 1 \leq s \leq S \quad \text{as } K \to \infty \qquad \text{(a3)}$$

where u_s^* is interpreted as a rescaled mean virtual waiting time at server s.

$$u_s^* \geq 0. \quad 1 \leq s \leq S \qquad (2.5)$$

Assumption (a1) is that mean queue lengths seen by arrivals do not differ significantly in percentage from time-average queue lengths. It is expected to hold when queue lengths are long and fluctuations relatively slow. Assumption (a2) holds when service times are bounded, which is the case in most practical systems. Assumption (a3) is justified by (2.3), which implies $W_{rs} \leq 1/\mu_{rs} + \sum_t K_t/\mu_{ts}$ so that $[W_{rs}/K]$ is uniformly bounded and possesses at least one cluster point as $K \to \infty$. This cluster point will be independent of r, as shown by (a1).

With these assumptions, (2.1) and (2.2) become

$$\lambda_{rs} \to \lambda^*_{rs} = \frac{\pi(r)_s \, f_r}{\sum_{t=1}^{S} \pi(r)_t \, u^*_t} \geq 0 \qquad 1 \leq r \leq R, \quad 1 \leq s \leq S, \quad K \to \infty \tag{2.6}$$

$$\frac{N_{rs}}{K} \to \lambda^*_{rs} \, u^*_s \qquad 1 \leq r \leq R, \quad 1 \leq s \leq S, \quad K \to \infty. \tag{2.7}$$

In addition, since server s operates at ≤100% utilization,

$$\sum_{r=1}^{R} \lambda^*_{rs}/\mu_{rs} \leq 1. \qquad 1 \leq s \leq S . \tag{2.8}$$

Finally, the <u>complementary slackness condition</u>

$$u^*_s > 0 \text{ implies (2.8) is equality} \tag{2.9}$$

holds because any server s with $u^*_s > 0$ has an infinite queue and therefore works at 100% utilization.

2.3 Computational Procedure

Relations (2.5), (2.6), (2.8), and (2.9) form a non-linear complementarity set for $\{u^*_s\}$ and $\{\lambda^*_{rs}\}$. They may be solved as follows.

Using complementary slackness, rewrite (2.9) as

$$u^*_s[1 - \sum_{r=1}^{R} \lambda^*_{rs}/\mu_{rs}] = 0 \qquad 1 \leq s \leq S$$

or

$$u^*_s = u^*_s \sum_{r=1}^{R} \frac{\pi(r)_s \, f_r}{\mu_{rs} \sum_{t=1}^{S} \pi(r)_t \, u^*_t} \qquad 1 \leq s \leq S . \tag{2.10}$$

Equations (2.10), S equations for $\{u^*_s\}_{s=1}^{S}$, is conveniently solved by successive substitution, starting with $u_s > 0$ for all s. In computational experiments by this author, this always converged, and the solution was unique (independent of the starting point).

Once a solution $\{u_s^*\}$ and therefore $\{\lambda_{rs}^*\}$ is available, the bottleneck servers are $\{s | \sum_r \lambda_{rs}^*/\mu_{rs} = 1\}$. The approximate behavior for large finite population K is given by

$$\lambda_{rs} \sim \lambda_{rs}^* \qquad (2.11a)$$

$$W_{rs} \sim K\, u_s^* + O(1) \qquad (2.11b)$$

$$N_{rs} \sim K\, \lambda_{rs}^*\, u_s^* + O(1) \qquad (2.11c)$$

which provides an initial guess for heuristic solutions of (2.1, 2.2, 2.3). In particular, if the approximation (2.4) is used, (2.11) are the leading terms in a Laurent expansion in powers of $\frac{1}{K}$. Next-order terms would be needed to fix $\{u_s^*\}$ in the rare case where the solution $\{u_s^*\}$ to (2.10) is <u>not unique</u>. This occurs, for example, if

$$f_r = \frac{1}{R} \qquad \pi(r)_t = \frac{1}{S} \qquad \mu_{rs} = \mu_0 \qquad 1 \le r \le R, \quad 1 \le s \le S$$

where the solution set to (2.10) is

$$\{(u_1^*, u_2^*, \ldots, u_s^*) |\ \text{each } u_i^* \ge 0 \text{ and } \sum_{i=1}^{S} u_i^* = (\mu_0)^{-1}\},$$

with (2.5), (2.6), (2.8), (2.9) holding with $\lambda_{rs}^* = \mu_0/R$.

2.4 Duality Interpretation

The above non-linear complementarity problem has the interpretation of being a <u>Kuhn-Tucker set</u> of a certain convex programming problem, i.e., of having a duality interpretation, provided the service times admit a <u>factorization</u>

$$\mu_{rs} = \frac{C_s}{J_r} = \frac{\text{service rate of server s}}{\text{work content imposed by a class-r job at any server}}$$

That is, a class-r job imposes the <u>same</u> work requirement J_r at every server it visits. This is met, for example, in telecommunication systems where

J_r = mean message length (in bits) of a class-r message

C_s = line speed of channel s, in bits per second.

To obtain this Kuhn-Tucker set, define

$$x_r \equiv \frac{f_r}{\sum_{s=1}^{S} \pi(r)_s \, u_s^*}, \quad 1 \leq r \leq R \text{ (so } \lambda_{rs}^* = \pi(r)_s \, x_r) \tag{2.12}$$

$$g(\underline{x})_s \equiv \sum_{r=1}^{R} x_r \, \pi(r)_s \, J_r - C_s \leq 0 \qquad 1 \leq s \leq S \tag{2.13}$$

(server s has utilization $\leq$100%).

Recall

$$u_s^* \geq 0 \qquad 1 \leq s \leq S \tag{2.14}$$

$$u_s^* \, g(\underline{x})_s = 0, \qquad 1 \leq s \leq S. \quad \text{(Complementary slackness)} \tag{2.15}$$

Relations (2.12, 2.13, 2.14, 2.15) are the Kuhn-Tucker set for the convex programming problem

$$\min \{f(\underline{x});\ x_r \geq 0 \text{ for } 1 \leq r \leq R,\ g(\underline{x})_s \leq 0 \text{ for } 1 \leq s \leq S\} \tag{2.16}$$

where $f(\underline{x}) \equiv -\sum_{r=1}^{R} f_r \, J_r \log x_r$. It is easily verified that

$0 = \frac{\partial}{\partial x_r} [f(\underline{x}) + \sum_{s=1}^{S} u_s^* \, g(\underline{x})_s]$ leads to (2.12).

This optimization problem does not seem to have any natural interpretation. The rescaled <u>virtual waiting time variables</u> u_s^* appear as <u>dual variables</u> (Lagrange multipliers) to the capacity constraints (2.13). The non-uniqueness of Lagrange multipliers $\{u_s^*\}$ illustrated above occurs when the response surface has cusps.

The dual problem to (2.16) is

$$\max_{\underline{u} \geq \underline{0}} L(\underline{u}) \equiv f(\underline{x}(\underline{u})) + \sum_{s=1}^{S} u_s \, g_s \, (\underline{x}(\underline{u})) \tag{2.17}$$

where, by (2.12), $x(\underline{u})_r \equiv f_r / [\sum_{s=1}^{S} \pi(r)_s \, u_s]$. One way to attempt solution of (2.17) is by the <u>large-step gradient method</u>

$$(u^{new})_s = u_s + (\frac{u_s}{C_s}) \frac{\partial L(u)}{\partial u_s} \qquad 1 \leq s \leq S \tag{2.18}$$

where $\frac{u_s}{C_s}$ = specific choice of step size along $s^{\underline{th}}$ axis.

Written out, this becomes

$$u_s^{new} = u_s \sum_{r=1}^{R} \frac{f_r \, \pi(r)_s \, J_r / C_s}{\sum_{t=1}^{S} \pi(r)_r \, u_t} \qquad 1 \leq s \leq S$$

and reduces to the successive substitution method used to solve (2.10). One can now compare objective functions in (2.16, 2.17) to provide upper and lower bounds on the optimum, and to suggest termination criterion. Convergence can always be assured since the step size in (2.18) can be successively halved until the objective function L(u) improves.

3. Bottleneck Determination in Open NOQ's with One Class of Customers

This problem is solvable by <u>finite</u> algorithms (linear programming, approximation in policy space) presented below. The case having multiple classes of customers, described in the next section, involves <u>simultaneous non-linear equations</u>, for which <u>no</u> finite algorithm exists.

Bottleneck determination again means identification of saturated servers: those having an infinite queue develop. The problem is non-trivial because of the <u>starving</u> of downstream nodes when upstream nodes saturate and have limited output. For example, consider servers in series with increasing service rates and external load only at the first server. Even if the external load to the first server is higher than every service rate, only the first server will saturate. The set of saturated servers cannot be found by the heuristic of computing the unsaturated flows induced by the external load, assuming no server saturates, and then identifying which servers would be saturated by such flows: the true set of saturated servers is a subset (generally a strict subset, as the tandem queue example illustrates) of the saturated servers found by this heuristic. (See (3.6)).

3.1 Notation and Assumptions

- S servers labelled $s = 1,2,\ldots,S$. Each server is FCFS with unlimited waiting room.
- a_s = external arrival rate to server s, $1 \leq s \leq S$ (customers per unit time)
- $\frac{1}{\mu_s}$ = mean service time of a customer at server s, $1 \leq s \leq S$
- $P = [P_{st}]$ = $S \times S$ routing matrix, $P^n \to 0$ as $n \to \infty$
- $1 - \sum_{t=1}^{S} P_{st}$ = Probability customer departs network after service completion at server s.

The S unknowns are $\lambda_s \equiv$ departure rate from server s, $1 \leq s \leq S$. These satisfy the S functional equations

$$\lambda_s = \min \left[\mu_s, a_s + \sum_{t=1}^{S} \lambda_t P_{ts}\right] \qquad 1 \leq s \leq S \tag{3.1}$$

or

$$\lambda = T\lambda \quad \text{(fixed point equation)} \tag{3.2}$$

which say the departure rate from server s is the minimum of the total arrival rate to s and the capacity (maximal work rate) of server s. That is, server s will serve all of the incoming load, up to his maximal work rate μ_s. The problem reduces to solving these functional equations for $\{\lambda_s\}$. The saturated servers are those with $\lambda_s = \mu_s$, and will have an ∞ queue.

Basic properties of these functional equations are

λ is unique (T is a contraction with unique fixed point) (3.3)

$\lambda_s \le \lambda_s^{un} \qquad 1 \le s \le S$ (3.4)

where $\lambda_s^{un} \equiv \sum_{t=1}^{S} a_t \, [I - P]^{-1}_{ts}$ = departure rate from s in <u>unsaturated</u> case

$\lambda = \lambda^{un}$ if and only if every $\lambda_s^{un} \le \mu_s$ (3.5)

$\{s | \lambda_s = \mu_s\} \subseteq \{s | \lambda_s^{un} \ge \mu_s\}$ (actually-saturated servers are a subset of servers that λ^{un} would saturate) (3.6)

$\lambda_s > 0$ if and only if $\lambda_s^{un} > 0$. (3.7)

3.2 Computational Procedures to Solve (3.1, 3.2) (à la dynamic programming)

(a) <u>Successive Substitution</u> (not finite algorithm)

$$\lambda^{(n+1)} = T\lambda^{(n)} \qquad n = 0,1,2,3,\ldots$$

This converges <u>geometrically</u> to λ, for any choice of $\lambda^{(0)}$, because T is a contraction operator. <u>Monotonicity</u> properties:

$$\lambda^{(0)} \ge [\mu_s] \text{ implies } \lambda^{(n)} \downarrow \lambda$$

$$0 \le \lambda^{(0)} \le [\min\{\mu_s, a_s\}] \text{ implies } \lambda^{(n)} \uparrow \lambda$$

(b) <u>Linear Programming</u> (finite algorithm)

The unique solution to the linear program

$$\max \sum_{s=1}^{S} \lambda_s$$

$$\lambda_s \le \mu_s \qquad 1 \le s \le S$$

$$\lambda_s \le a_s + \sum_{t=1}^{S} \lambda_t P_{ts} \qquad 1 \le s \le S$$

$$\lambda_s \ge 0 \qquad 1 \le s \le S$$

is the unique solution to (3.1).

(c) Approximation in Policy Space (finite algorithm)

The policy iteration algorithm is:

<u>Initialization</u>

enter with $A \subseteq \{1,2,\ldots,S\}$, A = your guess for set of saturated servers

<u>policy evaluation step</u>

Given A, solve the set of S linear equations

$$\lambda(A)_s = \begin{cases} \mu_s & s \in A \\ a_s + \sum_{t=1}^{S} \lambda(A)_t P_{ts} & s \notin A \end{cases}$$

for $\lambda(A)_s$, $1 \leq s \leq S$

<u>policy improvement step</u>

Let $B \equiv \{s | a_s + \sum_{t=1}^{S} \lambda(A)_t P_{ts} \geq \mu_s\}$

= new set of saturated servers

If B = A, exit with $\lambda(A) = \lambda$.

Otherwise, replace A by B and return to policy evaluation step. The main properties of this algorithm are

- <u>finite convergence</u> to λ, the unique solution to (3.1).
- <u>monotonicity</u>: if B succeeds A, then $\lambda(B)_s \leq \lambda(A)_s$ for all s.
- <u>variational characterization of optimum</u>:

$$\lambda_s = \min_{A \subseteq \{1,2,\ldots,S\}} \lambda(A)_s \qquad 1 \leq s \leq S$$

- the policy iteration algorithm is a <u>variant of the simplex algorithm</u> for solving the dual to the linear programming formulation (b) of the problem, in which one does block pivoting on the basis.

4. Bottleneck Determination in Open NOQ's with Multiple Classes of Customers

4.1 Notation and Assumptions

- R = number of customer classes, labeled $r = 1,2,\ldots,R$
- S = number of servers, labeled $s = 1,2,\ldots,S$. Each server is FCFS with unlimited waiting room.
- a_{rs} = external mean arrival rate of class-r customers to server s (customer per unit time)
- $\frac{1}{\mu_{rs}}$ = mean service time of a class-r customer at server s
- $P(r)_{st}$ = routing matrix for class r. $P(r)^n \to 0$ as $n \to \infty$.

The RS unknowns are

- λ_{rs} = mean departure rate of class-r customers from server s.

They satisfy the RS functional equations $\lambda = T\lambda$:

$$\lambda_{rs} = \frac{a_{rs} + \sum_{t=1}^{S} \lambda_{rt} P(r)_{ts}}{D(\lambda)_s} \qquad 1 \le r \le R, \quad 1 \le s \le S \tag{4.1}$$

where

$$D(\lambda)_s \equiv \max \left[1, \sum_{r=1}^{R} \frac{1}{\mu_{rs}} \left[a_{rs} + \sum_{t=1}^{S} \lambda_{rt} P(r)_{ts}\right]\right] \tag{4.2}$$

Equation (4.1) says departure rate from s = arrival rate to s divided by a scale factor D_s. This ensures that the proportions of customer classes in the input and output streams at s agree. The scale factor is chosen so that $D_s = 1$ (output = input) if the arrival streams would not oversaturate server s. If the arrival streams would oversaturate server s, then they are scaled down so that the server operates at exactly 100% utilization: $D_s > 1$ implies $\sum_r \frac{1}{\mu_{rs}} \lambda_{rs} = 1$. When $R = 1$, (4.1, 4.2) reduces to (3.1).

4.2 Properties of the Functional Equations $\lambda = T\lambda$

(a) Solution exists to (4.1, 4.2), by Brouwer fixed-point mapping theorem.

(b) Empirically, solution is unique.

(c) $\lambda_{rs} \leq \lambda^{un}_{rs} \equiv \sum_{t=1}^{S} a_{rt} \, [I-P(r)]^{-1}_{ts}$ (compare (3.4)).

(d) $\lambda = \lambda^{un}$ if and only if $\sum_{r=1}^{R} \lambda^{un}_{rs}/\mu_{rs} \leq 1$ all s (compare (3.5)).

(e) $\{s|D(\lambda)_s > 1\} \subseteq \{s| \sum_{r=1}^{R} \lambda^{un}_{rs}/\mu_{rs} > 1\}$ (compare (3.6)).

(f) $\lambda_{rs} > 0$ if and only if $\lambda^{un}_{rs} > 0$ (compare (3.7)).

(g) No finite algorithm exists for solving functional equations (simultaneous quadratic equations).

(h) T is monotone for variables in same customer class, antimonotone for variables in distinct customer classes (due to contention by distinct classes for the server's time).

4.3 Computational Procedures

(a) Successive substitutions on $\lambda = T\lambda$ always converged geometrically to same solution. Modest computer times, e.g., 20-50 iterations to get 8-figure accuracy.

(b) Can get upper and lower bounds on λ which move monotonically inward to λ as iterations continue. Due to (a) and (b), computations may be considered routine.

(c) The linear programming and policy interation algorithms from Section 3 do not generalize because T apparently loses the properties of monotonicity and contraction -- interclass contention for resources makes insights very difficult to obtain.

5. Acknowledgement

Professor Arie Hodrijk of the University of Leiden pointed out to me that (3.1) is the fixed point of a contraction operator.

6. References

[1] Bard, Y., "Some Extensions to Multiclass Queueing Network Analysis," Performance of Computer Systems, M. Arato, A. Butrimenko and E. Gelenbe (eds.), North Holland, Amsterdam, (1979).

[2] Bellman, R., Adaptive Control Processes: A Guided Tour, Princeton University Press, Princeton, New Jersey (1961).

[3] Derman, C., Finite State Markovian Decision Processes, Academic Press, New York (1970).

[4] Howard, R. A., Dynamic Programming and Markov Processes, Wiley, New York (1960).

[5] Lavenberg, S. S., "Closed Multichain Product Form Queueing Networks with Large Population Sizes," IBM Watson Research Center, Yorktown Heights, New York, RC 8496 (September 1980).

[6] Pittel, B., "Closed Exponential Networks of Queues with Saturation: The Jackson-Type Stationary Distribution and its Asymptotic Analysis," Math. Oper. Res., 4, (1979) pp. 357-378.

[7] Reiser, M., "A Queueing Network Analysis of Computer Communication Networks with Window Flow Control," IEEE Trans. Comm., COM-27, (1979) pp. 1199-1209.

[8] Reiser, M. and S. S. Lavenberg, "Mean-Value Analysis of Closed Multichain Queueing Networks," J. Assoc. Comput. Mach., 27, (1980) pp. 312-322.

[9] Schweitzer, P., Unpublished manuscript, IBM Watson Research Center, Yorktown Heights, New York (1977).

[10] Schweitzer, P., "Approximate Analysis of Multiclass Closed Networks of Queues," Proceedings of the International Conference on Stochastic Control and Optimization, Free University, Amsterdam, Netherlands, April 5-6, 1979.

[11] Schweitzer, P. J., "First Moment Analysis of Multiclass Closed Networks of Queues," (in preparation).

[12] Schweitzer, P. J. and S. Agnihothri, "Bottleneck Determination in Open Networks of Queues," (in preparation).

Graduate School of Management, The University of Rochester, Rochester, New York 14627

PROBABILISTIC ANALYSIS OF ALGORITHMS
Dave Liu, Chairman

G. Lueker

ON THE AVERAGE DIFFERENCE BETWEEN THE SOLUTIONS TO LINEAR AND INTEGER KNAPSACK PROBLEMS†

George S. Lueker††

Abstract

We analyze the expected difference between the solutions to the integer and linear versions of the 0-1 Knapsack Problem. This difference is of interest partly because it may help understand the efficiency of a well-known fast backtracking algorithm for the integer 0-1 Knapsack Problem. We show that, under a fairly reasonable input distribution, the expected difference is $O(\log^2 n/n)$; for a somewhat more restricted subclass of input distribution, we also show that the expected difference is $\Omega(1/n)$.

1. Introduction

The following optimization problem is known as the <u>0-1</u> <u>knapsack</u> <u>problem</u>:

$$\text{maximize} \quad \sum_{i=1}^{N} z_i a_i$$

$$\text{subject to} \quad \sum_{i=1}^{N} z_i b_i \leq B,$$

where a_i, b_i, and B are given and the z_i are to be either 0 or 1. This problem is known to be NP-complete [K72]. Sometimes we will refer to a version in which each z_i may be any real in the interval [0,1]; this will be called the <u>relaxed</u> version, as opposed to the <u>integer</u> version above, and may easily be solved exactly in O(N log N) time by a greedy algorithm[1]. See, for example, [HS78]. Because of the importance and simple structure of the 0-1 integer knapsack problem, it has been the subject of extensive investigation. For example, it is known [IK75] that it admits a fully polynomial time approximation scheme [GJ79]; that is, we may obtain a worst-case relative error of ε, for any $\varepsilon > 0$, by an algorithm whose time is bounded by a polynomial in N and ε^{-1}. See [A78] for an analysis of an algorithm which works well on the average under certain assumptions about the input distribution. The problem also lends itself readily to solution by a backtracking approach; the search tree can be pruned whenever the solution obtained by using the items not yet considered according to the relaxed constraint is not as good as the best integer solution seen previously. See [HS78] for a detailed discussion of this approach. When applied to randomly generated data, this approach, which always yields the exact optimum, seems to run very rapidly even for large values of N; in fact, it seems possible that its expected time is polynomial in N. A proof of this would be very interesting, but probably difficult. A first step towards such a proof might be to obtain a better understanding of the difference between the optimum solutions to the integer and relaxed versions of the problem. (In general, determining the quality of the heuristics that guide a search is useful for understanding the quality of the search algorithm; see, for example, [G77].) This is the goal of this paper.

We will assume that the a_i and b_i are chosen uniformly from the interval [0,1]. Thus the selection of the parameters of the N items can be viewed as the placement of N points at random in the unit square. In order to simplify the analysis, we will assume that N is drawn from a Poisson distribution with parameter n; this will cause the number of points in disjoint parts of the square to be completely independent.

(For large n, N will tend to be nearly equal to n.) We will assume that the items are numbered so that the profit density (a_i/b_i) is nonincreasing. In order to try to cause a constant fraction of the items to be used in the solution as n becomes large, we will assume that for some fixed β, $B=\beta n$.

Observation. If β is larger than 1/2, then <u>all</u> items will fit into the knapsack except with a probability which vanishes as $n\to\infty$.

Proof sketch. Define ε by $\beta=(1+2\varepsilon)/2$. Note that if not all N items fit into the knapsack, then either

a) $N > (1+\varepsilon)n$, or

b) $N \leq (1+\varepsilon)n$ and the sum of the N b_i values exceeds βn.

Now (a) has an exponentially small probability; intuitively this is because the expected value of N is only n. Also, (b) has an exponentially small probability; intuitively this is because the expected value of b_i is 1/2, so the expected value of the sum of $(1+\varepsilon)n$ such values would be only $(1+\varepsilon)n/2 < (1+2\varepsilon)n/2 = \beta n$. □

Thus we henceforth assume β is an element of the open interval (0,1/2). For a given n and β, the random structure created this way will be denoted $P_{n,\beta}$; since we will generally assume β is fixed but $n\to\infty$, we will usually abbreviate this as P_n. The greedy method can be visualized by imagining a ray, which we shall call the <u>profit density ray</u>, which passes through the origin and rotates clockwise; as this ray rotates from pointing up to pointing to the right, it intersects the points in the order in which they are considered. Let $\bar{m}$ be the limit as $n\to\infty$ of the average slope of this ray at the point when the greedy method for the relaxed version fills the knapsack. It is not difficult to show that $\bar{m}$ is such that

$$\beta = \int\int_A x \, dx \, dy,$$

where A is the area shown in Figure 1.

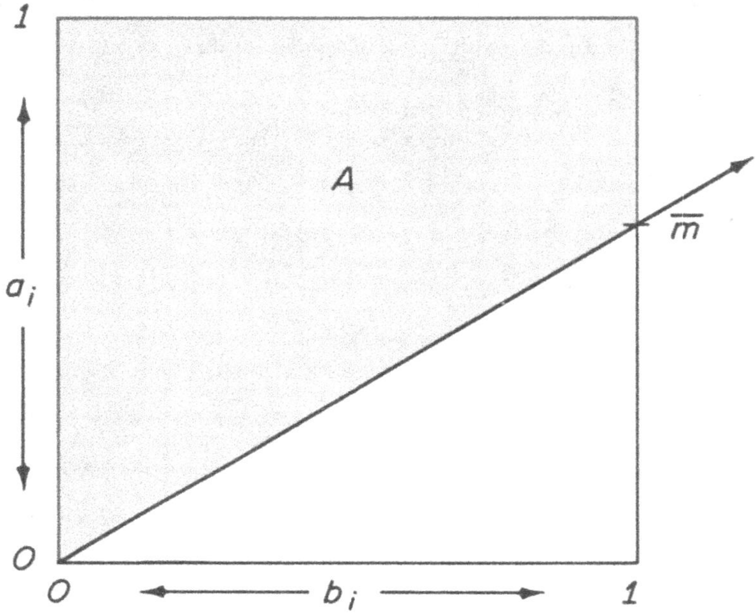

Figure 1. The profit density ray

(As a sketch of a justification, note that ($n \int\int_A x\, dx\, dy$) is the expected value of the sum of the b_i in the region A. If $\bar{m}$ is such that this value is less than βn, then except with exponentially small probability there will be $\Omega(n)$ space left in the knapsack when the ray advances to $\bar{m}$. On the other hand if $\bar{m}$ is such that this value exceeds n, then except with exponentially small probability all the space will be used up before the ray has advanced to $\bar{m}$.) Then if we let

$$\alpha = \int\int_A y\, dx\, dy,$$

it can be shown that the average optimum, to the integer or relaxed version, is asymptotic to αn. By our assumption on β, $\bar{m}$ is in the open interval $(0,\infty)$.

Since the linear and integer solutions are asymptotic to each other, it might not seem interesting to compare them. To obtain an interesting problem, we will look not at the ratios of the results, but rather at their differences. In [BZ77] it is observed empirically that for certain data this difference decreases as N increases; this is attributed to the fact that as N increases, more variables tend to lie in a region of small profit density change, which increases the chances of finding an integer solution with a value close to the relaxed optimum. The results presented in this paper formally establish that for our input distribution the average difference is $O((\log n)^2/n)$ for $\beta \in (0,1/2)$, and is $\Omega(1/n)$ for $\beta \in (1/6,1/2)$. (We strongly suspect that the bound of $\Omega(1/n)$ also holds for $\beta \in [0,1/6]$, but do not prove that here.)

2. A Theorem About Sums of Subsets

Before investigating the knapsack problem further, it is useful to consider the following problem about sums of subsets of random variables. We are given 2k random variables, and we wish to find a subset whose sum is as close as possible to some target x_k. How close can we hope to come? (See [AP80] for an analysis of an algorithm for a related subset sum problem. The method to be used below is nonconstructive, and gives an exponentially tighter bound.)

The following theorem provides a partial answer to this question. Since it appears to be of interest in its own right, we state it in a more general form than is needed for section 3.

Theorem 1. Let g be the piecewise continuous pdf of a variable which assumes values in $[-a,a]$. Suppose g is bounded and has mean 0 and variance 1. Let x_k be a real sequence with $x_k = o(k^{1/2})$. Suppose we draw 2k variables $X_1,\ldots,X_{2k}$ according to g. Then for large enough k, the probability that some subset of k of the 2k variables has a sum in $[x_k-\varepsilon, x_k+\varepsilon]$ is at least 1/2, provided $\varepsilon=7k\ 4^{-k}$.

Proof. A bit of notation is useful. Let G be the cumulative distribution function corresponding to g. Let G_n (resp. g_n) be the cumulative distribution (resp. density) function for the sum of n variables drawn according to g. Let F_n (resp. f_n) be the cumulative distribution (resp. density) function for the sum of n unit normal variables. Hence

$$f_n = C_n e^{-\frac{x^2}{2n}}, \tag{2.1a}$$

where

$$C_n = \frac{1}{\sqrt{2\pi n}} . \tag{2.1b}$$

Let Y_k be the random variable which tells the number of distinct subsets of size k whose sums lie in $[x_k-\varepsilon, x_k+\varepsilon]$. We seek to show $P\{Y_k > 0\} \geq 1/2$, provided $\varepsilon=7k4^{-k}$.

First note that the expectation of Y_k, as k becomes large, is

$$\begin{aligned} E[Y_k] &= \binom{2k}{k} [G_k(x_k+\varepsilon) - G_k(x_k-\varepsilon)] \\ &\sim \binom{2k}{k} f_k(x_k)\ 2\varepsilon \\ &\sim 2\binom{2k}{k} C_k\ \varepsilon \\ &= 2\binom{2k}{k} \varepsilon/\sqrt{2\pi k}, \end{aligned}$$

where we have employed [F66, Theorem 1, page 506] and the fact that $x_k/\sqrt{k} \to 0$. (A simple asymptotic analysis of the right hand term shows that it is about 3 for ε as in the lemma. This in itself, however, gives us no proof that the probability that Y is zero is small.) Now, as in [ES74, ER60, BE76, M70], we use the following well-known corollary of Chebyshev's inequality, which holds for an arbitrary random variable Y:

$$P\{Y=0\} \leq \frac{E[Y^2]}{(E[Y])^2} - 1. \tag{2.2}$$

The computation of $E[Y_k^2]$ is a bit messy, and is omitted from this paper; it can be shown that

$$E[Y_k^2] \sim \frac{\binom{2k}{k} 2\varepsilon}{\sqrt{2\pi k}} + \frac{\binom{2k}{k}^2 4\varepsilon^2}{\sqrt{3\pi k}} .$$

Hence

$$\frac{E[Y_k^{\ 2}]}{E[Y_k]^2} \sim \frac{4\sqrt{2}\ \binom{2k}{k}\varepsilon + 2\sqrt{3\pi k}}{2\sqrt{6}\ \binom{2k}{k}\varepsilon} .$$

Some asymptotic analysis, using the fact that

$$\binom{2k}{k} \sim 4^k/\sqrt{\pi k} ,$$

shows that if ε grows as $\alpha k 4^{-k}$, this ratio approaches

$$\sqrt{\frac{4}{3}} + \frac{\pi}{\sqrt{2}\alpha} . \tag{2.3}$$

Letting $\alpha=7$ causes this expression to achieve a value just under 1.5, which, in view of (2.2), establishes the result. □

It is interesting to note that letting α become very large does not cause (2.3) to approach 1; rather, it approaches $\sqrt{4/3}$. Thus, to show that a large ε gives a very small probability of failure, some different argument would be needed. Note, on the other hand, that if $\varepsilon=o(k4^{-k})$, one easily shows that $E[Y_k]$ approaches 0, so the probability of finding the desired subset of cardinality k approaches 0; thus in some sense the Theorem is tight up to constant factors.

A similar theorem could be obtained for a more general class of density functions, but we will not pursue that further here.

(Shepp [S81] has pointed out to me that a stronger form of inequality (2.2) can easily be demonstrated, namely,

$$P\{Y=0\} \leq 1 - \frac{(E[Y])^2}{E[Y^2]} .$$

He has used this in [S72a, S72b] and attributes the idea to Billard and Kahane [K68].)

3. An Upper Bound on the Asymptotic Average Difference

Let P_n denote a random problem generated as explained in the introduction; let INTEGER(P_n) and RELAXED(P_n) denote the value of the optimum solutions to the integer and relaxed versions of this problem. In order to bound the difference between these solutions, we will employ a procedure, named APPROX, which constructs a feasible solution to the integer version; it appears below. APPROX works as follows. Let $k = \lfloor \log_4 n \rfloor$. First APPROX advances the profit density ray and packs items until the remaining knapsack capacity is about 2k/3. Then it repeatedly considers successive sets of 2k items, trying to find a subset of k items which use up almost all of the remaining capacity, without regard to profit. For comparison, we have also presented the greedy procedure which solves the relaxed problem exactly.

<u>Theorem 2</u>. $E[\text{RELAXED}(P_n) - \text{INTEGER}(P_n)] = O(\log^2 n/n)$.

Proof. For simplicity, we first assume $\beta \in (1/6, 1/2)$. Then, by (1.1), $\bar{m} \in (0,1)$; thus, except with exponentially small probability, the profit density ray will be intersecting the right edge of the square when the knapsack becomes full. Since APPROX gives a lower bound on the true optimum, we may bound the difference between INTEGER and RELAXED by that between APPROX and RELAXED. Now the deviation between APPROX and RELAXED is attributable to two causes:

a) we do not completely fill the knapsack during APPROX, and

b) the part we do fill may be filled with items of a lower profit density.

For part (a), note that if the branch to OUT is taken in APPROX, the unused part of the knapsack has size at most 2ε, which is $O(\log n/n)$; the probability that the branch to OUT is never taken can be shown to be $O(n^{-e})$ for any positive integer e.

```
procedure APPROX;
begin
    BB := B; i :=1; A := 0;
    k := ⌊log_4 n⌋; ε := (7/√18)k4^(-k);
    while BB ≥ 2k/3 and i ≤ N do
        begin
            BB := BB - b_i;
            A := A + b_i;
            i := i + 1;
        end;
    comment at this point 2k/3-1 ≤ BB ≤ 2k/3;
    while i + 2k ≤ N do
        begin
            for all subsets S of {i+1,i+2,...,i+2k} do
                begin
                    if the sum of the b_j values over all j in S lies in
                        [BB-2ε,BB] then go to OUT;
                end;
            i := i + 2k;
        end;
    S := the empty set;
    OUT: A := A + the sum of the a_j values over all j in S;
    return A;
end;
```

```
procedure RELAXED;
begin
   BB := B; i := 1; A := 0;
   while BB > b_i and i ≤ N do
      begin
         BB := BB - b_i;
         A := A + a_i;
         i := i + 1;
      end
   if i≤N then A:= A + a_i * (BB/b_i);
   return A;
end;
```

Next consider part (b). Now it is not hard to see that the b_i values for the points intersected as the profit density ray advances are distributed according to the pdf 2x, for $0 \leq x \leq 1$. (Intuitively, this is because the rate at which a small segment of the profit density ray sweeps out area is directly proportional to the x coordinate of the segment, for $x \in (0,1)$; see Figure 1.) Thus, they have mean 2/3 and variance 1/18. Hence, by an appropriately scaled version of Theorem 1, the probability of success on a single iteration of the second while-loop is at least 1/2.

Since successive iterations are independent, the expected number of iterations is O(1). Now the extent to which the profit density ray advances at each iteration is independent of the values of the b_i, and can readily be seen to have an expectation of $O(\log n/n)$; this change in density applies to a portion of the knapsack whose capacity is at most $2k/3=O(\log n)$ so the expected contribution due to part (b) above is $O(\log^2 n/n)$.

This argument may be extended to cover the case in which $\beta < 1/6$; we omit the details here. (The constants hidden in the O-notation may depend on β.) □

Since we observed earlier that the $E[\text{INTEGER}(P_n)] \sim \alpha n$, we now see that for our distribution the difference between the integer and relaxed solutions asymptotically tends to be a very small fraction of the solution.

4. A Lower Bound on the Asymptotic Average Difference

An interesting question is whether the bound on the expected difference between the integer and relaxed solutions stated in Theorem 2 is tight. Although we have not been able to answer that question, we have established the following lower bound.

Theorem 3. If $\beta \in (1/6,1/2)$, then

$$E[\text{RELAXED}(P_n) - \text{INTEGER}(P_n)] = \Omega(1/n).$$

Proof Sketch. Intuitively, the idea of the proof is as follows. When we have advanced the profit density ray to the point where the remaining knapsack capacity is only 1 or 2, we are likely to find that no subset of the next few items can fit into the knapsack without leaving a substantial part of the remaining capacity unused. Then if we wish to try to fill this remaining capacity, we must either use items of lower profit density or consider combinations of items which involve removing some of the previously packed items of high profit density. In either case, we are likely to come up with a solution significantly different from $\text{RELAXED}(P_n)$ because of the lowering of profit density.

We now will make this more precise by describing a boolean procedure TEST with the following two properties:

a) It returns *true* with probability of at least 1/4 for large n, and

b) if it returns *true*, then for this problem instance the relaxed and integer solutions differ by $\Omega(1/n)$.

From this the Theorem follows readily.

TEST proceeds as follows. First it fills the knapsack as in RELAXED until the remaining capacity BB satisfies

$$1 < BB \leq 2. \tag{4.1}$$

Henceforth in the proof we fix BB at this value; let $\hat{p}_0$ be the profit density of the last item used. The procedure rejects (i.e., returns *false*) if the condition (4.1) cannot be met; since all b_i are in [0,1], rejection occurs here only if we run out of items to use in the knapsack, and this occurs with exponentially small probability. TEST also rejects if the profit density ray has not yet advanced past the upper right corner of the square, which again can be seen to have exponentially small probability under our model of input distribution.

Next we look at the next four points as the profit density ray advances. (The probability that fewer than this number remain is again exponentially small.) Call their profits, costs, and densities $\hat{a}_i$, $\hat{b}_i$, and $\hat{p}_i$, respectively, for $i = 1,\ldots,4$. Reject unless all three of the following hold:

$$\hat{p}_0 - \hat{p}_1 \geq \frac{1}{10n} \tag{4.2a}$$

$$\hat{p}_3 - \hat{p}_4 \geq \frac{1}{10n} \tag{4.2b}$$

$$\hat{p}_4 \geq \bar{m}/2 \ . \tag{4.3}$$

Note that the movement of the profit density ray between these items has an exponential distribution with mean $2/n$; hence the probability of rejection in (4.2a) or (4.2b) is less than (1/20) each, for a total of 1/10. The probability of a violation of (4.3) can be seen to be exponentially small, since it would mean that we had gone far beyond the $\bar{m}$ of Figure 1.

Next we impose some restrictions on the $\hat{b}_i$ values. (Note that each has, independently, a density function of $2x$ for $x \in [0,1]$.)

Reject if <u>any</u> subset of these four values has a sum in the range $BB \pm \frac{1}{600}$. Note that the sum of any fixed nonempty subset of the b_i has a density function uniformly bounded by 2; hence, for any such subset, the probability that its sum lies in the indicated range is at most $\frac{1}{150}$. On the other hand, there are only 15 nonempty subsets, so the probability of rejection here is at most $\frac{1}{10}$. Finally, reject unless

$$\hat{b}_1 < BB < \hat{b}_1 + \hat{b}_2 + \hat{b}_3 . \tag{4.4}$$

The left inequality is always true since $1 < BB$. The right inequality has probability greater than 1/2; this can be seen by noting that $BB < 2$, and performing a tedious but straightforward computation involving convolution of the densities of the $\hat{b}_i$. At this point the description of TEST is complete.

Now since the probability of the union of several events is bounded by the sum of their probabilities, we see that

$$P\{\mathrm{TEST}(P_n) = \underline{\text{false}}\} \le \frac{1}{10} + \frac{1}{10} + \frac{1}{2} + o(1),$$

which is smaller than 0.75 for large enough n, so condition (a) holds.

Next we establish condition (b). Assume that TEST returns <u>true</u>. Then we know from (4.4) that the procedure RELAXED fills the knapsack when the profit density ray is lying in the area labeled β_2 in Figure 2. Let B_α, B_β, and B_γ be the total knapsack capacity used in the relaxed solution by items lying in regions α, $\beta_1 \cup \beta_2 \cup \beta_3$, and γ (respectively). Note that $B_\beta = BB$, and $B_\gamma = 0$. Now consider the optimum solution to the integer problem; define $\widetilde{B}_\alpha$, $\widetilde{B}_\beta$, and $\widetilde{B}_\gamma$ for this solution analogously to B_α, B_β, and B_γ. Now by the restriction TEST imposed on sums of subsets of the $\hat{b}_i$, we know that $|B_\beta - \widetilde{B}_\beta| \ge \frac{1}{600}$. Hence it can be seen that at least one of the following three conditions must hold:

$$\widetilde{B}_\alpha + \widetilde{B}_\beta + \widetilde{B}_\gamma \le B_\alpha + B_\beta + B_\gamma - \frac{1/3}{600} \tag{4.5}$$

$$\tilde{B}_\alpha \leq B_\alpha - \frac{1/3}{600} \tag{4.6}$$

$$\tilde{B}_\gamma \geq B_\gamma + \frac{1/3}{600} \tag{4.7}$$

If (4.5) holds, at least $\frac{1}{1800}$ units of the knapsack is being wasted, from which it is not hard to see that the difference between the two solutions is $\Omega(1)$. If (4.6) holds, then since in the relaxed solution all items in $\alpha \cup \beta_1$ were used, at least $\frac{1}{1800}$ units of capacity has been shifted by the integer solution from α to $\beta_2 \cup \beta_3 \cup \gamma$, and hence by (4.2a) experienced a decrease of $\frac{1}{10n}$ in profit density; thus the integer solution is worse by $\Omega(1/n)$. A similar argument holds for case (4.7). □

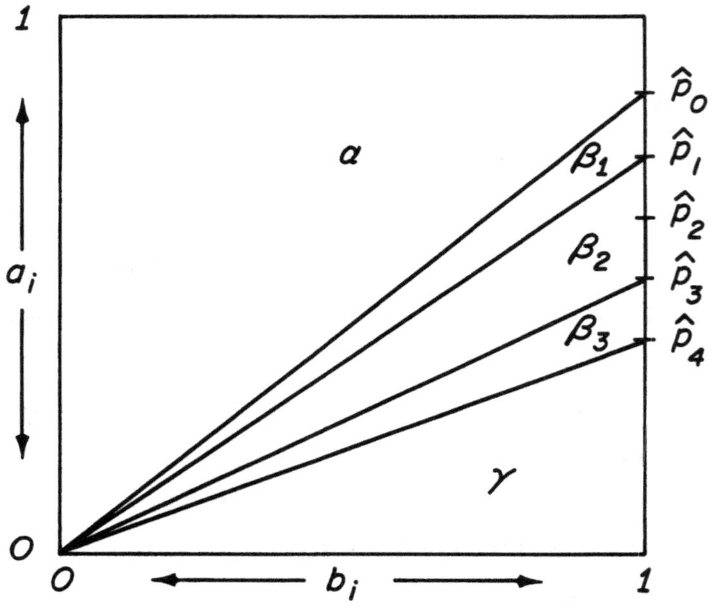

Figure 2. Illustration for the lower bound on the difference. Regions are to include the segment bounding them from below, but not that bounding them above.

5. References

[A78] G. d'Atri, "Probabilistic Analysis of the Knapsack Problem," Technical Report No. 7, Groupe de Recherche 22, Centre National de la Recherche Scientifique, Paris, October 1978.

[AP80] G. d'Atri and C. Puech, "Probabilistic Analysis of the Subset-Sum Problem," Technical Report No. 1 (1980), Dipartimento di Matematica, Universita' Della Calabria, Italy, March 1980.

[BZ77] Egon Balas and Eitan Zemel, "Solving Large Zero-One Knapsack Problems," Management Sciences Research Report No. 408, Carnegie-Mellon University, July 1977.

[BE76] B. Bollobás and P. Erdős, "Cliques in Random Graphs," Math. Proc. Cambridge Philos. Soc., 80 (1976), pp. 419-427.

[ES74] P. Erdős and J. Spencer, Probabilistic Methods in Combinatorics, Academic Press, New York, 1974.

[ER60] P. Erdős and A. Rényi, "On the Evolution of Random Graphs," Publ. Math. Inst. Hung. Acad. Sci. 5A, (1960), pp. 17-61.

[F66] William Feller, An Introduction to Probability Theory and Its Applications, Volume II, John Wiley and Sons, New York, 1966.

[GJ79] M. R. Garey and D. S. Johnson, Computers and Intractability: A Guide to the Theory of NP-Completeness, W. H. Freeman and Company, San Francisco, 1979.

[G77] John Gaschnig, "Exactly How Good are Heuristics?: Toward a Realistic Predictive Theory of Best-First Search," Proc. Intl. Joint Conf. on Artificial Intelligence, Cambridge, Mass., August 1977.

[HS78] E. Horowitz and S. Sahni, Fundamentals of Computer Algorithms, Computer Science Press, Potomac, Maryland, 1978.

[IK75] O. H. Ibarra and C. E. Kim, "Fast Approximation Algorithms for the Knapsack and Sum of Subset Problems," J. Assoc. Comput. Mach., 22, (October 1975), pp. 463-468.

[K68] J. P. Kahane, Some Random Series of Functions, D. C. Heath and Company, Lexington, Massachusetts, 1968.

[K72] R. M. Karp, "Reducibility among Combinatorial Problems," in R. E. Miller and J. W. Thatcher, eds., Complexity of Computer Computations, Plenum Press, New York, 1972, pp. 85-104.

[K76] D. E. Knuth, "Big Omicron and Big Omega and Big Theta," SIGACT News, 8 (April-June, 1976), pp. 18-24.

[M70] L. Moser, "The Second Moment Method in Combinatorial Analysis," in Combinatorial Structures and Their Applications, Gordon and Breach, New York, 1970.

[P77] Nicholas Pippenger, "An Information-Theoretic Method in Combinatorial Theory," J. Combin. Theory Ser. A, 23, (July 1977), pp. 99-104.

[S72a] L. A. Shepp, "Covering the Line with Random Intervals," Z. Wahrsch. Verw. Gebiete, 23 (1972), pp. 163-170.

[S72b] L. A. Shepp, "Covering the Circle with Random Arcs," Israel J. Math., 11 (1972), pp. 328-345.

[S81] L. A. Shepp, private communication, April 1981.

6. Endnotes

[1]By $f(n) = O(g(n))$, we mean that for some C and n_o, $|f(n)| \leq Cg(n)$ for all $n \geq n_o$. By $f(n) = \Omega(g(n))$, we mean that for some C and n_o, $f(n) \geq Cg(n)$ for all $n \geq n_o$. See [K76].

†This work was facilitated by the use of MACSYMA, a large symbolic Manipulation program developed at the MIT Laboratory for Computer Science and supported by the National Aeronautics and Space Administration under grant NSG 1323, by the Office of Naval Research under grant N00014-77-C-0641, by the U.S. Department of Energy under grant ET-78-C-02-4687, and by the U.S. Air Force under grant F49620-79-C-020.

††Supported by the National Science Foundation under grant MCS79-04997.

Department of Information and Computer Science, University of California, Irvine, Irvine, CA 92717.

www.ingramcontent.com/pod-product-compliance
Lightning Source LLC
Chambersburg PA
CBHW071355050326
40689CB00010B/1647
* 9 7 8 0 8 1 7 6 3 1 1 6 1 *